Recent Advances in Solar Cells

Online at: https://doi.org/10.1088/978-0-7503-5994-8

Recent Advances in Solar Cells

N M Ravindra
New Jersey Institute of Technology, Newark, NJ, USA

Leqi Lin
Shanghai Jiao Tong University, Shanghai, China

Priyanka Singh
CSIR-National Physical Laboratory, New Delhi, India

IOP Publishing, Bristol, UK

ISBN 978-0-7503-5994-8 (ebook)
ISBN 978-0-7503-5992-4 (print)
ISBN 978-0-7503-5995-5 (myPrint)
ISBN 978-0-7503-5993-1 (mobi)

DOI 10.1088/978-0-7503-5994-8

Version: 20250301

IOP ebooks

British Library Cataloguing-in-Publication Data: A catalogue record for this book is available from the British Library.

Published by IOP Publishing, wholly owned by The Institute of Physics, London

IOP Publishing, No.2 The Distillery, Glassfields, Avon Street, Bristol, BS2 0GR, UK

US Office: IOP Publishing, Inc., 190 North Independence Mall West, Suite 601, Philadelphia, PA 19106, USA

To the Guru—the eternal teacher.

Contents

Preface

My journey into solar cells began in 1977. As a graduate student in the Department of Physics at the University of Roorkee (Indian Institute of Technology, Roorkee—since 2001), I was introduced to solar cells [1, 2] by reading books such as *Semiconductors and Semimetals. Vol. 11, Solar cells* [3], *Solar cells* (IEEE Press selected reprint series) by Charles E Backus [4] and *Festkörperprobleme 18* [5]. The idea of microwave transmission of space-based solar power [6], fictionalized by Isaac Asimov in 1941 [7] and patented by Peter Glaser in 1968 [8] seemed to address all the energy needs of the modern world—without the need to fight a war or conquer a country in order to control the natural energy producing assets and resources. The transition to Montpellier [9, 10], Paris [11, 12], Trieste [13] and Torino [14, 15], with a focus on amorphous silicon, during the years 1982–1985, was a significant opportunity to understand the role of hydrogenation in amorphous silicon and its impact on enhanced optical absorption of the incident solar radiation in silicon. The assignments at the Microelectronics Center of North Carolina and collaboration with Oak Ridge National Laboratory and the North Carolina State University [16, 17], provided an insight into the gains for silicon photovoltaics from traditional silicon microelectronics.

The subsequent move to the New Jersey Institute of Technology (NJIT) in 1987 [18] and the research opportunities funded by various US DOD agencies—DARPA [19] and DURIP [20], combined with initiatives from SEMATECH and Semiconductor Research Corporation (SRC) [21, 22], provided a common focus for research in the form of rapid thermal processing and the need to reduce the thermal budget in silicon microelectronics manufacturing. The impact of DARPA's Microelectronics Manufacturing Science and Technology (MMST) program [23, 24] on silicon solar photovoltaics manufacturing was significant. This resulted in notable research efforts in the development of belt furnaces for silicon PV manufacturing for various process steps [25]. This was in the context of the fact that the manufacture of silicon solar cells requires various thermal steps such as in-line diffusion, metallization, and anti-reflection coatings. The metallization process includes solar cell drying and firing. With a view to reduce the thermal budget, there was an immediate need for the development of non-contact sensors for process monitoring and control. This led to significant investments in understanding the emissivity of silicon and silicon-related materials as function of wavelength and temperature and its subsequent applications in the development of radiation thermometry [26, 27] and a variety of rapid thermal processing systems [28–34]. The subsequent collaborations between this research group and the National Renewable Energy Laboratory, from 1996 to 2016, resulted in a significant understanding of the various aspects of silicon solar cell processes [35–44] and the utilization of solar energy for hydrogen production via photoelectrochemical reactions [45].

Over the years, the author and his team have been working on problems related to the thermal management of solar cells [46–50]. A summary of problems of current interest in photovoltaics is presented in table P.1.

Table P.1. Summary of some of the focus areas in photovoltaics.

Topic	Areas of focus
Need for standards	As silicon wafers for solar cell manufacturing are getting larger, there is a need for standardization of wafer size and panel size at the global level.
Cost-effective materials	New materials that are stable with improved spectral match and high absorption coefficients; preferably light weight; should be completely scalable.
Use of silver	Due to its intrinsic property of excellent electrical conductivity and thermal conductivity, silver is a standard material today to form the conducting layer on the front and back of the silicon solar cells; the quantity of silver used is directly proportional to the cell efficiency; Bloomberg NEF estimates that within the next six years, i.e., by 2030, solar panels will consume 20% of the world's silver; by 2050, the World Bank expects that more than half of the world's silver will be consumed by the energy sector. As an illustration of the seriousness of the problem, the solar PV industry utilized 100 million ounces of silver in 2023—this represents ~14% of the global silver demand. A typical solar panel uses ~20 grams of silver. From cost considerations, it is estimated that silver accounts for ~10% of the total cost of a standard silicon PV module.
Material stability	Minimal degradation due to exposure to heat and humidity; long-term stability.
Perovskites	While high efficiency solar cells, made from perovskites, have been demonstrated, their stability and scalability for manufacture need to be addressed. The promise of the ability of perovskites to exhibit a better solar spectral match than most materials will continue to drive the research focus on these materials. The immediate challenges are their stability and reliability from the perspective of life expectancy of commercial perovskites-based solar panels. The US Department of Energy and agencies throughout the world are supporting R&D projects to expedite the commercialization of hybrid organic–inorganic perovskite solar technologies, while focusing on their increased

	efficiency and lifetime and reducing manufacturing costs.
Solar photons—material interactions	There is a continued interest in developing methodologies for light trapping [51]—particularly in thin film solar cells; panels that can operate efficiently in low-light level is also of significant interest.
Grid integration and energy storage	Especially in utilizing large scale solar farms, the integration with the grid as well energy storage continue to be the major challenges; in many countries, including the US, the Grid needs to be upgraded and modernized, with increased capacity, to accommodate additional input from renewable energy sources such as wind and solar.
Solar panels—use of glass	The industry is moving towards glass on glass panels; while this will create issues such as the added weight and thus the increased cost of transportation, the impact of hail on glass etc, there is a need to formulate global standards for guidance for manufacturers and customers alike. The continued rise in the use of solar panels has resulted in increased demand for glass, resulting in the shortage of glass. This is also due to supply-chain related issues and enhanced costs of raw materials. Glass-free solar panels, including transparent solar panels are, in general, less efficient. They find use in specific applications that involve small areas, such as cell-phone touch screens, and windshields in vehicles.
End-of-life management	With increased use and implementation of solar photovoltaics, the industry needs to develop long-term strategy for sustainable processes for recycling of solar panels in order to minimize its impact on the environment; if this is not addressed in a timely manner, it may result in reduced use of solar photovoltaics.
Energy storage	Energy storage systems that can be efficiently integrated with solar panels in order to address the intermittent nature of sunlight and provide continuous and reliable power during periods of low solar intensity/low power generation, due to the incident solar energy, are required.

(*Continued*)

Table P.1. (*Continued*)

Topic	Areas of focus
	In general, the increased use of electric vehicles is anticipated to reduce the costs of energy storage. Significant research is being performed to address the materials as well as the thermal management of rechargeable batteries.
Implementation of artificial intelligence (AI) and machine learning (ML) techniques	Potential applications of AI/ML include the following: cost reduction, defect identification and analysis, design optimization, performance prediction—including efficiency and reliability, predictive maintenance, and site selection. Examples of recent studies, in the context of perovskites, are several [52–54].
Incentives	Historically, the use of renewable energy sources, such as wind and solar, has been dependent on incentives at the local, state and federal level. This has been particularly due to the availability of solar panels at reduced prices in some countries. While imposing additional tariffs on such products may be a solution, in reality, the additional costs are simply passed on to the consumer. In general, implementation of tariffs, from the perspective of the consumer, may become a liability over the long term due to enhanced product costs and the unavailability of the panels, sometimes, at the local consumer level.

An illustration of the world market share of Czochralski single crystal silicon wafers, of various dimensions, is presented in figure P.1 [55]. At least for the next few years, until 2032, M10 (182 mm) and G12 (210 mm) will continue to dominate the market share of wafers, i.e., $> 50\%$, used in silicon photovoltaics.

The convergence of the Inflation Reduction Act (IRA), the geo-political problems and the associated concerns with the global supply chain, have led to enhanced opportunities for the solar panel manufacturing capacity in the United States. In fact, it has quadrupled from 8.3 GW in August 2021 to over 27 GW in 2024 [56, 57]. One such major player, DYCM Power, a collaborative venture between Das & Company and APC Holdings (APCH), is in the process of building a $800 million, 2 GW annual capacity silicon (n-TopCon), bifacial, M10 (182 mm × 182 mm) wafer-to-cell-to-module manufacturing in the United States [58]. An illustration of the US Solar Capacity, during the years 2014–24, is presented in figure P.2 [59]. As can be

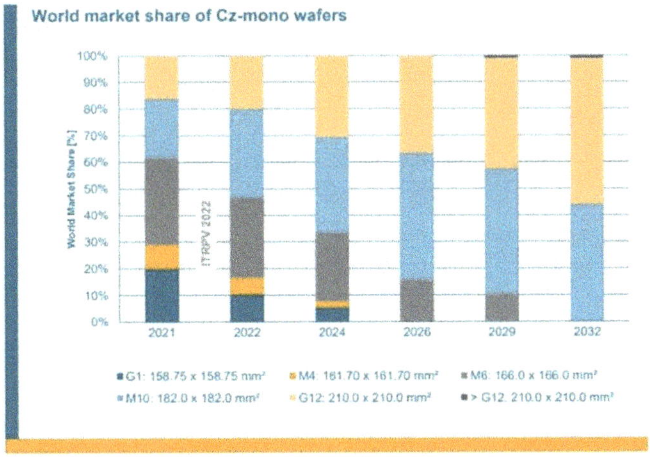

Figure P.1. Illustration of world market share of Cz-monocrystalline silicon wafers as function of wafer size. Reproduced with permission from [55], copyright VDMA 2021–2024.

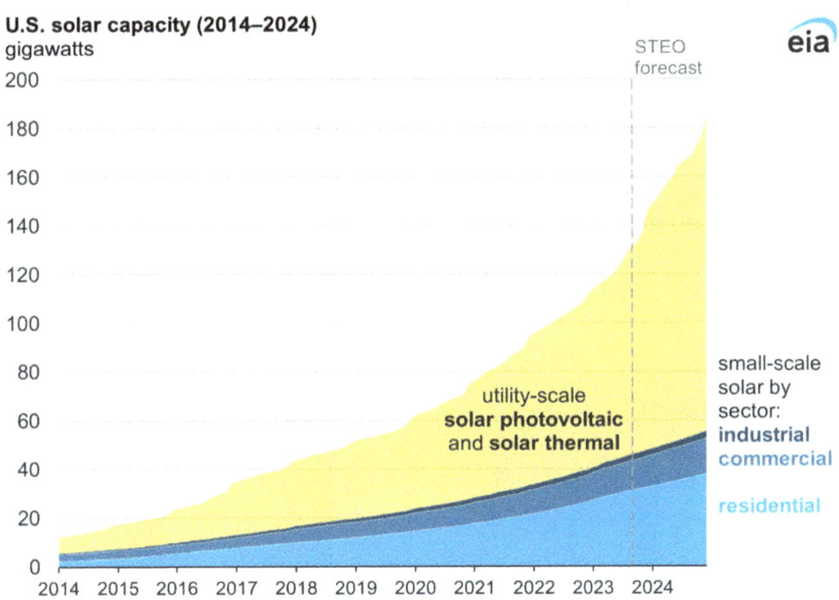

Figure P.2. Illustration of the growth of US solar capacity during the last ten years. **Source:** U.S. Energy Information Administration (2023) *Short-Term Energy Outlook*. **Data values:** U.S. regional electricity generating capacity, small-scale solar. Reproduced with permission from [59] credit: U.S. Energy Information Administration (Dec 2024).

seen in this figure, the solar capacity has grown over 18 times, from 10 GW to >180 GW in less than ten years (2014–23).

A comparison of solar energy appropriations (1950–78) [60] and the allocated budget to energy efficiency and renewable energy in 2024 [61] is presented in

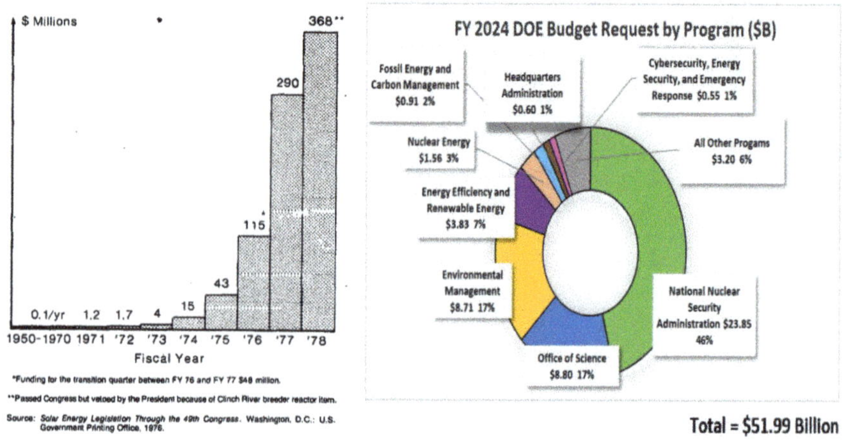

Figure P.3. Solar energy appropriations, 1950–78 [60] credit: USA Department of Energy; energy efficiency and renewable energy—$3.83 billion, 2024 [61].

figure P.3. In 1978, $368 million was invested towards solar energy; in 2024, energy efficiency and renewable energy represents $3.83 billion, an increase of > ten- fold, in the US Department of Energy Budget.

August 16 2024 marks the second anniversary since the Inflation Reduction Act (IRA) was implemented. From a historical and global perspective, the IRA represents the single largest investment in climate action and clean energy [62]. With the anticipation that some form of Inflation Reduction Act is firmly in place, coupled with the emphasis on United States-based manufacture and strong commitments to Work Force Development, the Solar Industry is expected to grow significantly over the next eight years.

According to the Climate Jobs National Resource Center, the existing income and registered apprenticeship requirements could also enhance the pay and job quality for clean energy workers across the United States [63].

Newark, NJ,
December 12, 2024

References

[1] Ravindra N M and Prasad B 1980 Saturation current in solar cells—an analysis *Sol. Cells* **2** 109–13

[2] Prasad B and Ravindra N M 1981 The dependence of solar cell active layer resistance on illumination *Int. J. Electron.* **50** 499–504

[3] Hovel H J, Willardson R K and Beer A C 1976 Semiconductors and semimetals *Solar Cells* **11** (New York: Academic) 1st edn

[4] Backus C E 1976 Solar cells *IEEE Press Selected Print Series*

[5] Fischer H 1978 Solar cells based on nonsingle crystalline silicon *Festkörperprobleme 18*Advances in Solid State Physics *ed J Treusch (Berlin: Springer)*

[6] James R A and Fork W E 2024 Microwave transmission of space-based solar power: the focus of new attention https://gravel2gavel.com/microwave-transmission-space-solar-power/
[7] Asimov I 1941 Reason *Astounding Science Fiction*
[8] Glaser P 1973 Method and apparatus for converting solar radiation to electrical power *US Patent* 3781647A
[9] Prasad B, Gopal K, Ravindra N M and Robin S 1983 Maximum consideration in lossy solar cells *Int. J. Electron.* **55** 681–4
[10] Ravindra N M, Narayan J, Ance C, Dechelle F and Ferraton J P 1986 Low temperature optical properties of hydrogenated amorphous silicon *Mater. Lett.* **4** 343–9
[11] Jouanne M, Kanehisa M A, Mhoronge J F, Ravindra N M and Balkanski M 1985 Bound state energy and line width due to the resonant interaction between optical phonon and electronic transitions in degenerate silicon *Solid-State Electron.* **28** 39–45
[12] Ravindra N M, Mhoronge J F and Jouanne M 1985 Optical properties of laser-induced heavily doped Si *Infrared Phys.* **25** 707–14
[13] Demichelis F, Minetti Mezzetti E, Tagliaferro A, Tresso E, Rava P and Ravindra N M 1986 Optical properties of hydrogenated amorphous silicon *J. Appl. Phys.* **59** 611–8
[14] Ravindra N M and Demichelis F 1985 Influence of electrical and optical losses on fundamental solar-cell parameters *Il Nuovo Cimento* D **6** 251–60
[15] Ravindra N M and Demichelis F 1985 Cody disorder: absorption-edge relationships in hydrogenated amorphous silicon *Phys. Rev.* B **32** 6591–5
[16] Ravindra N M, Narayan J, Fathy D, Srivastava J K and Irene E A 1987 Silicon oxidation and Si-SiO_2 interface of thin oxides *J. Mater. Res.* **2** 216–21
[17] Ravindra N M, Narayan J, Fathy D, Srivastava J K and Irene E A 1986 Two step oxidation processes in silicon *Mater. Lett.* **4** 337–42
[18] Ravindra N M and Zhao J 1992 Fowler-Nordheim tunneling in thin SiO_2 films *Smart Mater. Struct.* **1** 197–201
[19] Kosonocky W F *et al* 1994 Multiwavelength imaging pyrometer *Proc. of SPIE—The International Society for Optical Engineering* ed E L Dereniak and R E Sampson (Society of Photo-Optical Instrumentation Engineers) pp 26–43
[20] Ravindra N M *et al* 1994 Development of emissivity models and induced transmission filters for multi-wavelength imaging pyrometry (M-WIP) *Proc. of SPIE—The International Society for Optical Engineering,* Thermosense XVI**2245** ed J R Snell Jr (SPIE) pp 304–18
[21] Ravindra N M *et al* 1998 Temperature-dependent emissivity of silicon-related materials and structures *IEEE Trans. Semicond. Manuf.* **11** 30–9
[22] Ravindra N M *et al* 1998 Radiative properties of SIMOX *IEEE Trans. Compon., Packag. Manuf. Technol.—Part* A **21** 441–9
[23] Doering R R 1993 The MMST program *Int. Symp. on Semiconductor Manufacturing (Austin, TX)* (Piscataway, NJ: IEEE)
[24] Ravindra N M, Marthi S R and Bañobre A 2017 *Radiative Properties of Semiconductors* (Morgan & Claypool Publishers—IOP Publishing)
[25] Solar Cell Manufacturing https://beltfurnaces.com/solarcellprocess.html (accessed 19 October 2024)
[26] Ravindra N M *et al* 2001 Emissivity measurements and modeling of silicon related materials —an overview *Int. J. Thermophys.* **22** 1593–611
[27] Ravindra N M 2021 *Microbolometers Fundamentals, Materials, and Recent Developments* 1st edn (Amsterdam: Elsevier)

[28] https://appliedmaterials.com/in/en/semiconductor/semiconductor-technologies/rapid-thermal-processing.html (accessed 19 October 2024)
[29] https://ecm-usa.com/ecm-lab-solutions/rtp-rta (accessed 19 October 2024)
[30] https://annealsys.com/products/rtp-and-rtcvd/as-one.html (accessed 19 October 2024)
[31] https://heraeus.com/en/hng/light_is_more/how_does_it_work/rapid_thermal_processes/rapid_thermal_processing.html (accessed 19 October 2024)
[32] https://mattson.com/technology/rapid-thermal-processing/ (accessed 19 October 2024)
[33] https://photonexport.com/rapid-thermal-processing/ (accessed 19 October 2024)
[34] https://allwin21.com/semiconductor-process-equipment/application-of-rapid-thermal-annealing-systems/ (accessed 19 October 2024)
[35] Chen W 2000 Modeling, design and fabrication of thin-film microcrystalline silicon solar cells *PhD Dissertation* New Jersey Institute of Technology https://digitalcommons.njit.edu/dissertations/397/
[36] Zhang Y 2001 Modeling hydrogen diffusion for solar cell passivation and process optimization *PhD Dissertation* New Jersey Institute of Technology https://core.ac.uk/download/pdf/232274734.pdf
[37] Appel J S 2008 Modeling edge effects of mesa diodes for silicon photovoltaics *PhD Dissertation* New Jersey Institute of Technology https://digitalcommons.njit.edu/dissertations/876/
[38] Li C 2009 Surface and bulk passivation of multicrystalline silicon solar cells by silicon nitride (H) layer: modeling and experiments *PhD Dissertation* New Jersey Institute of Technology https://digitalcommons.njit.edu/dissertations/892
[39] Mehta V R 2010 Formation of screen-printed contacts on multicrystalline silicon (mc-Si) solar cells *PhD Dissertation* New Jersey Institute of Technology https://digitalcommons.njit.edu/dissertations/215/
[40] Budhraja V 2012 Influence of defects and impurities on solar cell performance *PhD Dissertation* New Jersey Institute of Technology https://digitalcommons.njit.edu/dissertations/294/
[41] Guhabiswas D 2013 Extension to PV optics to include front electrode design in solar cells *PhD Dissertation* New Jersey Institute of Technology https://digitalcommons.njit.edu/dissertations/343/
[42] Basnyat P M 2013 Investigations into b-o defect formation-dissociation in cz-silicon and their effect on solar cell performance *PhD Dissertation* New Jersey Institute of Technology https://digitalcommons.njit.edu/dissertations/359/
[43] Sahoo S 2013 Influence of SiNx/Si interface states on Si solar cells *PhD Dissertation* New Jersey Institute of Technology https://digitalcommons.njit.edu/dissertations/378/
[44] Devayajanam S 2016 Experimental and theoretical evaluation of in-depth damage distribution in sawn silicon wafers *PhD Dissertation* New Jersey Institute of Technology https://digitalcommons.njit.edu/dissertations/1411/
[45] Shet S 2010 Synthesis and characterization of metal oxide semiconductors for photoelectrochemical hydrogen production *PhD Dissertation* New Jersey Institute of Technology https://digitalcommons.njit.edu/cgi/viewcontent.cgi?article=1258&context=dissertations
[46] Singh P and Ravindra N M 2012 Temperature dependence of solar cell performance—an analysis *Sol. Energy Mater. Sol. Cells* **101** 36–45

[47] Kaufman P and Ravindra N M 2024 Protecting solar panels from damage due to overheating *US Patent* 11,876, 484 B2

[48] Lin L and Ravindra N M 2022 CIGS and perovskite solar cells—an overview *Emerg. Mater. Res.* **9** 812–24

[49] Lin L and Ravindra N M 2020 Temperature dependence of CIGS and perovskite solar cell performance—an overview *SN Appl. Sci.* **2** 1361

[50] Tarifa A, Lee E S and Ravindra N M 2024 Opto-electro-thermal simulation of heat transfer in monocrystalline silicon solar cells *Eur. Phys. J. Spec. Top.* **233** 2303–24

[51] Sopori B L, Chen W, Abedrabbo S and Ravindra N M 1998 Modeling emissivity of rough and textured silicon wafers *J. Electron. Mater.* **27** 1341–6

[52] Liu Y *et al* 2023 Machine learning for perovskite solar cells and component materials: key technologies and prospects *Adv. Funct. Mater.* **33** 2214271

[53] Yilmaz B and Yildirim R 2021 Critical review of machine learning applications in perovskite solar research *Nano Energy* **80** 105546

[54] Liu Y *et al* 2022 How machine learning predicts and explains the performance of perovskite solar cells *Sol. RRL* **6** 2101100

[55] International Technology Roadmap for Photovoltaic (ITRPV) 2023 https://vdma.org/international-technology-roadmap-photovoltaic (accessed 19 October 2024); Pickerel K (2023) Downstream players adapt to irregular panel sizes entering all markets *Trends in Solar, Solar Power World* https://solarpowerworldonline.com/2023/01/downstream-players-adapt-to-irregular-panel-sizes-entering-all-markets/

[56] American solar panel manufacturing capacity increases 71% in Q1 2024 as industry reaches 200-Gigawatt milestone https://seia.org/news/american-solar-panel-manufacturing-capacity-increases-71-q1–2024-industry-reaches-200-gigawatt/ (accessed 18 October 2024)

[57] Kennedy R 2023 IREC National solar jobs census, in U.S. solar workforce needs to double in less than a decade *SEIA Analysis Based on Data from Wood Mackenzie* https://pv-magazine-usa.com/2023/03/02/u-s-solar-workforce-to-double-in-less-than-a-decade/, pv magazine

[58] Kennedy R 2024 DYCM Power Announces $800 Million US Solar Cell, Module Factory. *PV Magazine* https://pv-magazine.com/2024/09/18/dycm-power-announces-800-million-us-solar-cell-module-factory/

[59] Short-Term Energy Outlook (2024, December 10), STEO Between the Lines: Small-scale solar accounts for about one-third of U.S. solar power capacity, https://eia.gov/outlooks/steo/report/BTL/2023/09-smallscalesolar/article.php

[60] Annual Review of Solar Energy 1978 Program Evaluation Branch, SERI, Department of Energy, Contract No. EG-77-C-01-4042, SERI/TR-54-066

[61] Department of Energy 2023 *FY 2024, Budget in Brief, FY 2024 Congressional Justification*

[62] *Fact Sheet: Celebrating Two Years of the Inflation Reduction Act* https://usda.gov/media/press-releases/2024/08/16/fact-sheet-celebrating-two-years-inflation-reduction-act (accessed 19 October 2024)

[63] *IRA's Labor Standard Requirements Could Create 3.9 M 'high quality' jobs: Report* https://utilitydive.com/news/inflation-reduction-act-ira-labor-standards-job-creation/724693/ (accessed 19 October 2024)

Acknowledgements

Ravi thanks his doctoral thesis advisors—the late Professors V K Srivastava and Sushil Auluck. He acknowledges with thanks the research groups at the laboratories in Montpellier, Paris and Torino and The Abdus Salam International Center for Theoretical Physics. His lasting association with Drs Bhushan L Sopori, Mowafak Al-Jasim and the research groups at the National Renewable Energy Laboratory (NREL) for over 20 years is acknowledged with sincere thanks and gratitude. His collaborations with Dr Basudev Prasad Saklani, formerly of Bharat Heavy Electricals Limited (BHEL) India; Dr Jochen Rentsch and his team at Fraunhofer-Institut für Solare Energiesysteme ISE, Freiburg, Germany; Dr Anis Jouini, Mr Yvan Trouillot, Mr Laurent Pelissier and team with the ECM Group, Grenoble, France; Dr Steve Rose, Mr Michael Cerone, Mr Todd Sorber and Ms Janet Albrecht at the Passaic County Community College, Paterson, New Jersey; Mr John Conte and his team at Renewable Energy Products Manufacturing, Collingswood, New Jersey; Mr Scott Daniel and Mr Sheldon Fereira at Materium Technologies, Summit, New Jersey; Dr Michael Jaffe at the New Jersey Innovation Institute, Newark, New Jersey; Dr Shuki Yeshurun at Nanopass, Ness Ziona, Israel; Dr John Magno of Magno Fibers, Philadelphia, Pennsylvania; the team at DYCM Power, New York—Mr Sriram Das, Mr Richard Powell, Mr Suryaprakash Jayaramu, Ms Rachel Lucero, Mr Steven Shapiro, Mr Jay Sharma, Ms Kim Monnar, Ms Michelle Hernandez and Mr Venkat Chandolu; Ms Judith Sheft at the New Jersey Commission of Science, Innovation and Technology, Trenton, New Jersey; Dr Anthony T Fiory of AT&T Bell Labs-Retired, Summit, New Jersey; Dr Sufian M Abedrabbo of Khalifa University, Abu Dhabi, UAE; Dr Oktay Gokce at NJIT, Newark, New Jersey; and the support of the administration at NJIT, Newark, New Jersey, Ms Christine Li, Ms Keyannah Watkins, Mr Sanjiv Chokshi, Ms Iris Pantoja, Drs Atam Dhawan, Kevin Belfield, Cristo Ernesto Yáñez León, and Andrew Gerrard; and the contribution of my students is acknowledged with thanks.

The author and his collaborators, are very thankful to Ms Bethany Hext, Ms Betty Barber, Ms Isabelle Defillion, Ms Caroline Mitchell and the publishing team at the Institute of Physics, London, UK.

Author biographies

N M Ravindra

Dr Nuggehalli M. Ravindra (Ravi) is a Professor of Physics at the New Jersey Institute of Technology (NJIT). Previously, Ravi was the Chair of the Department of Physics and Director of the Interdisciplinary Program in Materials Science & Engineering at NJIT. A coauthor of more than 350 papers and seven books, Ravi's research interests are in the areas of education, electronic materials, devices and structures, energy storage, renewable energy, and semiconductors.

Leqi Lin

Ms Leqi Lin is a senior PhD student at Shanghai Jiao Tong University in China. Previously, Leqi obtained her Master's degree in civil and environmental engineering at NJIT and mechanical engineering at State University of New York - Buffalo. Leqi is a coauthor of more than 12 papers in the areas of devices and structures, semiconductors, pharmaceutical manufacturing and machine learning.

Priyanka Singh

Dr Priyanka Singh is a Woman Scientist with the National Physical Laboratory, New Delhi, India. Previously, she worked as a Research Scientist, with Professor N M Ravindra, in the Department of Physics at New Jersey Institute of Technology (NJIT). Priyanka is a coauthor of more than 13 papers in the field of solar cells and semiconductors.

IOP Publishing

N M Ravindra, Leqi Lin and Priyanka Singh

Chapter 1

Introduction

A historical timeline of the discoveries, inventions and applications of solar energy, in a variety of forms is illustrated in this chapter. A brief introduction to the solar radiation, solar thermal energy, photovoltaics, solar cells and the Shockley–Queisser limit is described. The solar cell fundamentals are presented.

1.1 Historical perspective

The utilization of renewable sources of energy is of significant interest today. This is particularly the case due to the growing interest and concern in addressing climate change, carbon footprint and the associated challenges for the environment across the globe, combined with the demands on reduction in dependence on fossil fuels. The various forms of renewable energy of interest are illustrated in figure 1.1 [1]. Additionally, thermoelectric generators, by their inherent ability to convert waste heat to clean electricity, are renewable energy sources [2]. Thermal energy storage via phase change materials is another example of a renewable energy source [3, 4]. Solar driven hydrogen production is yet another case of green synthesis [5]. The US Department of Energy (USDOE) anticipates that, by 2025, US solar energy generation will increase by 75% and wind by 11% [1].

According to a recent research study performed by Rocky Mountain Institute (RMI), in collaboration with Bezos Earth Fund, amongst the various forms of renewable energy, wind, solar and battery deployment are anticipated to produce over a third of global power by 2030 [6].

Since the inception of the ability to utilize natural resources of energy, mankind has looked up to the Sun to utilize its radiation. It is estimated that 'the amount of sunlight arriving at the top of Earth's atmosphere is only one-fourth of the total solar irradiance, or ~340 watts per square meter' [7]. A brief overview of some of the discoveries, practices and methodologies, relating to the use of solar energy, are summarized in table 1.1 [8–27].

doi:10.1088/978-0-7503-5994-8ch1

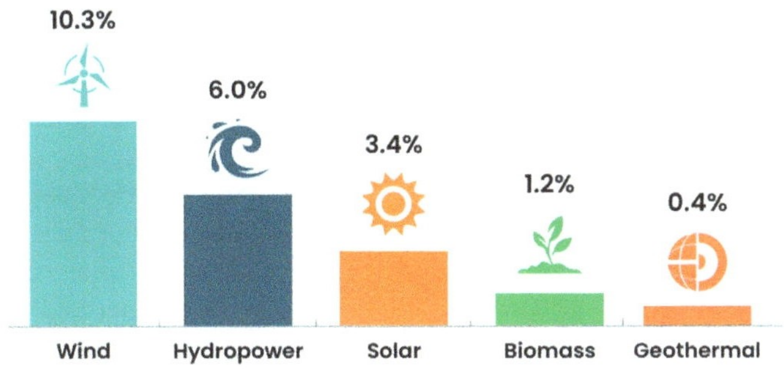

Figure 1.1. Illustration of the various forms of renewable energy. Percentages are based on 2022 data. Reuse with permission from the US Department of Energy [1] credit: NREL.

Table 1.1. Summary of a brief overview of the discoveries and inventions relating to the uses of the solar energy [8–27].

Year	Event	Application
7th Century BC	Use of magnifying glass	Create fire and burn ants
3rd Century BC	Use of burning mirrors	Light torches
2nd Century BC	Use of reflective properties of bronze shields to focus sunlight	Set fire to wooden ships
20 AD	Use of burning mirrors	Light torches for religious purposes
1st–4th Century AD	Use of south-facing windows	Facilitate sun's warmth
6th Century AD	Use of sunrooms	Justinian Code—sun rights to ensure individual access to sun.
1200 AD	Use of winter sun in south-facing cliff dwellings	Anasazi, a member of an ancient North American people of southwestern US, and the present-day Pueblo culture, practiced this methodology
1699	Hot air engine	Guillaume Amontons demonstrated a 'Fire Mill' along the lines of a new thermodynamic cycle
1767	World's first solar oven	Horace-Benedict de Saussure's invention turned light into heat; it could reach temperatures of 230 °F
1816	Invention of the 'Economiser'	Letters Patent No. 4081 of 1816, which later on became the Stirling Engine—that cools the heated air and sends it back to the hot sink to be heated again, in a closed loop system
1839	Discovery of the photovoltaic effect	Edmond Becquerel's demonstration of the photovoltaic effect while experimenting with an electrolytic cell made up of two metal

		electrodes that were placed in an electricity-conducting solution—electricity-generation increased when exposed to light
1866	Construction of the first solar powered engines	Auguste Mouchet proposed the concept of the solar powered steam engines using parabolic troughs to heat water and produce steam. After 20 years, he and his assistant, Abel Pifre, constructed the first solar powered engines and used them for a variety of applications. These engines became the predecessors of the modern parabolic dish collectors
1873	Photoconductivity of selenium	Willoughby Smith's discovery of photoconductivity of selenium catalyzed research in selenium
1876	Discovery of production of electricity in selenium when exposed to light	William Grylls Adams and Richard Evans Day's discovery was instrumental in the demonstration of production of electricity without the need for heat or moving parts
1880	Invention of the bolometer	Samuel P Langley's invention demonstrated the ability to measure light from distant stars and planets by the change in electrical resistance in a wire
1883	Selenium solar cells	Charles Fritts reported the results of the first solar cells made from selenium
1887	Discovery of alteration of the lowest sparking voltage between two metal electrodes	Heinrich Hertz's discovery of the influence of ultraviolet radiation on the sparking voltage
1891	First commercial solar water heater	In US Patent 451,384 Clarence Kemp described the Climax System—the first commercial passive solar water heater
1904	Photosensitivity of copper and cuprous oxide	Wilhelm Hallwachs's discovery of interaction of light with a combination of copper and cuprous oxide
1905	Photoelectric effect	On March 18th 1905, Albert Einstein submitted his paper 'Über einen die Erzeugung *u*nd Verwandlung

(*Continued*)

Table 1.1. (*Continued*)

Year	Event	Application
		des Lichtes betreffenden heuristischen Gesichttspunkt' ('On a Heuristic Viewpoint Concerning the Production and Transformation of Light') to *Annalen der Physik.* For his services to theoretical physics, and especially this paper, he won the 1921 Nobel Prize in Physics
1908	Invention of the solar collector	William Bailey's design of the solar collector had copper coils and an insulated box—it is almost identical to today's design
1914	Existence of a barrier layer in photoelectric devices	Goldman and Brodsky's report of the correlation of the photoelectric effect with the existence of a barrier to current flow at a semiconductor-metal interface helped to provide critical insights that were necessary for building practical photovoltaic devices.
1916	Experimental proof of the photoelectric effect	Robert Millikan's experiments provided the evidence for the photoelectric effect
1917	First experimental reports of phase change materials	Alan T Waterman discovered phase change materials [27]. During his studies of thermionic emission of certain salts, at high temperatures, Waterman noted certain peculiarities in the electrical conductivity of molybdenite
1918	Growth of single-crystal silicon	Jan Czochralski developed a process to grow single-crystal silicon
1932	Photovoltaic effect in CdS	Audobert and Stora discovered photovoltaic effect in CdS
1941	Creation of the solar cell design	Russel Ohl created the solar cell design—U.S. Patent 2 402 662, 'Light sensitive device' that is utilized in many of the modern solar panels
1953	First theoretical calculations of efficiencies of solar cells made from various materials	Dan Trivich reports the first theoretical calculations of efficiencies of solar cells made from various materials—based on their energy gaps

1954	First solar cell Patent—US2780765A—Solar energy converting apparatus	Daryl Chapin, Calvin Fuller, and Gerald Pearson develop the silicon solar cell at Bell Labs with a 4% efficiency—figure 1.3
1955	Commercial licenses	Western Electric began to sell commercial licenses for silicon photovoltaic technologies
1957–60	Hoffman electronics	Development of 8% (1957); 9% (1958); 10% (1959); 14% (1960) efficient cells
1962	First commercial telecommunication satellite	Telstar, launched by Bell Telephone Labs, was the first commercial use of solar cells in space
1966	First astronomical observatory	1 kW photovoltaic array launched by NASA
1976	First amorphous silicon solar cell; U.S. Patent—4 064 521	David Carlson and Christopher Wronski developed the world's first cell with an efficiency of 1.1%
1977	Formation of Solar Energy Research Institute (SERI)	The U.S. Department of Energy launches the Solar Energy Research Institute
1978	World's first village-based solar photovoltaic system	For residents of Schuchuli, Arizona installed with a capacity of 3.5 kW—activated by NASA's Lewis Research Center
1981	Fraunhofer Institute for Solar Energy Systems (ISE)	Fraunhofer ISE is founded by Adolf Goetzberger in Freiburg, Germany. It was the first non-university establishment for applied solar energy research in Europe.
1982	World's first megawatt-sized solar photovoltaic project	Developed by ARCO Solar, began its operation. Covering 20 acres of land and employing dual-axis solar trackers—at Lugo near Hesperia, California
1986	World's largest solar thermal facility, located in Kramer Junction, California, was commissioned.	The solar field contained rows of mirrors that concentrated the sun's energy onto a system of pipes circulating a heat transfer fluid. The heat transfer fluid was used to produce steam, which powered a conventional turbine to generate electricity.

(*Continued*)

Table 1.1. (*Continued*)

Year	Event	Application
1990	Demonstration of >20% efficiency silicon solar cells	University of South Wales reported this result for cells under space illumination
1991	1000 roofs program	Germany's success of this program proved to be a catalyst for increased commitments to solar installations throughout the world
1991	SERI is elevated to NREL	On Sept. 16, President George H W Bush elevates SERI to national laboratory status and renames it the National Renewable Energy Laboratory (NREL) to more accurately reflect its range of work
1992	15.9% efficient CdTe thin film solar cells	University of South Florida reported these results—breaking the efficiency limit of 15%
1994	NREL's first two-terminal monolithic cell to exceed 30% efficiency limit	The cells were made of gallium indium phosphide and gallium arsenide. The efficiency peaked at 30.2% in a concentration range of 140–180 suns
1994	Solar energy research facility	NREL completes the construction of its Solar Energy Research Facility —it was recognized as the most energy-efficient of all U.S. government buildings worldwide. It features not only solar electric system, but also a passive solar design
1994	Solar dish generator tied to a utility grid in Kerman, California	7.5—kilowatt—First solar dish generator, using a free-piston Stirling engine, is tied to a utility grid

1.2 Solar radiation fundamentals

The solar radiation spectrum for direct light at both the top of the Earth's atmosphere (represented by the area in yellow) and at sea level (area in red) is illustrated in figure 1.2 [28]. Approximately 49% of the solar radiation is in the infrared range of wavelengths—700 nm to 1 mm; ~7% is in the ultraviolet range of 100–400 nm; less than 1% represents emission as x-rays, gamma rays and radio waves; 42%–43% lies in the visible range of wavelengths—i.e., 400–700 nm.

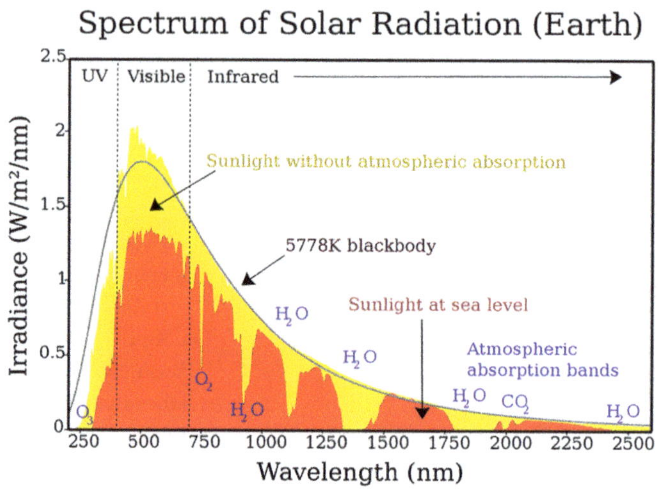

Figure 1.2. The solar radiation spectrum for direct light at both the top of the Earth's atmosphere (represented by the area in yellow) and at sea level (area in red). Reuse with permission from Wikimedia [28]. This Solar Spectrum.png has been obtained by the authors from the Wikimedia website, where it is stated to have been released into the public domain. It is included within this article on that basis.

Table 1.2. A summary of the wavelengths of solar radiation [29].

Short-wave	Visible	400–780 nm	Visible light
Long-wave (IR)	Near infrared	780 nm–3 μm	Heat radiation from the Sun: Violet, Indigo, Blue, Green, Yellow, Orange, Red
Long-wave (IR)	Far infrared	3–50 μm	Heat radiation from the atmosphere, clouds, earth and surroundings

Due to the significant component of solar radiation in the infrared range of wavelengths, numerous applications, that take advantage of this heat, have been developed over several centuries. These include solar thermal collectors, solar space heating and cooling as well as electric power generation by utilizing high temperature collectors that operate using mirrors or lenses. Agriculture and water desalination are beneficiaries of solar thermal energy. In this context, it must be noted that thermoelectrics and phase change materials offer significant potential for efficient power generation. A summary of the wavelengths of radiation is presented in table 1.2 [29].

1.3 Photovoltaic effect

The photovoltaic (PV) effect is the process of direct conversion from light to electricity in a solar cell system. The photovoltaic effect was first discovered in

1839 by the French physicist, Edmond Becquerel, while investigating the electronic behavior of metal electrodes in electrolytes. The process includes the generation of charge carriers from the absorption of photons in the materials that form a junction, the separation of the photogenerated charge carriers in the junction and the collection of the photogenerated charge carriers at the terminals of the junction [30].

In recent years, with the advancement in technologies in manufacturing, characterization techniques and the advent of nanomaterials, several solar cells based on various semiconductor materials, such as gallium arsenide (GaAs), indium phosphide (InP), copper indium gallium sulfur selenide (CIGS) and cadmium telluride (CdTe), have been commercialized and are being utilized in various applications. In typical p-n (silicon) homojunction solar cells or the first-generation solar cells, they can be further categorized into single, multi, micro and poly crystalline cells. Thin film solar cells including CIGS, GaAs, CdTe and amorphous silicon are termed as the second generation. Among them, amorphous silicon solar cells hold significant potential for rapid growth due to their low cost of production, light weight and scalability. Currently, the third generation PV cells such as perovskite-based, organic materials such as polymers, dye-sensitive solar cells are appealing to researchers and companies due to the high power conversion efficiency (*PCE*) in addressing the future energy needs and environmental concerns as well as the use of flexible substrates for applications in portable electronics [31]. The materials that would reduce the production costs are termed as the third-generation solar cells.

Since the report of the first solar cell in 1954 by Daryl Chapin, Calvin Fuller, and Gerald Pearson at Bell Laboratories [32] (figure 1.3), the solar cell industry has seen an exponential growth. An illustration of the growth of global solar installation is presented in figure 1.4 [33].

Figure 1.3. The first practical p-n silicon solar cell demonstration by a research team working at Bell Laboratories. Reuse with permission from Bell Laboratories [32]. Copyright MCMXCVII - MMXXIII AGB 1847, Inc. All rights reserved.

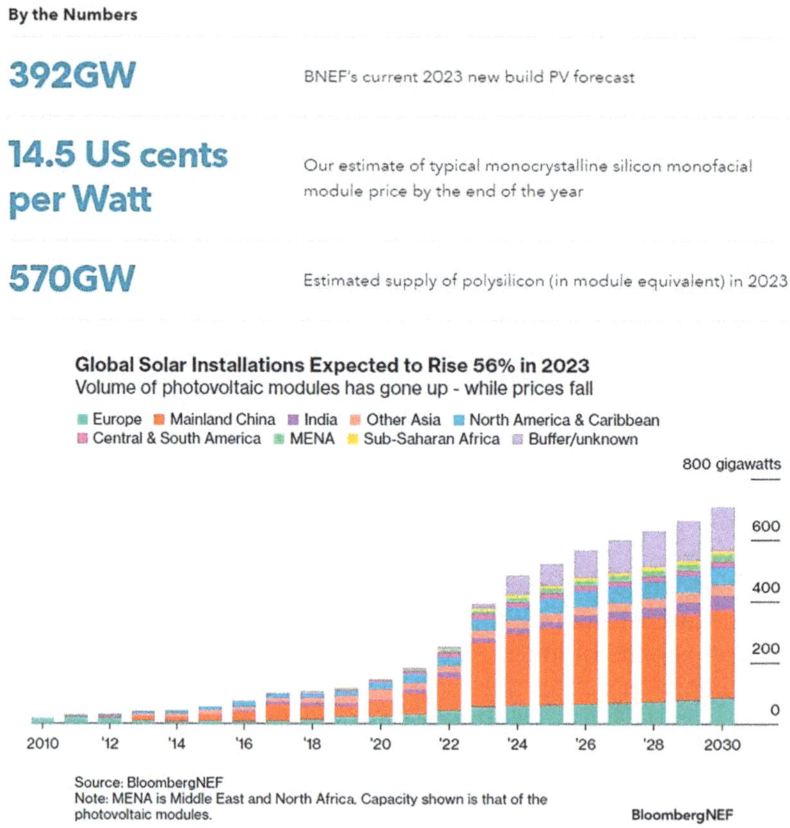

Figure 1.4. Global growth of photovoltaics. Reuse with permission from BNEF [33]. Copyright 2024 Bloomberg Finance L.P. All rights reserved.

1.4 Solar cell fundamentals

Solar cells convert sunlight into electricity via the PV effect. The generation of voltage and current occurs upon the absorption of photons in the semiconductor diode.

The basic structure of a solar cell is a p-n junction, illustrated in figure 1.5.

A p-n junction is formed in a semiconductor by processes such as diffusion or ion-implantation, followed by subsequent annealing for dopant activation. A group IV semiconductor, such as silicon or germanium, becomes p-type when a trivalent impurity, such as boron or gallium, is added (doped) to (into) an intrinsic or pure semiconductor. Similarly, n-type semiconductor is formed by diffusing an impurity from group V of the Periodic Table such as phosphorous or arsenic. When sunlight interacts with the cell, nearly free electrons that are present in the valence band (VB) will absorb energy and get excited to the conduction band (CB), leaving behind a hole in the VB, as shown in figure 1.5. This hole can also move, but in the opposite direction to the p-side, which generates an electromotive force and an electric current, and thus some of the light energy is converted into electric energy.

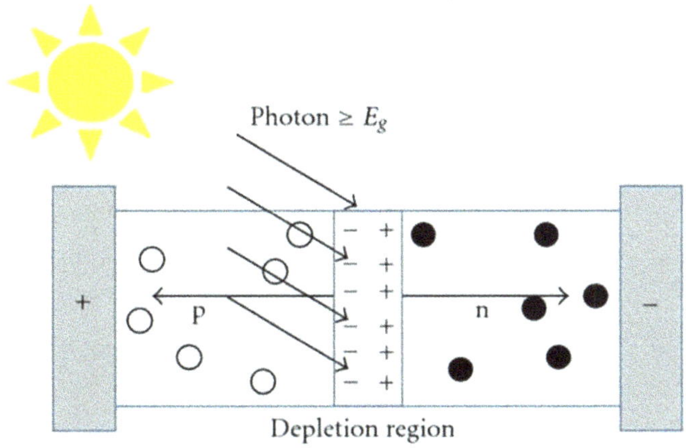

Figure 1.5. Illustration of a p-n junction diode through which the current flows and a voltage is generated when photons interact with the p-n junction. Reuse with permission from Hindwai [34] CC BY 4.0.

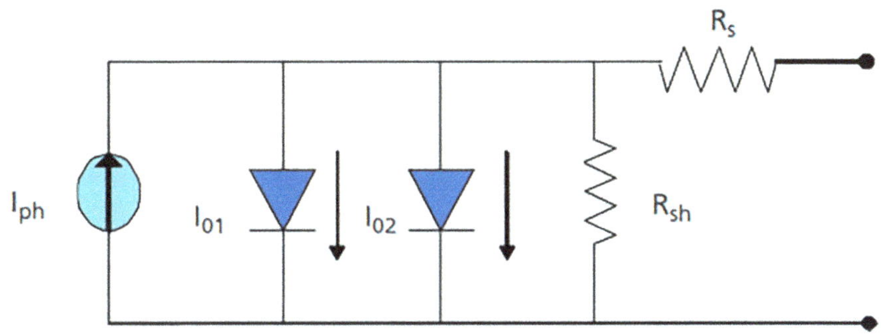

Figure 1.6. Schematic diagram of a solar cell described by a two-diode model. I_{ph} is the photogenerated current, I_{01} represents the diffusion limited current, I_{02} is the generation-recombination term, R_s is the series resistance, R_{sh} is the shunt resistance. Reuse with permission from N M Ravindra [35].

The amount of energy produced by the PV cells depend on the incident light energy; this will determine the conversion efficiency of the device. The incident light induced electron–hole pairs are separated via a load that is connected to the cell, leading to the generation of power in the cell via the PV effect.

Schematic diagram of a solar cell, described by a two-diode model, is presented in figure 1.6 [35].

Examples of the current–voltage characteristics of some silicon solar cells are presented in figure 1.7 [35].

The fundamental equations governing the performance of solar cells have been described in a large number of studies in the literature [35–42]. In a previous study, Singh and Ravindra [36] have investigated the temperature dependence of Si, Ge, GaAs, InP, CdTe and CdS solar cells. The first detailed studies of the comparison of solar cell efficiency η versus energy gap E_g for various semiconductors, in the temperature range of 273 K–673 K, by Wysocki and Rappaport, was reported over

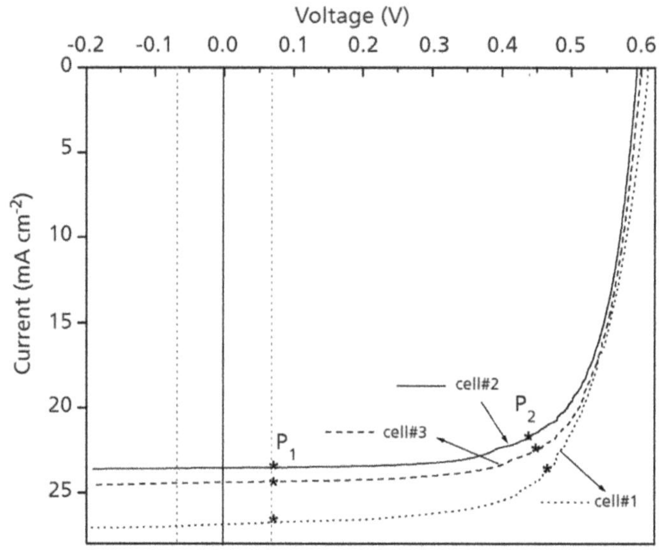

Figure 1.7. Current (I)–Voltage (V) curves without bias (fourth quadrant) and with reverse bias (third quadrant) at room temperature (25 °C) under a simulated air mass 1.5 solar irradiance of 110 mWcm^{-2} intensity for the cells (Cell1, Cell 2 and Cell 3). Reuse with permission from N M Ravindra [35].

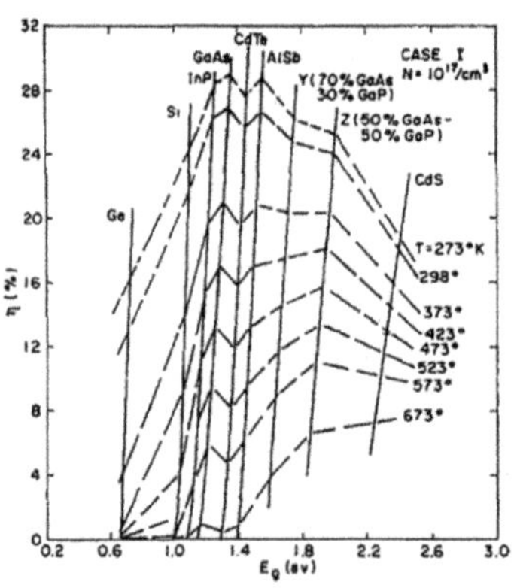

Figure 1.8. Efficiency versus E_g for various semiconductors in the temperature range of 273 K–673 K. Reprinted with permission from AIP Publishing [44].

60 years ago (figure 1.8) [44]. It is interesting and relevant to note that while there have been significant improvements in solar cell materials and device technology, not much has changed in relation to the absolute solar cell efficiencies as function of temperature.

In general, during real-time utility and applications in the field, solar cells/solar panels operate at lower or higher temperature than in the laboratory under standard temperature conditions (STC), and predominantly under varying light intensity; therefore, it is necessary to realize the relationship between solar cell parameters and temperature. The performance of solar cells is determined by the key PV parameters, for example, short-circuit current density (J_{sc}), open circuit voltage (V_{oc}), fill factor (FF) and PCE. These parameters are temperature dependent and hence affect the performance of solar cells [36, 44–47]. V_{oc} decreases with increasing temperature whereas J_{sc} increases slightly with increasing temperature [36, 43, 45–48]. Both FF and PCE decrease with increasing temperature and efficiency degradation is mainly due to the decrease in V_{oc} [36, 44–48]. The variation of R_s and R_{sh} with temperature slightly affects the efficiency [36, 44–48], while with increasing temperature, J_o increases exponentially and V_{oc} decreases rapidly. Therefore, J_o is a critical PV parameter in determining the PCE of solar cells.

The temperature dependence of the bandgap (E_g) of semiconductors is described by the well-known Varshni relation [48] in equation (1.1),

$$E_g(T) = E_g(0) - \frac{\alpha T^2}{(T + \beta)} \tag{1.1}$$

where, $E_g(T)$ is the bandgap of the semiconductor at temperature T, E_g (0) is its value at 0 K, and α and β are constants that are characteristic of the material.

As shown in equation (1.2), J_{sc} depends on the solar spectral irradiance; the value of J_{sc} may be limited by reflection losses (series and shunt resistance), shadowing losses (front metal coverage) and recombination losses [36].

$$J_{sc} = q\int_{hv=E_g}^{\infty} \frac{dN_{ph}}{dhv}\, d(hv) \tag{1.2}$$

V_{oc} is the maximum voltage of a solar cell; by setting $J = 0$, the expression for V_{oc} is given by equation (1.3). The temperature dependence of V_{oc} can be further obtained from equation (1.3) as shown in equation (1.4).

$$V_{oc} = \frac{kT}{q}\ln\left(\frac{J_{sc}}{J_o} + 1\right) \tag{1.3}$$

where, (kT/q) is the thermal voltage and J_o is the reverse saturation current density; it is a measure of the leakage (or recombination) of minority carriers across the solar cell in reverse bias.

$$\frac{dV_{oc}}{dT} = \frac{V_{oc}}{T} + V_{th}\left(\ln\frac{1}{J_{sc}}\frac{dJ_{sc}}{dT} - \frac{1}{J_o}\frac{dJ_o}{dT}\right) \tag{1.4}$$

where, $V_{th} = (kT/q)$.

For a solar cell, J_o has been modeled [49, 50] as in equation (1.5).

$$J_o = q\left(\frac{D_n}{L_n N_A} + \frac{D_p}{L_p N_D}\right)n_i^{\,2} \tag{1.5}$$

where, n_i is the intrinsic carrier density, N_A and N_D are the densities of acceptor and donor atoms, D_n and D_p are diffusion constants of minority carriers in p and n regions, respectively. As seen from equation (1.5), J_o is strongly determined by the proportionality to $\sim n_i^2$ and n_i can be represented as in equation (1.6).

$$n_i{}^2 = N_c N_v \exp\left(-\frac{E_g}{kT}\right) = 4 \times \left(\frac{2\pi kT}{h^2}\right)^3 m_e \tfrac{3}{2} m_h \tfrac{3}{2} \exp\left(-\frac{E_g}{kT}\right) \quad (1.6)$$

where, N_c, N_v are the effective density of states in the conduction band, valence band and m_e, m_h are the effective mass of electron, hole, respectively. By combining equation (1.5) and equation (1.6), the expression for J_o can be written in terms of temperature and bandgap energy [51] as in equation (1.7).

$$J_o = C \times T^3 \exp\left(-\frac{E_g}{kT}\right) \quad (1.7)$$

FF is defined as the ratio of the maximum power output ($P_{\max}$) at the maximum power point to product of V_{oc} and J_{sc} and can be expressed as in equation (1.8).

$$FF = \frac{P_{\max}}{V_{oc} J_{sc}} \quad (1.8)$$

The efficiency of a solar cell is the ratio of the power output, corresponding to the maximum power point, to the power input and is represented as in equation (1.9).

$$PCE = \frac{V_{oc} J_{sc} FF}{P_{in}} \quad (1.9)$$

where, P_{in} is the intensity of the incident radiation. At each temperature, corresponding to the calculated V_{oc}, J_{sc} and FF, PCE is calculated using equation (1.9). Under the influence of solar concentrators, the open circuit voltage V'_{oc} in a solar cell is given by:

$$V'_{oc} = \frac{nkT}{q} \ln\left(\frac{XI_{sc}}{I_o}\right) = V_{oc} + \frac{nkT}{q} \ln X \quad (1.10)$$

where X is the concentration of sunlight [52]. The open circuit voltage, under illumination, can be described by the equation:

$$V_{oc} \approx V_{ocn} + \frac{nkT}{q} \ln\left(\frac{E}{E_n}\right) \quad (1.11)$$

where, V_{ocn} and E_n are the open circuit voltage and the irradiation under nominal conditions, respectively [53].

As shown in figure 1.9, the National Renewable Energy Laboratory (NREL) is one of the most well established laboratories in the world. It provides certification and calibration for the PV devices with the latest record of PV systems annually [54]. Up to this year, 2023, 47.6% of PCE has been the highest record achieved with the structure of four junction cells with concentrator from the Fraunhofer ISE and 35.5% of PCE in two junction cells with concentrator. A 27.6% of PCE has been

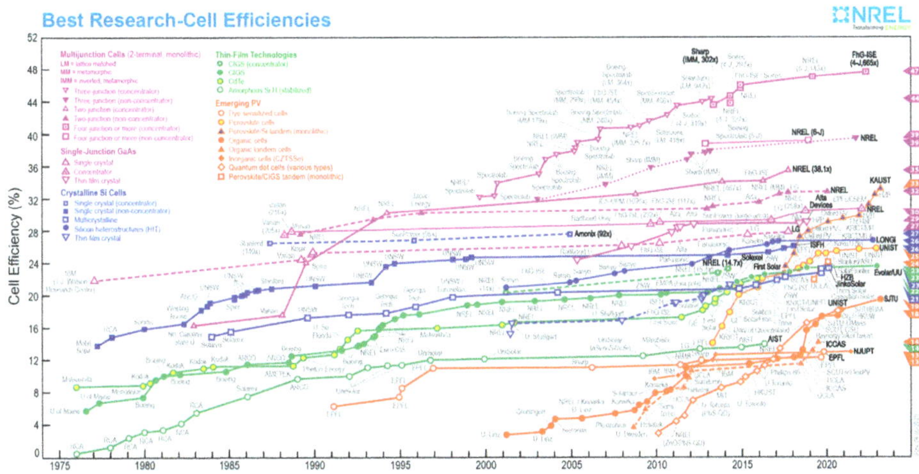

Figure 1.9. The latest best research-cell efficiencies from research labs and companies from around the world in 2023—reported from NREL. Rprinted with permission from [54]. Credit: NREL. This plot is courtesy of the National Renewable Energy Laboratory, Golden, CO.

achieved with the structure of crystalline silicon cells with concentrator from LONGi, China. In thin-film types of solar cells, 23.6% of *PCE* has been achieved with the structure of CIGS without concentrator from the Evolar/UU, 22.1% of *PCE* in CdTe from First Solar and 14.0% of *PCE* for amorphous Si from AIST. In emerging PV, with the most rapid development, 25.8% of *PCE* is achieved in perovskite cells from UNIST, 19.2% of *PCE* in organic cells from SJTU and 33.2% of *PCE* in perovskite/Si tandem solar cells from KAUST. It may be noted from figure 1.9 that cell efficiencies have grown from less than 10% of *PCE* in the early stage to now close to 50% of *PCE* due to novel smart materials, advancements in nanotechnologies and high-precision, repeatability and reproducibility in manufacturing.

The *PCE* of solar cells is very sensitive to the environmental factors such as temperature, humidity, irradiance, and mechanical stimulation from the surroundings, which may lead to disparity from the reported values. Most of the PV cells are reporting less than 40% of conversion efficiency from sunlight after losses due to heat [55]. In order to overcome this low efficiency, instead of directly utilizing the solar to electricity by using PV panels, concentrated solar power (CSP) concentrates sunlight to produce electricity. However, the heat dissipation is still the bottleneck problem affecting the material and the cell, which further restricts large-scale development. With more focus from both industry and universities and comprehensive policies towards the PV and CSP systems, it is foreseeable to have over 50% of *PCE* commercially available in the near future to solve the energy and environmental issues around the world. The merits of solar cells are obvious with the advantage of being a quiet, clean and renewable source of energy. Moreover, under the low-carbon emission policy in many countries, the efficient and widespread use of solar cells is imperative and necessary. Ideally, the PV and CSP systems would employ solar panels with very low production costs, high efficiency and a good design, enabling easy and low-cost installation [56].

1.5 Shockley–Queisser limit

The conversion of solar energy to electricity for solar cells assuming complete utilization of sunlight, which is similar to the Carnot efficiency, is described by equation (1.12). This consideration is expanded to include solar radiation and the cell circuitry as well as to take into account some empirical quantities (such as the charge carrier lifetime, for instance) [57]. In 1961, the radiative efficiency limit (also known as the detailed balance limit) was defined as the ultimate efficiency for any p-n junction using PV process wherein the main dominant factor is the radiative recombination. This limit depends on the energy bandgap of semiconductors and certain geometrical factors of devices such as the angle subtended by the sun, the angle of incidence towards the radiation, and certain degrading factors such as the absorption coefficient for solar energy striking the surface [58]. The first calculation in 1961 with silicon (bandgap of 1.1 eV) with 6000 K black-body spectrum and 300 K as the temperature of the cell, yielded a maximum 30% efficiency [59]. It is worth mentioning that this Shockley–Queisser (S–Q) limit is applicable to single junction only. Comparison of the 'semiempirical limit' of efficiency of solar cells with the 'detailed balance limit', derived by Shockley and Queisser, is presented in figure 1.10.

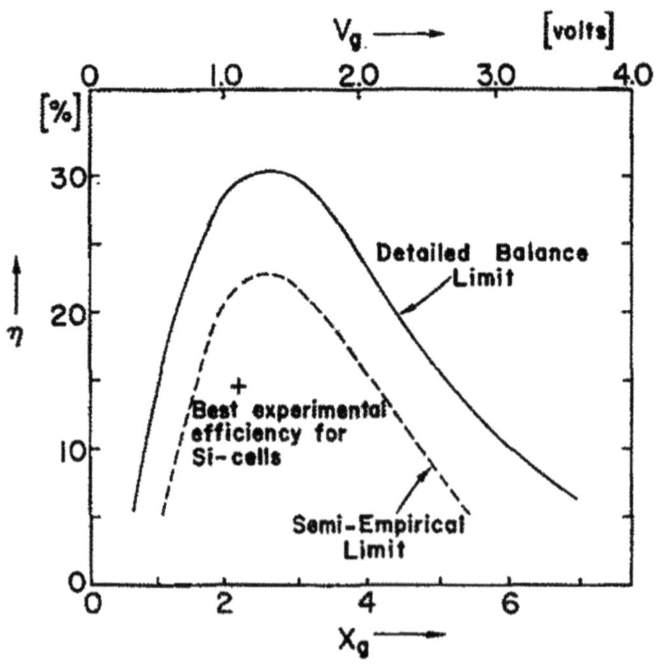

Figure 1.10. Comparison of the 'semiempirical limit' of efficiency of solar cells with the 'detailed balance limit', derived by Shockley and Queisser. + represents the 'best efficiency to date' for silicon cells. Reprinted with permission from [59] AIP Publishing.

The Shockley–Queisser detailed balance limit simulation proceeds as follows:

$$\frac{T_s - T_c}{T_c} \tag{1.12}$$

where T_s is the temperature of solar and T_c is the temperature of solar cell.

To a very good estimation, this upper limit efficiency is a function of some variables such as temperature, voltages or frequencies and the efficiency of radiative recombination (a fraction f_c of all the recombination). Here the relationship of these variables and energy is discussed.

$$kT_s = qV_s \tag{1.13}$$

$$kT_c = qV_c \tag{1.14}$$

$$E_g = hv = qV_g \tag{1.15}$$

where, k is the Boltzmann's constant, q is the electronic charge and h is Planck's constant.

Thus, the efficiencies are calculated by utilizing the following methodology:

$$x_g = \frac{E_g}{kT_s} \tag{1.16}$$

$$x_c = \frac{T_c}{T_s} \tag{1.17}$$

The efficiency also depends on t_s, which is defined as the probability that a photon with $hv > E_g$ that is incident on the surface will produce an electron–hole pair [59].

The four most important parameters that define the efficiency of PV cells are: the short-circuit current I_{sc} (corresponding to the maximum electric current generated by the solar cell), the open circuit voltage V_{oc} (maximum voltage of the cell), the fill factor FF (ratio between the maximum power $P_{\max}$, and the efficiency η (defined as the ratio of the electrical power output to the total incoming sunlight power P_{Sun}), in which these parameters are characterized and determined from the experiments under standard solar radiation condition.

Figure 1.11(a) provides the Shockley–Queisser limit of several semiconductor PV cells and the record efficiency in 2016–20 [60]. Figure 1.11(b) presents a summary of the current density relative to the maximum possible current density, under standardized AM1.5 illumination conditions, of several semiconductor PV cells.

1.6 Conclusions

An overview of the historical developments relating to the use of solar energy was presented. In this context, some of the applications of solar thermal energy were

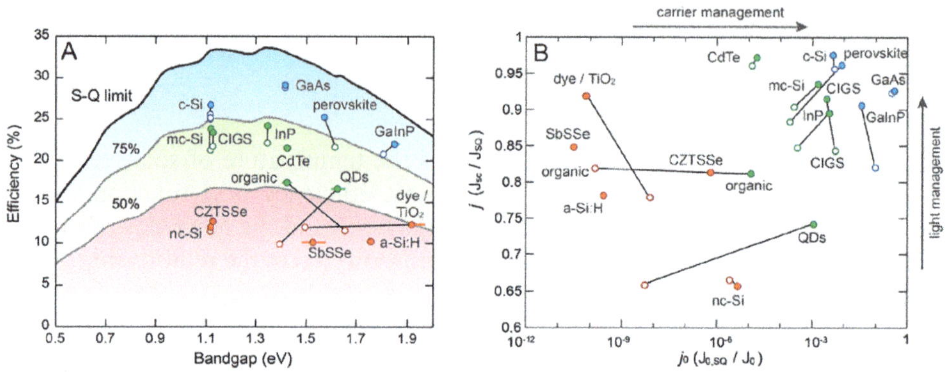

Figure 1.11. (a) Record efficiency of solar cells of different semiconductors with respect to their bandgaps, in comparison to the SQ limit (top solid line). (b) Current density relative to the maximum possible current density, under standardized AM1.5 illumination conditions, versus minimum dark recombination current density relative to the recombination current derived for the record cells in panel A. The open symbols show the record efficiency in April 2016, the solid symbols show the numbers in July 2020. Reprinted with permission from [60], copyright (2020) American Chemical Society.

highlighted. The fundamentals of photovoltaics and the equations governing the operation of solar cells were summarized.

This chapter was reproduced with permission from [42].

Notation

J_{sc}	Short-circuit current density
FF	Fill factor
V_{oc}	Open circuit voltage
PCE	Power conversion efficiency
R_s	Series resistance
R_{sh}	Shunt resistance
α	Constant, characteristic of material
β	Constant, characteristic of material
T	Temperature
k	Boltzmann constant
J_o	Reverse saturation current density
J_{ph}	Photogenerated current density
q	Electric charge
D_n	Diffusion constant of minority carriers in n- region
D_p	Diffusion constant of minority carriers in p- region
L_n	Diffusion length of minority carriers in n- region
L_p	Diffusion length of minority carriers in p- region
N_A	Density of acceptor atoms
N_D	Density of donor atoms
n_i	Intrinsic carrier density
V_{th}	$V_{th} = kT/q$
N_c	Effective density of states in conduction band

N_v	Effective density of states in the valence band
m_e	Effective mass of electron
m_h	Effective mass of hole
C	Constant of combined doping and materials parameters of solar cells
$P_{\max}$	Maximum power output at maximum power point
P_{in}	Intensity of incident radiation
η	Power conversion efficiency
X	Concentration of sunlight
E	Photon energy
E_n	Irradiation under nominal condition
$E_g(T)$	Energy bandgap of semiconductor at temperature T
V_{ocn}	Open circuit voltage under nominal condition
STC	Standard test condition
h	Planck's constant
v	Frequency of light in vacuum

References

[1] https://energy.gov/eere/renewable-energy (Accessed 17 December 2024)

[2] Ravindra N M, Jariwala B, Bañobre A and Maske A 2019 *Thermoelectrics Fundamentals, Materials Selection, Properties and Performance* (New York: Springer Briefs in Materials): Springer)

[3] Barbi S *et al* 2022 Phase change material evolution in thermal energy storage systems for the building sector, with a focus on ground-coupled heat pumps *Polymers* **14** 620

[4] Jo H, Joo Y and Kim D 2023 Thermal design of solar thermoelectric generator with phase change material for timely and efficient power generation *Energy* **263** 125604

[5] Fehr A M K *et al* 2023 Integrated halide perovskite photoelectrochemical cells with solar-driven water-splitting efficiency of 20.8% *Nat. Commun.* **14** 3797

[6] https://rmi.org/press-release/renewable-energy-deployment-puts-global-power-system-on-track-for-ambitious-net-zero-pathway/ (Accessed 14 October 2023)

[7] https://earthobservatory.nasa.gov/features/EnergyBalance/page2.php (Accessed 15 October 2023)

[8] https://www1.eere.energy.gov/solar/pdfs/solar_timeline.pdf (Accessed 15 October 2023)

[9] https://urbansolar.com/the-history-of-solar-power/ (Accessed 15 October 2023)

[10] https://en.wikipedia.org/wiki/Guillaume_Amontons (Accessed 15 October 2023)

[11] https://jstor.org/stable/4025200 (Accessed 15 October 2023)

[12] https://lindahall.org/about/news/scientist-of-the-day/robert-stirling/ (Accessed 15 October 2023)

[13] http://hotairengines.org/closed-cycle-engine/stirling-1816 (Accessed 15 October 2023)

[14] https://google.com/search ? channel = ftr&client = firefox-b-1-d&q = Clarence + Kemp + patent + water + heater (Accessed 15 October 2023)

[15] https://purevolt.ie/resources/solar/history-of-solar.php (Accessed 15 October 2023)

[16] https://en.wikipedia.org/wiki/Russell_Ohl (Accessed 15 October 2023)

[17] https://jlanka.com/solar-photovoltaics-the-best-solution-to-reverse-the-climate-change/

[18] Raval N and Gupta A K 2015 Historic developments, current technologies and potential of nanotechnology to develop next generation solar cells with improved efficiency *Int. J. Renew. Energy Dev.* **4** 77–93

[19] https://nasa.gov/history/nasa-lewis-designed-the-first-solar-electric-village/ (Accessed 15 October 2023)

[20] https://google.com/search ? channel = ftr&client = firefox-b-1-d&q = World%E2%80%99s + first + megawatt-sized + solar + photovoltaic + project (Accessed 15 October 2023)

[21] https://google.com/search ? channel = ftr&client = firefox-b-1-d&q = The + 1000-roof + program + in + Germany + was + launched + (Accessed 15 October 2023)

[22] https://kfw.de/kfw.de.html (Accessed 15 October 2023)

[23] https://patents.google.com/patent/US2780765A/en (Accessed 15 October 2023)

[24] https://ae-solar.com/history-of-solar-module/ (Accessed 15 October 2023)

[25] https://nrel.gov/about/history.html (Accessed 15 October 2023)

[26] https://ise.fraunhofer.de/en.html (Accessed 15 October 2023)

[27] Waterman A T 1917 On the positive ionization from certain hot salts, together with some observations on the electrical properties of molybdenite at high temperatures *Philos. Mag. J. Sci.* **33** 225

[28] Rohde R A 2008 https://commons.wikimedia.org/wiki/File:Solar_Spectrum.png

[29] https://kippzonen.com/Knowledge-Center/Theoretical-info/Solar-Radiation (Accessed 15 October 2023

[30] landis G A *et al* 2004 High-temperature solar cell development *19th European Photovoltaic Science and Engineering Conf. (Paris, June 7–11)*

[31] Dambhare M V, Butey B and Moharil S V 2021 Solar photovoltaic technology: a review of different types of solar cells and its future trends *J. Phys. Conf. Ser.* **1913** 012053

[32] Bell Laboratories B 1954 *Bell Labs Engineer Testing Solar Battery in 1954* (Nokia Bell Labs)

[33] https://about.bnef.com/blog/3q-2023-global-pv-market-outlook/ (Accessed 14 October 2023)

[34] Gupta N, Alapatt G, Podila R, Singh R and Poole K F 2009 Prospects of nanostructure-based solar cells for manufacturing future generations of photovoltaic modules *Int. J. Photoenergy* **2009** 154059

[35] Singh P and Ravindra N M 2011 Analysis of series and shunt resistance in silicon solar cells using single and double exponential models *Emerg. Mater. Res.* **1** 33–8

[36] Singh P and Ravindra N M 2012 Temperature dependence of solar cell performance—an analysis *Sol. Energy Mater. Sol. Cells* **101** 36–45

[37] Riesen Y, Stuckelberger M, Haug F J, Ballif C and Wyrsch N 2016 Temperature dependence of hydrogenated amorphous silicon solar cell performances *J. Appl. Phys.* **119** 044505

[38] Mesrane A, Mahrane A, Rahmoune F and Oulebsir A 2017 Temperature dependence of InGaN dual-junction solar cell *J. Electron. Mater.* **46** 2451–9

[39] Steinkemper H, Geisemeyer I, Schubert M C, Warta W and Glunz S W 2017 Temperature-dependent modeling of silicon solar cells—eg, N I, recombination, and VOC *IEEE J. Photovolt.* **7** 450–7

[40] Bandara T M W J, Fernando H D N S, Furlani M, Albinsson I, Dissanayake M A K L, Ratnasekera J L *et al* 2017 Dependence of solar cell performance on the nature of alkaline counterion in gel polymer electrolytes containing binary iodides *J. Solid State Electrochem.* **21** 1571–8

[41] Lin L and Ravindra N M 2020 Temperature dependence of CIGS and perovskite solar cell performance; an overview *SN Appl. Sci.* **2** 1361

[42] Lin L and Ravindra N M 2020 CIGS and perovskite solar cells—an overview *Emerg. Mater. Res.* **9** 812–24

[43] Fortunato E, Gaspar D, Duarte P, Pereira L, Águas H, Vicente A *et al* 2016 Optoelectronic devices from bacterial nanocellulose *Bacterial Nanocellulose* ed M Gama, F Dourado and S Bielecki (Amsterdam: Elsevier) ch 11 pp 179–97

[44] Wysocki J J and Rappaport P 1960 Effect of temperature on photovoltaic solar energy conversion *J. Appl. Phys.* **31** 571–8

[45] Fan J C C 1986 Theoretical temperature dependence of solar cell parameters *Sol. Cells* **17** 309–15

[46] Singh P, Singh S N, Lal M and Husain M 2008 Temperature dependence of I–V characteristics and performance parameters of silicon solar cell *Sol. Energy Mater. Sol. Cells* **92** 1611–6

[47] Nakada T 2003 ZnO/ZnS(O,OH)/Cu(In,Ga)Se_2/Mo solar cell with 18.6% efficiency *Material Science: Proc. of the Third World Conf. of Photovoltaic Energy Conversion* M Hongo

[48] Varshni Y P 1967 Temperature dependence of the energy gap in semiconductors *Physica* **34** 149–54

[49] Hu C and White R M 1983 *Solar Cells* (New York: McGraw-Hill) 21

[50] Nell M E and Barnett A M 1987 The spectral p-n junction model for tandem solar-cell design *IEEE Trans. Electron Devices* **34** 257–66

[51] Chen W S *et al* 1993 Thin film $CuIn_{1-x}Ga_xSe$ cell development *23rd IEEE Photovoltaic Conf. (Louisville, KY)*

[52] Wang X, Li S S, Kim W K, Yoon S, Craciun V, Howard J M *et al* 2006 Investigation of rapid thermal annealing on Cu(In,Ga)Se_2 films and solar cells *Sol. Energy Mater. Sol. Cells* **90** 2855–66

[53] Repins I, Contreras M A, Egaas B, DeHart C, Scharf J, Perkins C L *et al* 2008 19·9%-efficient ZnO/CdS/CuInGaSe_2 solar cell with 81·2% fill factor *Prog. Photovolt. Res. Appl.* **16** 235–9

[54] *The National Renewable Energy Laboratory (NREL)* 2023 *Is Operated for the U.S. Department of Energy (DOE) by Alliance for Sustainable Energy* (L. A. Best Research-Cell Efficiencies) https://nrel.gov/pv/module-efficiency.html

[55] Haoliang B A I, W. C and Jing L U 2023 Solar cell heat dissipation technology and development status of concentrating photovoltaic system *Chem. Ind. Eng. Prog.* **42** 159–77

[56] Edoff M 2012 Thin film solar cells: research in an industrial perspective *Ambio* **41** 112–8

[57] Zanatta A R 2022 The Shockley–Queisser limit and the conversion efficiency of silicon-based solar cells *Res. Optics* **9** 100320

[58] Katagiri H, Jimbo K, Yamada S, Kamimura T and Motohiro T 2008 Detailed balance limit of efficiency of p-n junction solar cells *Appl. Phys. Express* **32** 510

[59] Shockley W and Queisser H J 1960 Detailed balance limit of efficiency of p-n junction solar cells *J. Appl. Phys.* **32** 510

[60] Ehrler B *et al* 2020 Photovoltaics reaching for the Shockley–Queisser limit *ACS Energy Lett.* **5** 3029–33

Chapter 2

Material considerations

The selection of material candidates for solar cell manufacture is the most fundamental and critical step in photovoltaics. It directly determines the efficiency and performance characteristics of the solar cell to be able to convert the incident sunlight into useful electricity. Materials with suitable bandgaps, high absorption coefficients, low reflectance, high carrier mobility, chemical, structural and thermal stability and minimal defects, that are cost-effective and abundantly available, are sought for enhanced device performance, reproducibility and reliability. Advancements in material options and bandgap engineering have led to breakthroughs in solar cell efficiency, with tandem and multi-junction/heterojunction cells offering improved performance by utilizing multiple materials that better absorb various regions of the solar spectrum.

2.1 Material fundamentals

The ability of a semiconductor to absorb the incident sunlight to create electron–hole pairs and the subsequent separation of the light-induced electrons and holes via a built-in electric field in a device, such as p-n junction, forms the basis of a solar cell. This ability of the semiconductor to absorb the incident solar photons is primarily determined by its energy gap and the nature of the energy gap (direct or indirect). Examples of semiconductors of interest to photovoltaics, and their respective energy gap, are presented in table 2.1 [1–5].

An illustration of the direct bandgap semiconductor, such as indium phosphide or gallium arsenide, and indirect bandgap semiconductor, such as germanium, silicon or gallium phosphide, is presented in figure 2.1 [6].

The absorption of light in a direct bandgap semiconductor can be summarized as follows:

$$hv = E_{g-\text{direct}} \tag{2.1}$$

where, $E_{g-\text{direct}}$ is the direct bandgap of the semiconductor, h is the Planck Constant ($h = 6.626 \times 10^{-34}$ J s^{-1}) and v is the frequency of the incident photon.

doi:10.1088/978-0-7503-5994-8ch2

Table 2.1. Examples of semiconductors of interest to photovoltaics [1–5].

Semiconductor	Energy gap (eV) 0 K	Energy gap (eV) (300 K)	Nature of bandgap
Ge	0.74	0.66	Indirect
Si	1.17	1.11	Indirect
InP	1.42	1.27	Direct
GaAs	1.52	1.43	Direct
GaP	2.32	2.25	Indirect
GaSb	0.73	0.68	Indirect
CdS	2.53	2.42	Direct
CdTe	1.61	1.44	Direct
CdSe	1.84	1.74	Direct
$CuIn_{1-x}Ga_xSe_2$ (CIGS)	—	1.0 ($CuInSe_2$)–1.7 ($CuGaSe_2$)	Direct
ZnSe	2.80	2.70	Direct
ZnO	3.37	3.20	Direct
PbS	0.29	0.42	Direct

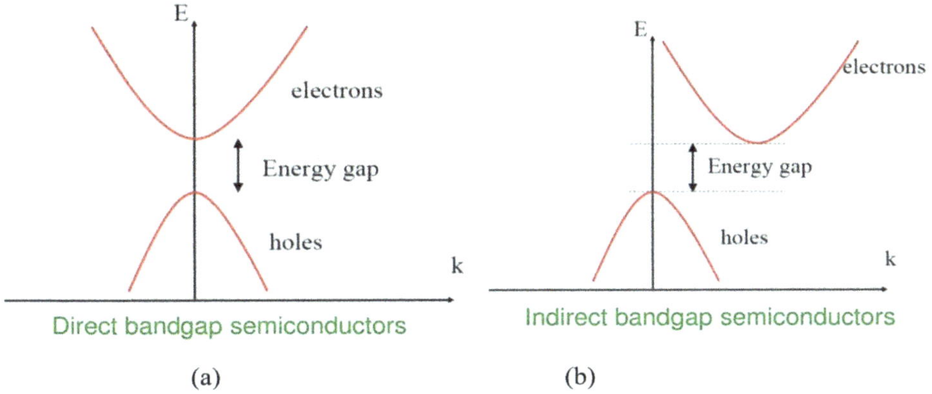

Figure 2.1. Illustration of (a) direct bandgap and (b) indirect bandgap semiconductors. Reproduced from [6] CC BY 4.0.

For an indirect bandgap semiconductor, the absorption of light in the semiconductor is given by:

$$hv = E_{g-\text{indirect}} + E_{\text{phonon}} \tag{2.2}$$

where, $E_{g-\text{indirect}}$ is the indirect bandgap of the semiconductor and E_{phonon} represents the phonon energy.

An example of the bandgap engineering in Si_xGe_{1-x} and $Al_xGa_{1-x}As$ is presented in figure 2.2(a) [7] and 2.2(b) [8]. As can be seen in figure 2.2(a), a desired energy gap that matches the wavelength in the solar spectrum can be conveniently obtained by choosing the corresponding composition of the semiconductors. Figure 2.2(b)

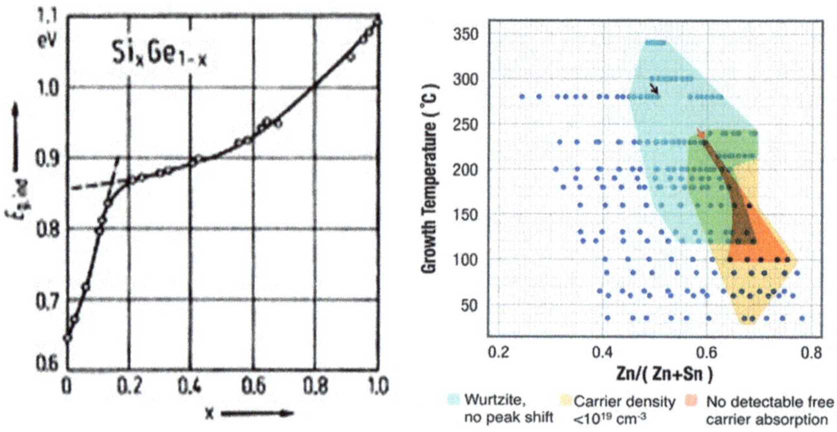

Figure 2.2. (a) Indirect energy gap versus Composition of Si_xGe_{1-x} at 296 K. Reprinted with permission from [7], copyright (1958) by the American Physical Society. (b) Examples of light absorbers. Reproduced from [8], credit: NREL.

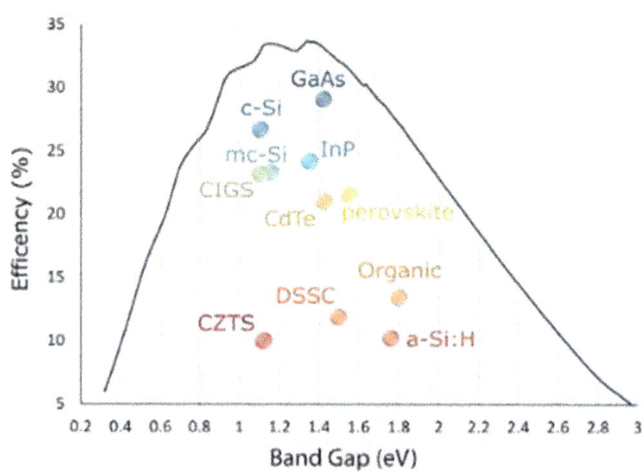

Figure 2.3. Efficiency as function of bandgap for various semiconductors. Reprinted from [9], copyright (2020), with permission from Elsevier.

illustrates the development of thin film light absorber materials for application in photovoltaics [8].

The refractive index of a material determines the reflectance of the material at that particular wavelength. The solar cell material of choice must have high absorption and low reflectance. In 'Solar materials find their bandgap', [9] Brandon Sutherland presents an analysis of the recent papers that have been published on the applications of machine learning techniques for the determination of the energy gap of materials of choice for photovoltaics [10, 11].

As can be seen in figure 2.3, GaAs has the maximum efficiency; it is a direct bandgap semiconductor with an energy gap of 1.43 eV at 300 K. However, from theoretical considerations, CdTe, with a direct bandgap of 1.44 eV at 300 K, has a better match with the average solar photon energy of 1.5 eV in the visible region of the solar spectrum [12]. From the point of view of its abundance, cost-considerations, chemical stability, environmental safety, maturity in technology and device performance, repeatability and reproducibility, crystalline silicon, in spite of its indirect bandgap, surpasses all the material candidates as the most dominant material for solar cell manufacturing. In this context, it must be emphasized that the silicon solar cell technology has benefited significantly from developments in the mainstream silicon semiconductor technology.

2.2 Amorphous silicon

From a historical perspective, the discovery of a mixture of silicon and iron, in 1810, is attributed to Jöns Jacob Berzelius. In 1823, Berzelius obtained iron-lean silicon by reducing SiF_4 with potassium metal. The first report of the commercial production of silicon in 1902 was in the form of ferrosilicon—an alloy of 25% iron and silicon [13]. Chemical vapor deposition (CVD) accounted for much of the production of silicon during the early years. The fabrication of polycrystalline silicon microwave diodes by CVD was reported by Teal and Storks in 1943 [13]. The first of the batches of crystalline silicon were grown by the Czochralski technique in the 1950s. Chittick, Alexander and Sterling reported their results of the preparation and properties of amorphous silicon in 1969 [14]. In 1970, Le Comber and Spear reported their studies on the electronic transport in amorphous silicon films [15]. The first amorphous silicon solar cell, with an efficiency of 2.4%, was reported by Carlson and Wronski in 1976 [16]. Catalano *et al* reported 10% conversion efficiency in amorphous silicon solar cells in 1982 [17]. While hydrogenated amorphous silicon exhibits direct bandgap and hence enhanced absorption, it suffers from light-induced metastable changes in its properties.

2.2.1 Staebler–Wronski effect

The Staebler–Wronski effect (S–W effect) refers to light-induced changes in the photoelectronic properties of hydrogenated amorphous silicon. It is illustrated in figure 2.4. The exposure to light causes an increase in the recombination current and

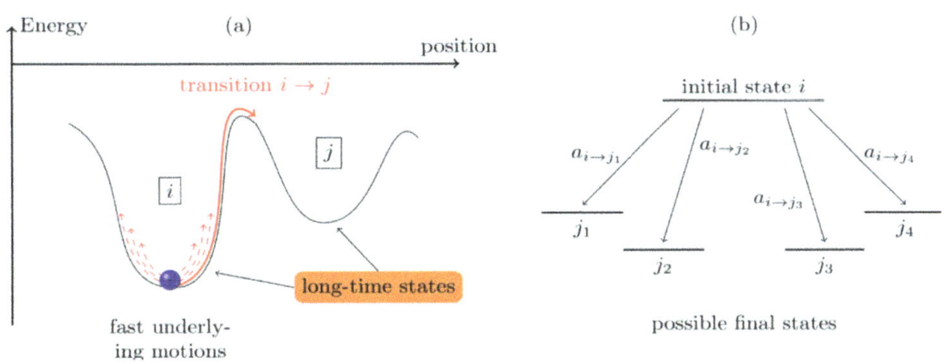

Figure 2.4. The sketch of the reaction process and energy band diagram of S–W effect. Reproduced from [26] CC BY 4.0.

degrades the efficiency of the conversion of sunlight into electricity [18, 19]. Originally, it was found that the large light-induced conductivity changes are a bulk phenomenon that occurs between what may be considered a thermally stable state A and a new metastable conductivity state B. Such effect was discovered in 1977 by David L Staebler and Christopher R Wronski [20]. Although the S–W effect caused degradation in the cell performance, Staebler and Wronski also found that the effect was perfectly reversible when the device was annealed at temperature over 150 °C and returned to room temperature. The performance of hydrogenated amorphous silicon solar cell strongly depends on the microstructure or the spatial relationship between incorporated hydrogen atoms and dangling bonds. Hydrogenated amorphous silicon plays an important role in the semiconductor industry for applications such as transistors, displays, batteries, hydrogen production and especially solar cells due to the low cost production [21, 22]. The variety of results that have been reported by using different experimental techniques clearly indicate that, from the aspect of the electronic properties, the photo-induced reversible changes depend on the fabrication conditions, doping, and impurities in different films [19].

The unknown and natural complexity of S–W effect remains ambiguous; thus, some models have been proposed for the kinetics of defect creation and annealing. One is the bond-breaking model [23], and the other is the metastable hydrogen collision model [24]. The first one, a weak bond adjacent to the Si–H bond is broken by the local phonon energy emitted after a direct recombination of photo-excited electron and hole and the later accounts quantitatively for the kinetics of light-induced defect creation, both near room temperature and at 4.2 K. Subsequently, A F Meftah *et al* [25] reported a model for the light-induced defect creation and annealing, which considered two cases of the illumination intensity G, the moderate and the very intense illuminations. This proposed model reproduces many experimental features of the S–W effect that have been reported in the literature.

2.3 Materials overview

Solar cells can be p-n homojunctions, metal–semiconductor contacts/Schottky barriers, tandem cells, metal-insulator–semiconductor/metal-oxide–semiconductor structures, heterojunctions, multijunctions or tandem cells. The material of choice can be a single crystal, polycrystalline/microcrystalline/multicrystalline, amorphous, or a thin film or a combination of structures. While each of them offer specific advantages, from a materials perspective, crystalline silicon is the most utilized material for the manufacture of solar cells in the world. A summary of the bBest module (blue), best cell (green), and Shockley–Queisser (SQ) efficiency (yellow) of major technologies is presented in figure 2.5 [27]. The material composition in a silicon solar panel is illustrated in figure 2.6 [28, 29].

2.3.1 Schottky barrier

Metal–semiconductor interfaces play a critical role in semiconductor devices. When a semiconductor is in intimate contact with a metal, the band aligned at its interface

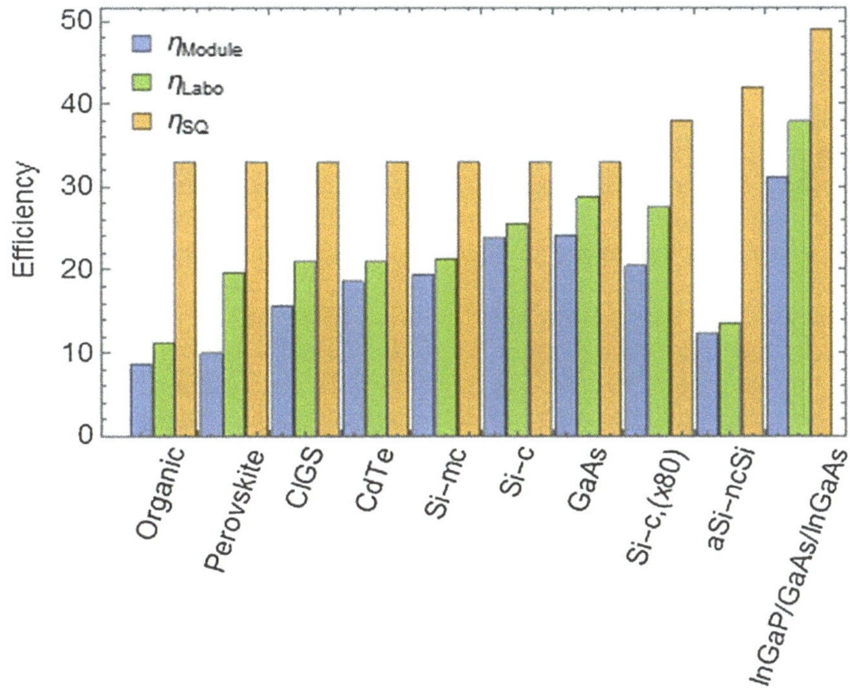

Figure 2.5. Summary of the best module (blue), best cell (green), and Shockley–Queisser (SQ) efficiency (yellow) of major solar cell technologies: Si–mc, Si–c, Si–c(x80), and a-Si–ncSi represent multicrystalline, crystalline, crystalline under 80 suns illumination and amorphous/nanocrystalline silicon, respectively. Reproduced from [27] CC BY 4.0.

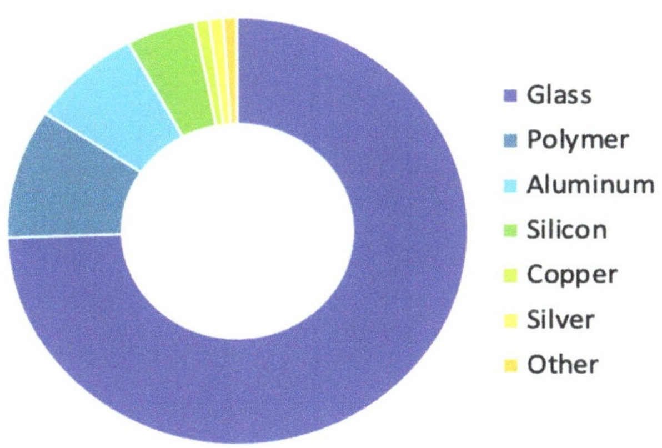

Figure 2.6. Illustration of the constituents in a typical crystalline silicon solar panel: 76% glass, 10% plastic polymer, 8% aluminum, 5% silicon, 1% copper, and less than 0.1% silver and other metals [28, 29]. What's in a solar panel? By weight, the typical crystalline silicon solar panel is made of about 76% glass, 10% plastic polymer, 8% aluminum, 5% silicon, 1% copper, and less than 0.1% silver and other metals, according to the Institute for Sustainable Futures. Graphic: Union of Concerned Scientists. Reproduced with permission from [29]. Copyright UTS 2019.

will establish a common chemical potential. This usually leads to a shift in the band. For example, in an n-type semiconductor, the band bending due to the upward shift of the conduction band edge, upon contact with the metal, forms a Schottky barrier. Eventually, the free electrons from the metal are prevented from entering into the semiconductor. Under forward bias conditions, the thermal population of majority carriers which can be transmitted over the top of the barrier is enhanced, and thus this results in current flow [30]. The ability to generate the current in the forward direction and not in the reverse direction leads to junctions which are rectifying in nature.

The energy band diagram for the formation of Schottky barrier between metal and semiconductor is shown in figure 2.7. First, the metal and semiconductor, before contact, are electrically neutral. Upon contact, the energy of an electron at rest outside the surfaces of the two solids is no longer the same; the Fermi level of the semiconductor shifts upwards to reach the Fermi level on the side of the metal. After equilibrium, the Fermi levels align with each other across the metal–semiconductor junction. The work function of the metal (Φ_m) and the electron affinity of the semiconductor (χ_s) remain unaltered, as shown in figure 2.7. The vacuum level in the metal will follow the same variations as E_c and gradually approach the semiconductor side to preserve continuity. The difference between the two vacuum levels is given by $qV_i = (\Phi_m - \Phi_s)$, where V_i is expressed in volts and is known as the contact potential difference or the built-in potential of the junction. qV_i is also the

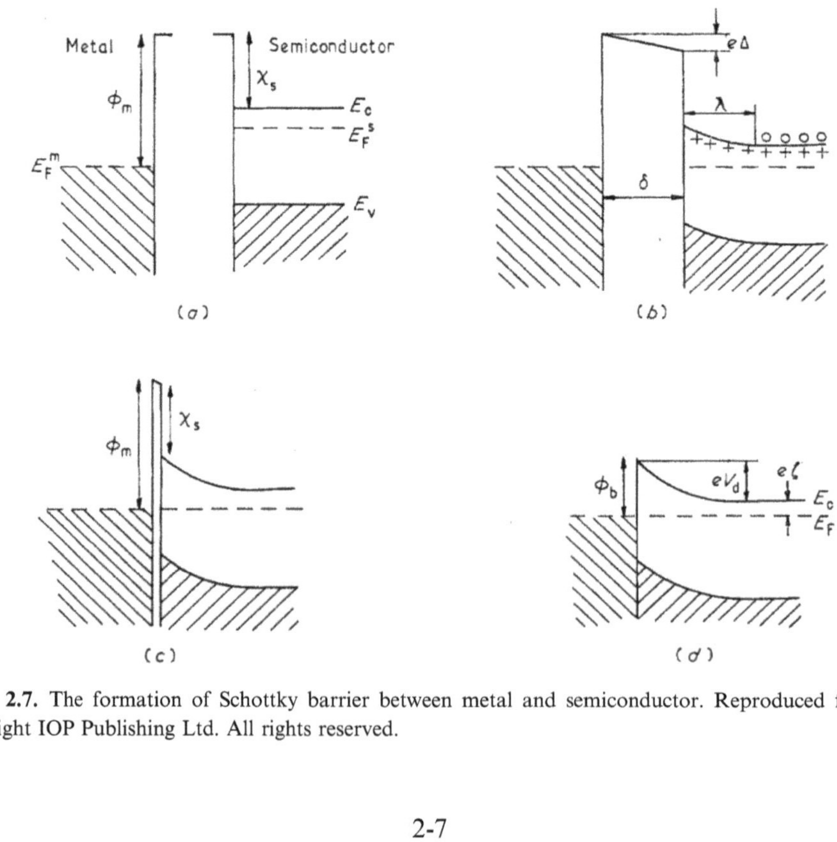

Figure 2.7. The formation of Schottky barrier between metal and semiconductor. Reproduced from [32]. Copyright IOP Publishing Ltd. All rights reserved.

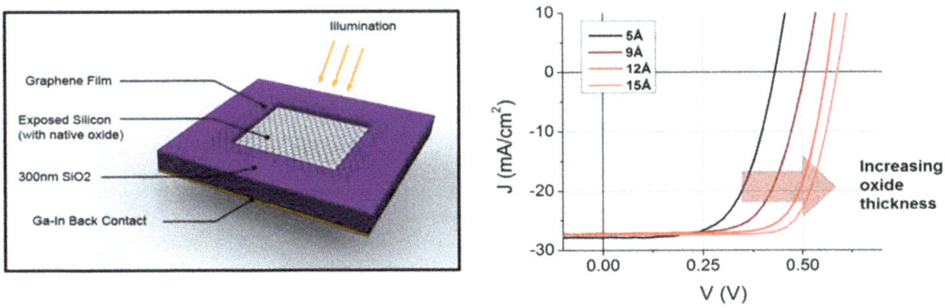

Figure 2.8. (a) Example of a graphene/n-Si device. (b) Current–voltage characteristics of the device—as a function of increasing oxide thickness. Reprinted with permission from [33], copyright (2015) American Chemical Society.

energy needed for an electron to move from the semiconductor side to metal side and the barrier is given by:

$$\Phi_B = \Phi_m - \chi_s \tag{2.3}$$

In most cases, the height Φ_B of the barrier is orders of magnitude larger than the thermal voltage (kT/q), and the space-charge region in the semiconductor becomes a high-resistivity depletion region devoid of mobile carriers [31].

In general, in Schottky barrier solar cells, the work function difference between the metal and the semiconductor is responsible for the space-charge electric field that is utilized to separate the electron–hole pairs.

With the advent of 2D semiconductors, there has been significant interest in Schottky barrier solar cells that take advantage of their properties such as high electrical conductivity and optical transmittance. An illustration of a high efficiency graphene—Schottky barrier solar cell is presented in figure 2.8 [33].

2.3.2 Experimental methods

An illustration of the vertically integrated solar panel manufacturing—from poly-silicon feedstock to ingot to wafer to cell to panel is presented in figure 2.9 [34].

Czochrslski and Float Zone are the most commonly used methods for the manufacture of silicon wafers. The following process steps are utilized for the fabrication of solar cells:

1. Pre-check and pretreatment;
2. Texturing;
3. Acid cleaning;
4. Diffusion;
5. Etching and edge isolation;
6. Post-etching washing;
7. Anti-reflective coating deposition;
8. Contact printing and drying;
9. Testing and cell sorting.

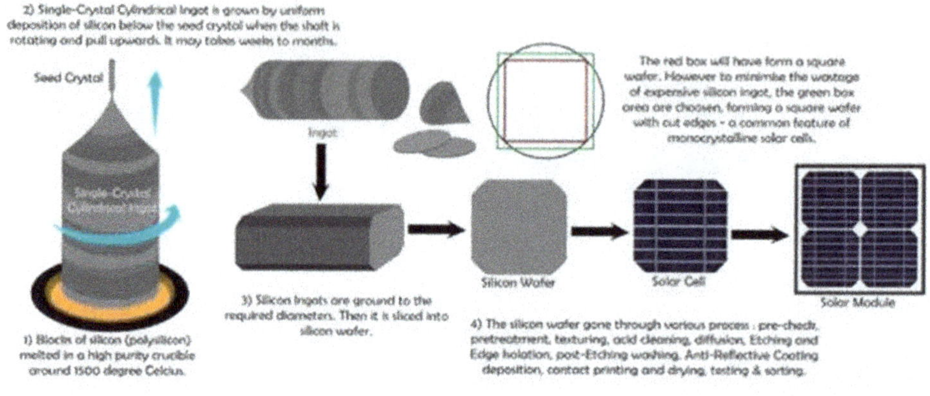

Figure 2.9. The manufacture of crystalline silicon modules involves fabricating silicon wafers, transforming the wafers into cells, and assembling cells into modules. Reproduced from [34], with permission from Springer Nature.

2.3.3 Wafer cleaning

In general, for silicon solar cell manufacturing, the wafers are cleaned through a series of wet chemical steps. However, in some cases, RCA (Radio Corporation of America) cleaning process (RCA clean) is utilized for wafer cleaning. A brief introduction to RCA clean is presented here.

The procedure for silicon wafer cleaning is very crucial for the successful fabrication of semiconductor devices. It ensures that the wafers are free from contaminants, particulates, and other impurities. Normally, the RCA clean is a widely used method in the silicon semiconductor-VLSI (very large-scale integration) industry. It is effective in removing organic and inorganic contaminants from the wafer surface. In table 2.2, two common RCA clean solutions: RCA-1 and RCA-2 are described. RCA-1 solution is prepared by mixing the DI water: hydrogen peroxide: ammonium hydroxide (5:1:1) and RCA-2 solution is prepared by DI water: hydrogen peroxide: ammonium hydroxide (6:1:1) [35, 36]. In solar cell manufacturing, the emphasis is on the reduced cost of manufacture, while guaranteeing cell performance, repeatability and reproducibility as well as process yield. Thus, in most solar cell manufacturing methods, a controlled working environment is implemented instead of the standard cleanroom practices that are rigorously followed in the VLSI industry.

2.4 Spectral response of material candidates for solar cells

The spectral response of a photovoltaic (PV) module is the fraction of the absorbed irradiance that can be detected by the photo-sensor that is converted into a current, which is a function of wavelength or frequency of the signal. The bandgaps of many commonly used semiconductor materials are listed in table 2.1. For example, the ideal spectral response of silicon solar cell is limited at long wavelengths. This is

Table 2.2. The procedure for silicon wafer cleaning.

RCA-1	Procedure
Safety precautions	Ensure a cleanroom or controlled working environment
Preparation	Clean container, pure high-quality deionized (DI) water and well-defined labels for chemicals
Handling wafer	Cleanroom-compatible tweezers for silicon wafer handling
RCA-1-step 1	Rinse with DI water to remove loose particles and contaminants
RCA-1-step 2	Place the wafer in RCA-1 solution for 5–10 min to remove organic contaminants
RCA-1-step 3	Rinse with DI water again to remove the residual of RCA-1 solution
RCA-2-step 1	Place the wafer in RCA-2 solution for 5–10 min to remove metal ions and inorganic contaminants
RCA-2-step 2	Rinse with DI water to remove residue RCA-2 solution
RCA-2-step 3	Final rinse with DI water to ensure that all chemicals and contaminants are removed
Drying	Dry the wafer using a spin rinse dryer or by blowing dry, particle-free nitrogen gas
Post-cleaning	Inspect the wafer for cleanliness

because of the fact that energies below the energy bandgap cannot be absorbed by the material. The mismatch in the ideal and measured spectral response is from the factors from the environment in actual use such as the irradiance angle, sensor aperture area, inhomogeneous illumination etc, which are typically quantified by a spectral mismatch factor [37]. Moreover, the spectral response obtained from the solar simulator and outdoor sunshine is different. The spectral response is a function of wavelength and is related to the quantum efficiency (QE):

$$SR = QE \times \lambda \frac{q}{hc} \tag{2.4}$$

where, SR is the spectral response, QE is the quantum efficiency, λ is the wavelength of the incident light, q is the electronic charge, h is the planck's constant and c is the speed of light. The solar cell spectral response depends on the solar cell materials, as shown in figure 2.10.

2.5 Types of device structures

A solar cell is typically composed of a p-n junction; when the light-induced electrons and holes are separated at the interface of the p- and n-type semiconductor, a voltage is generated. A typical schematic diagram of a solar cell is shown in figure 2.11. The electrical current generated at the interface of p- and n-type semiconductors is extracted by the front and back contacts. The top layer, normally made of glass or transparent conducting oxide, allows the penetration of light. Antireflection coating on top of the n-type semiconductor is used to minimize the reflection of light in order

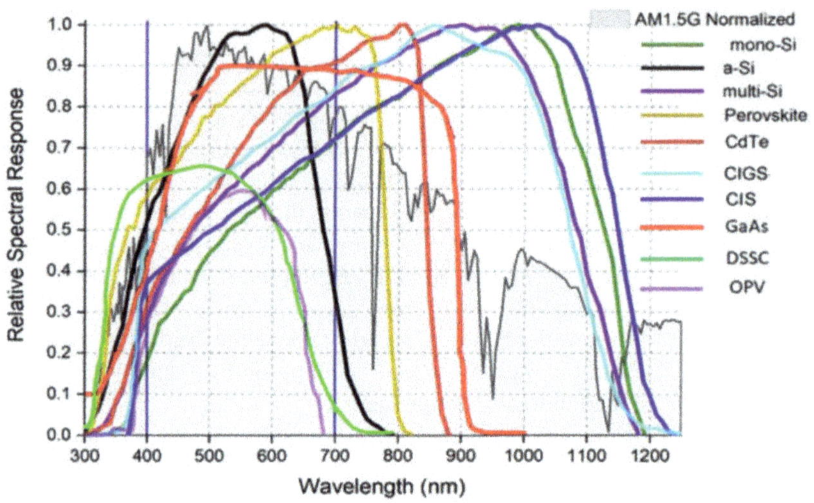

Figure 2.10. Spectral response of solar cells of several different materials under AM 1.5. Reproduced from [38] CC BY 3.0.

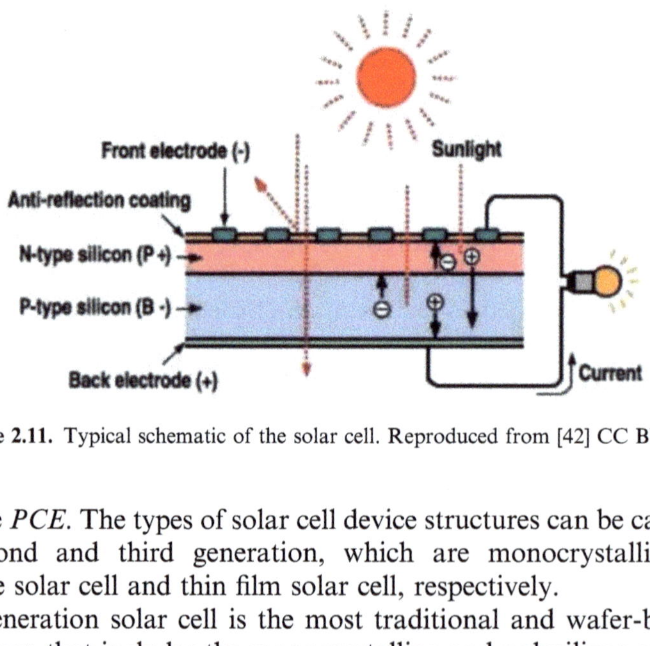

Figure 2.11. Typical schematic of the solar cell. Reproduced from [42] CC BY 4.0.

to increase the *PCE*. The types of solar cell device structures can be categorized into the first, second and third generation, which are monocrystalline solar cell, polycrystalline solar cell and thin film solar cell, respectively.

The first-generation solar cell is the most traditional and wafer-based made of crystalline silicon that includes the monocrystalline and polysilicon solar cells. The second-generation solar cells include amorphous silicon (a-Si), CdTe, GaAs and copper indium gallium diselenide (CIGS), as shown in figure 2.12. In comparison to the conventional wafer-based crystalline silicon solar cell, a-Si solar cell utilizes randomized structure which has higher bandgap of 1.7 eV with stronger absorption and an improved spectral response. CIGS achieves the highest *PCE* in thin film solar cells due to its tunable bandgap (1.0–2.4 eV) in the semiconductor by varying the

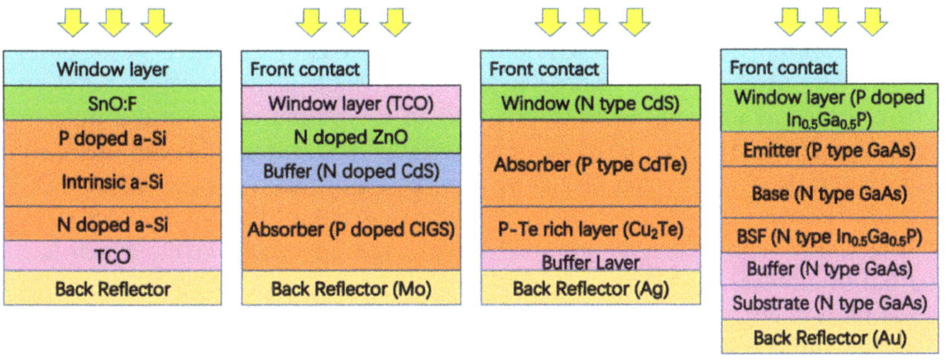

Figure 2.12. The structure of second-generation solar cells, from the left to right are: amorphous silicon (a-Si), copper–indium–gallium diselenide (CIGS), CdTe and GaAs solar cell respectively [TCO: transparent conducting oxide].

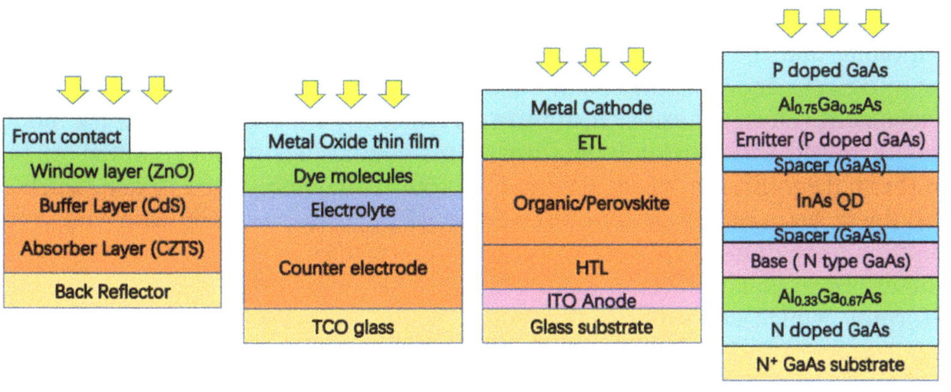

Figure 2.13. The structure of third-generation solar cell, from the left to right are: CZTS, CIGS, dye-sensitized, organic/perovskite and quantum dot (QD) solar cells, respectively [TCO: transparent conducting oxide; ITO: indium tin oxide; ETL: electron transport layer; HTL: hole transport layer].

ratios of In:Ga and Se:S. CdTe solar cells achieve similar *PCE* as CIGS cells. But with the capability of low-temperature processing, the use of CdTe allows flexible and affordable production among the second-generation solar cells. GaAs achieves the highest *PCE* among the single-junction solar cells with E_g of 1.42 eV; however, the manufacturing process of a GaAs-based solar cell is expensive and complicated in comparison with other second-generation solar cells.

As shown in figure 2.13, the third-generation solar cells, that are potentially able to overcome the Shockley–Queisser limit of 31%–41% power conversion efficiency for single bandgap solar cells, include copper zinc tin sulfide (CZTS), dye-sensitized, organic, perovskite and quantum dot (QD) solar cells. CZTS is emerging as a non-toxic and relatively abundant semiconductor with a tetragonal structure; kesterites belong to the A2BCX4 family. Dye-sensitized solar cells (DSSCs) have arisen as a technically and economically credible alternative to the p-n junction photovoltaic devices [39]. Halide perovskite solar cells (PSCs) are usually fabricated using methylammonium lead halide (CH3NH3PbX3 (MAPbI3), where X = halogen: chlorine (Cl), bromine (Br) and/or iodine (I)), which is sandwiched between an electron transport material (ETM)—that is, an n-type material—and a hole transport material (HTM)—that is, a p-type material [40]. Organic solar cell uses carbon-based materials and organic electronics instead of silicon as the semiconductor material to convert the incident light to electricity. A quantum dot solar cell (QDSC) is a solar cell design that uses quantum dots as the absorbing layer, which has the potential to increase the maximum attainable thermodynamic conversion efficiency of solar photon conversion up to about 66% by utilizing hot photogenerated carriers to produce higher photovoltages or higher photocurrents [41].

2.5.1 Thermal management of solar cells

Thermal management refers to the maintenance of the system temperature of the PV cell/module within the allowed operating temperature since the performance of solar cell usually degrades when the temperature exceeds 80 °C [43]. It is worth mentioning that the solar cell can only convert less than 30%–40% of the incident light into electricity while the rest of the energy from light will dissipate as heat to the environment. Thus, the excess heat must be efficiently expelled from the PV module to avoid any adverse impact in the form of loss of efficiency/power from the cell [44]. For thermal equilibrium of the PV module, four main energy fluxes exist, which are as follows: solar radiation into the system (P_{solar}), thermal emission (P_{emi}) and radiation dissipation (P_{diss}) out of the system and electricity out of the system (P_{elec}), as described in equation (2.5). Solar cells also suffer from significant thermal non-uniformity, which can be caused by shadows, micro-cracks, thermal strains, and other factors [45]. In addition to developing effective techniques to decrease the waste-heat generation, many studies have been conducted in the literature to seek good thermal management solutions for PV systems using two main approaches such as passive or active convection cooling [46].

$$P_{\text{solar}} - P_{\text{emi}} - P_{\text{diss}} - P_{\text{elec}} = 0 \tag{2.5}$$

Most solar cells utilize passive cooling with flowing medium such as heat sinks, heat pipes, and microchannels, due to their reliability and low cost. Heat sinks are made of high thermal conductivity metals such as Al and Cu, and the cooling efficiency can be adjusted by the geometrical shape [47]. Heat pipes are metal pipes with a small amount of water that transfer heat by evaporating and condensing continuously [48]. Microchannel cooling is made of high thermal conductivity material in which the inner fluid thermal conductivity is inversely proportional to the channel width [49]. In 2023, Poudel *et al* [50] utilized a porous nanochannel device on the back face of the PV panel to reduce the PV surface temperature significantly with an average cooling of 31.5 °C. Radiative cooling, as the general solution to passive cooling the PV system, involves the addition of a spectrally selective filter on top of the solar cell. The radiative cooling of solar cell can be performed in two ways: the chosen filter needs to be highly transparent to ensure that photons (below the

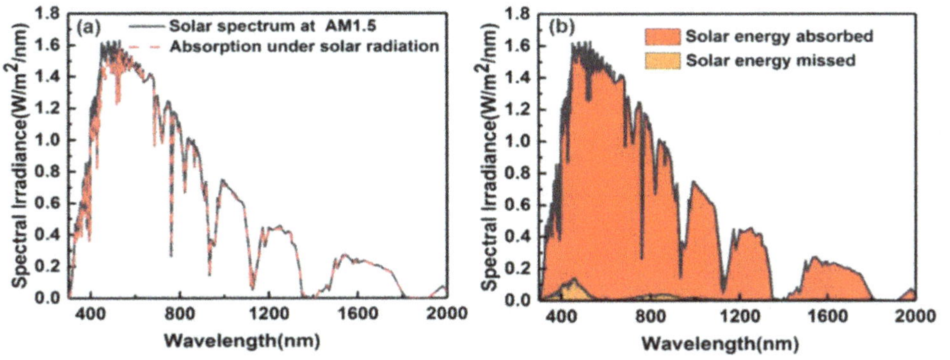

Figure 2.14. Illustration of effective wavelength for PV module conversion under normalized AM 1.5. Reproduced from [53] CC BY 4.0.

effective wavelength) with high energy can reach the solar cell and be partly converted into electricity, and the filter needs to exhibit strong thermal emission to maximally dissipate the solar cells' waste heat into the cold sky. For wavelengths above the allowed range, the filter can be highly reflective and reflect the unused photons back to the atmosphere; this is close to the ideal filter [51]. Wang *et al* designed a new method of coupling radiative cooling with a flat heat sink on the PV module, effectively reducing the operating temperature of concentrator photovoltaic (CPV) system by 36 °C under a heat load of 6.1 W, which also increased the open-circuit voltage and lifetime of the CPV system [52]. Over the years, a variety of new materials and structures have been proposed to enhance the radiative passive cooling; these developments have been well documented in recent literature [18]. An illustration of the effective wavelength for PV module conversion is presented in figure 2.14 [53].

2.5.2 Cost considerations

The cost analysis of the utilization of energy via photovoltaics is extremely complicated and must include various factors including its associated geo-politics, supply chain, the availability of raw materials, the physical infrastructure such as power and water, governmental incentives, labor costs, transportation, warranties, energy storage, inverters, connection to the grid and recycling. Further, the dynamics, variation and sometimes the unpredictability of the weather on a particular day, at a particular time, as function of location complicates the immediate dependency on electricity produced from solar panels. The obtained direct current output from the solar panel must be: (a) connected to an energy storage device such as a battery or a supercapacitor via a solar charge controller; (b) converted into alternate current via the use of inverter before connecting it to the grid. In several countries, the grid cannot support the load of the power generated by solar farms and needs to be upgraded.

In spite of all these challenges, the cost of solar electricity is decreasing rapidly and the demand is growing exponentially. As an illustration, the price of solar electricity has decreased from $2.36 USD/peak watt (Wp) in 2010 [21] to $0.35 USD/Wp in 2020 [54]; predictions are that module prices can be expected to halve again

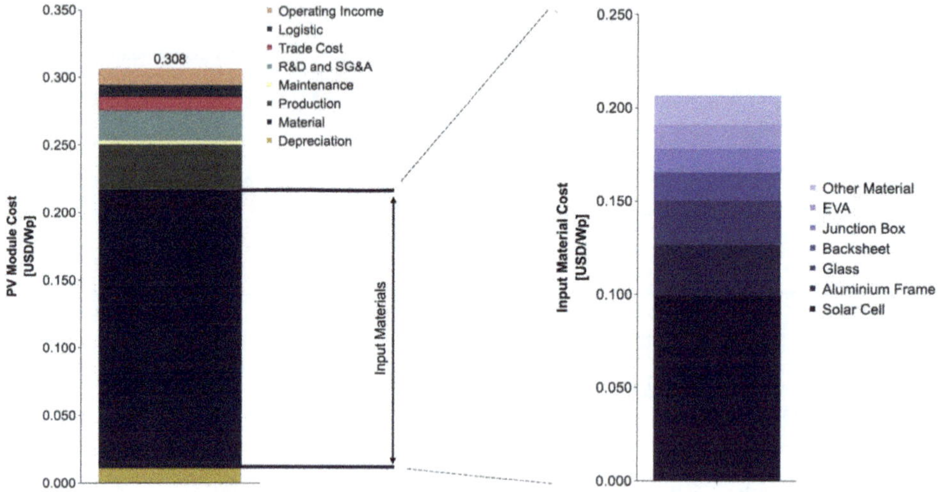

Figure 2.15. The cost structure of PV module. Reprinted from [57], copyright (2022), with permission from Elsevier.

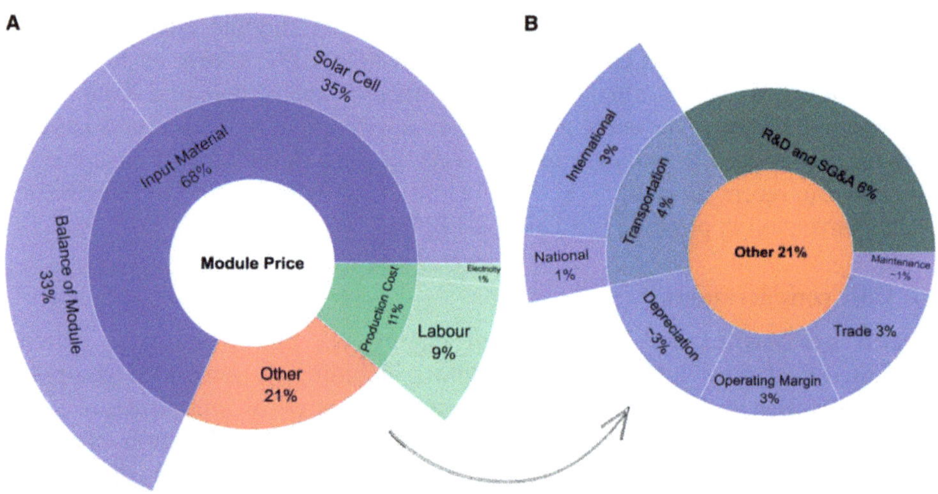

Figure 2.16. (a) The cost percentage of PV module components; (b) details of other cost components. Reprinted from [57], copyright (2022), with permission from Elsevier.

before 2030 [55]. The cost consideration analyses are mainly based on bottom-up cost models along the supply chain with key outputs of minimum sustainable prices, total manufacturing costs, trade costs and possible ways of cost reduction [56]. The proposed cost structure, along the manufacturing processes, is summarized in figure 2.15. The cost consideration is first location dependency and size dependency; taking Australia as an example, figure 2.16 shows that 68% of the PV module price is allocated to input materials, while production costs are responsible for 11% and other costs account for the remaining 21%. Key cost drivers for local module assembly are input materials followed by production costs, principally labor and electricity [57]. It must be emphasized that the continued growth in PV technologies can be attributed to the following considerations: (a) solar energy, in general, is a renewable source of energy; (b) the use of solar cells/panels for production of electricity does not produce any pollutants; (c) there is a growing global concern about climate change; and (d) the continued decrease in the cost of solar electricity, coupled with the decrease in cost of energy storage and support of local, private and governmental entities, has been a major driver for its growth. An improved education and understanding of solar cell technology, its implementation and utilization and approaches to cost reduction as well as methods of recycling have been some of the contributing factors for its overall acceptance throughout the world.

Recently, perovskites-based solar cells have become one of the most well investigated and rapidly developing technologies in the PV industry. Researchers have been analyzing the designs and costs that are associated with various perovskite materials for utilization in PV cells and modules. In this context, it may be noted that the lowest price corresponds to option 3 [SnO_2-perovskite-spiro-Cr/Cu] with SnO_2 as electron transport layer (ETL) and Cr/Cu as the electrode [58].

2.6 Architecture of silicon solar cells

Several technologies have been developed to improve the solar cell efficiency and stability; a summary of these technologies is presented in table 2.3. The passive emitter and rear contact (PERC) solar cells have seen rapid increase in the PV industry due to their better efficiency from the architecture that leads to very low surface recombination and higher carrier generation. When the photons strike the solar cell, the unabsorbed light is absorbed by the addition of the rear passivated

Table 2.3. Summary of highly efficient silicon-based single-junction solar cells [61].

Cell	Max. efficiency	Structure
PERC	22.8% [60]	Silicon wafer with passivation layer that helps to absorb more light
Bifacial PERC	Up to 25% [76]	Double-sided PERC cell
PERL	24.5% [77]	Addition of P+ passivation layer to reduce the rear surface recombination rate
HIT	26% [67]	Intrinsic a-Si layer is inserted between p-type a-Si and n-type a-Si for passivation
HJ-IBC	26.7% [67]	HIT cell with interdigitated back contacts that has the collection region and contact on the rear side
TOPCon	26.1% [61]	Tunnel oxide layer is inserted followed by a highly doped p-type or n-type polysilicon layer

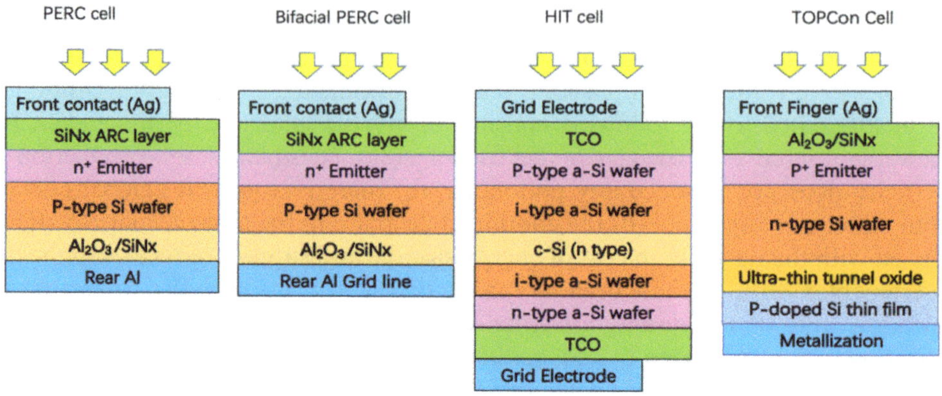

Figure 2.17. Architecture of (a) PERC cell, (b) Bifacial PERC cell, (c) HIT cell and (d) TOPCon cell.

layer, which improves the cell efficiency [59]. The first PERC solar cells were reported in 1989, with a *PCE* of 22.8% [60]; the basic architecture is composed of front contact, antireflection coating layer, emitter, Si-wafer, passivated oxide layer, SiNx capping layer and rear contact from top to bottom, as shown in figure 2.17(a). The bifacial solar cell structure can be used for PERC solar cell to further improve the efficiency, which achieves up to 25% with the structure of double-sided PERC cell, as shown in figure 2.17(b). However, the passivated emitter with rear locally diffused (PERL) solar cell is limited by the fact that the rear contact areas with Al directly in contact with the silicon wafer increase the recombination rate [61]. This gave birth to the modified PERL solar cells in 1992, which has a p+ passivation layer underneath the rear metal contact area that helps to lower the recombination rate, thereby improving the efficiency [62].

Heterojunction with intrinsic thin layer (HIT) solar cell is another type of silicon solar cells; they were first proposed in 2000 with structure of a very thin-layer of intrinsic a-Si inserted in-between the p-type a-Si and n-type c-Si, as shown in figure 2.17(c). According to the latest *International Technology Roadmap for Photovoltaics* (13th edn 2022), HIT solar cells are expanding their growth rapidly, with anticipated market share of more than 75% by 2032. Conventionally, HJT solar cells use a-Si:H (i/p) and a-Si:H (i/n) stacks to passivate dangling bonds on two surfaces of a wafer and form a passivating-contact, which is the most successful passivating-contact structure [63, 64]. HIT (HJT) solar cell is one of the most promising structures with higher conversion rate due to its low production costs and low production temperature, which has caught the attention of the PV industry. For improvement in efficiency of HIT solar cells, methods in improving the quality of grid electrode (resistance loss suppression) and reducing the absorption rate in the a-Si and TCO layers are widely studied [65, 66]. Currently, the highest efficiency for a single-junction silicon-based solar cell is the HJ solar cell with interdigitated back contacts (IBC) with a *PCE* of 26.7% [67]. The screen printing technology is the most commonly used metallization method for HIT solar cell fabrication [68]. HJT solar cells have an excellent V_{oc} of up to 750 mV and more than 85% *FF* value due to the superior interface passivation and perfect passivating-contact architecture. However, J_{sc} is lower than that of other silicon solar cells and even that of conventional aluminum back surface field (Al-BSF) solar cells [68].

Tunnel oxide passivated contact (TOPCon) solar cells have gained significant interest as an upgraded and advanced version of the PERC/PERT (passivated emitter and rear totally diffused technology) with *PCE* of 26.1% achieved by JinkoSolar at the end of 2022 [67]. TOPCon solar cells can be made with structure comprising of p-type or n-type Si solar cell, in which n-type is confirmed to be more efficient and has low impurities. The present approach of integration of TOPCon on p-PERC structure as an evolutionary upgrade may be an attractive option, since the TOPCon-upgraded cell benefits from the ongoing mainstream development of the p-PERC technology [69]. However, the production cost of TOPCon solar cells is comparably higher than PERC/PERT technology. The market trends have shifted in the last decade from back surface field (BSF) solar cells to PERC solar cell technology, and now it seems to be shifting to TOPCon solar cells [70]. Messmer *et al* [69] have reported detailed studies of the performance evaluation, via numerical device simulation, to improve the efficiency of the new generation of TOPCon-upgraded p-PERC technologies. The loss analysis of the final TOPCon-upgraded p-PERC cell reveals that its efficiency is strongly limited, by PERC features such as the phosphorus diffused front emitter and p-type bulk, to below 24%.

The efficient collection of charge carriers with the correct polarity at the contact is essential for silicon solar cells [71]. However, the Fermi level is often pinned at the metal–semiconductor interface due to interfacial defects such as dangling bonds, interfacial dipoles, or metal-induced gap states, which is called Fermi level pinning (FLP) effect [72]. As illustrated in figure 2.18(a), FLP effect is detrimental because it generates significant Schottky barrier at the Si interface. One strategy to mitigate the FLP effect is to introduce heavy doping of the c-Si, which reduces the depletion layer and increases the probability of quantum tunneling, as shown in figure 2.18(b) [73]. Another strategy is to de-pin the Fermi level by introducing a thin oxide layer in-between metal and c-Si, as shown in figure 2.18(c) [74, 75]

2.7 Bandgap engineering

2.7.1 Material selection

In PV devices, it is crucial to match the bandgap of the semiconductor material to the solar spectrum to maximize energy conversion. As single-junction solar cells are limited to 30%–32% conversion efficiency under 1-sun, multi-junction or tandem solar cells are expected to overcome this limitation to 36%–42% of *PCE* [79]. Binary and ternary semiconductors of II–VI group (ZnS, ZnSe, ZnTe, CdS, CdSe, CdTe, HgTe, HgS, HgSe, HgCdTe, CdZnTe, CdSSe, and HgZnTe) are very important as they provide optimization of energy conversion for a variety of applications, as shown in figure 2.19. Most materials of group II–VI are semiconductors with a direct bandgap and high optical absorption and emission coefficients [80]. The efficiencies of II–VI compounds, chalcopyrites, and kesterite-based tandem solar cells are 27.2% with III–V/CIGS 3-junction, 24.6% with perovskite/CIGS 2-junction [79], 16.8%

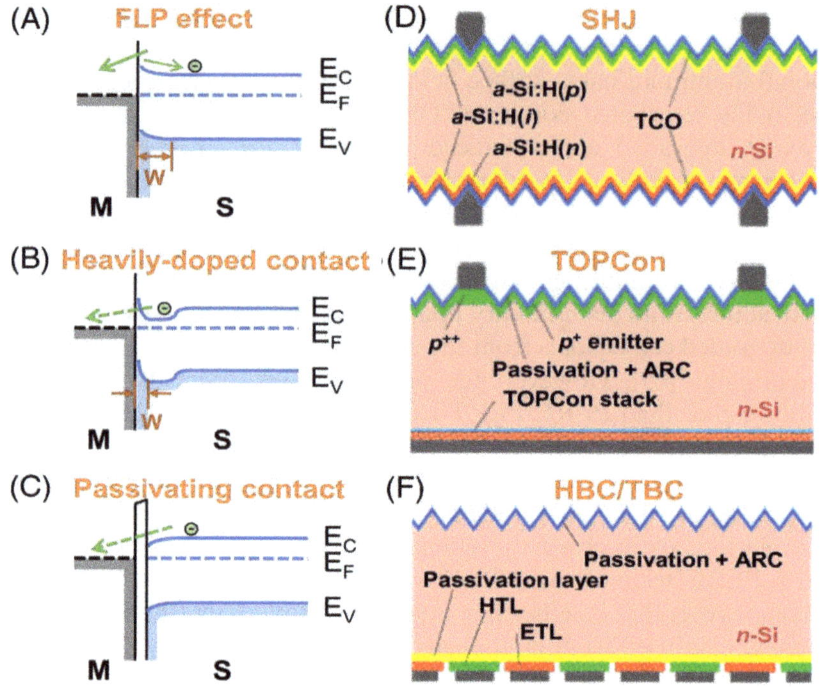

Figure 2.18. Energy band diagram of (a) FLP effect, (b) strategies of heavily-doped contact and (c) addition of passivating contact. (d)–(f) Structure of HIJ, TOPCon and HBC/TBC solar cell. Reprinted from [78] with permission from Springer Nature.

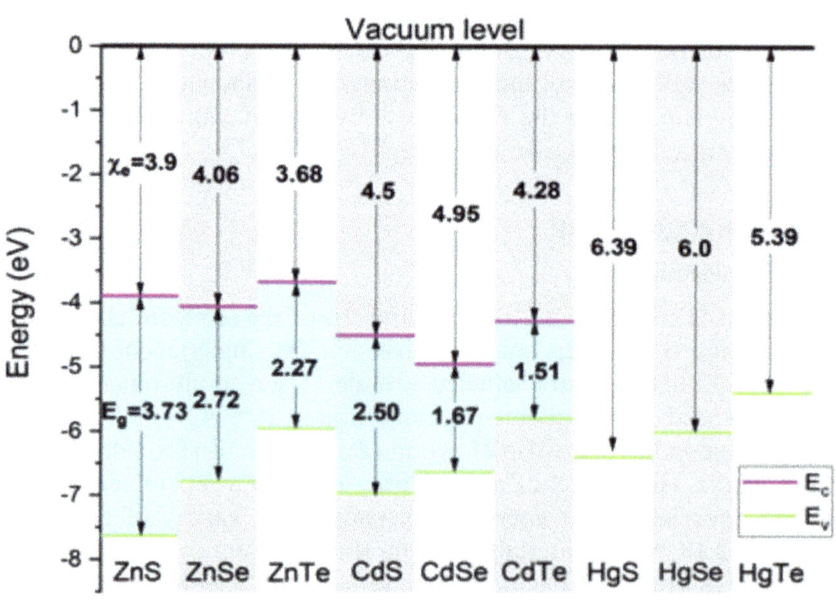

Figure 2.19. Band gap alignment for II–VI compounds. Reprinted from [80] with permission from Springer Nature.

with CdZnTe/Si 2-junction [81], 15.3% with CdTe/CISe 2-junction [82] 8.5% with $CuGaSe_2$/$CIGSe_2$ 2-junction [83], 3.5% with CZTS/Si 2-junction tandem solar cells [84]; these are lower compared to 39.5% with III–V 3-junction and 35.9% with III–V/Si 3-junction tandem solar cells [85].

III–V solar cells are composed of elements from Group III (typically boron, aluminum, gallium, or indium) and Group V (typically nitrogen, phosphorus, arsenic, or antimony); they commonly include semiconductors such as gallium arsenide (GaAs), indium gallium arsenide (InGaAs), and gallium antimonide (GaSb). They are also known for their high efficiency and being used widely in space applications and concentrator photovoltaics. III–V materials offer a wide range of bandgap options, which make them suitable for capturing different portions of the solar spectrum. In 2020, 47.1% *PCE* was achieved by utilizing a monolithic, series-connected, six-junction inverted metamorphic structure operating under the direct spectrum at 143 Suns concentration, which is the highest record efficiency reported in the literature [86]. In 2022, the *PCE* of GaSb/InSb and GaAs/InAs solar cells were found to attain 22.88% and 20.65%, respectively [87]. Despite the significant fabrication costs associated with solar cells, based on these III–V compound semiconductors, ongoing efforts are continually optimizing their structure, consistently achieving new *PCE* records. Their associated critical advantages include tolerance to particle radiations as well as high temperatures that make them uniquely important and desirable for space applications. Furthermore, the technology for lattice-matched GaInP/GaAs/Ge triple-junction solar cells is maturing, enabling large-scale mass production while maintaining a remarkable conversion efficiency of over 30% [88].

2.7.2 Applications

The primary threat for solar cell performance and longevity in an orbital environment is the space radiation environment. High-energy particles, including electrons and protons, cause displacement damage in various parts of the solar cell structure. The principal factor for the degradation of space solar cells is that the radiation-induced displacement damage creates non-radiative recombination centers. These centers diminish the minority carrier lifetime, consequently causing a decrease in the electrical and spectral performance of the solar cells. The high radiation exposure in orbits around the earth requires solar cells with high radiation tolerance and device stability. InP based solar cells provide the highest power, more than 30% in comparison with GaAS-based cells [89]. Moreover, as the technologies are constantly improved, new solar cell structures are being proposed to meet the demands of operation of solar cells in space. GaInP/GaAs/Ge (1.82/1.42/0.67 eV) lattice-matched (LM) triple-junction cells are well established with efficiencies of over 30% and fulfill many device performance requirements for space applications [88]. In particular, InGaP/InGaAs/Ge and AlInGaP/AlInGaAs/InGaAs/Ge solar cells, produced by several companies such as, Azur Space, Spectrolab, SolAero, and CESI, are nowadays the standard in aerospace applications as they offer better performance than the other PV technologies [90–92]. As shown in figure 2.20, the performance of the solar cell diminishes with increasing particle fluences and

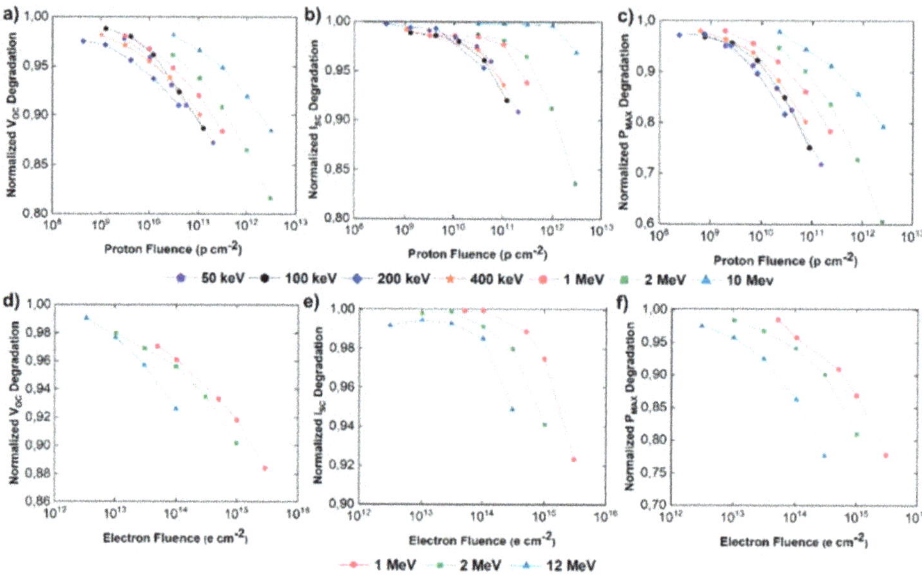

Figure 2.20. Degradation of V_{OC}, I_{SC}, and $P_{\max}$ of lattice matched (LM) GaInP/GaAs/Ge as a function of proton and electron fluence for various particles energies. Reproduced with permission from Wiley [96] CC BY 4.0.

ionizing radiations. Regarding the influence of electron bombardment, the degradation of solar cell performance becomes more pronounced as the energy of the electrons increases. In contrast, when it comes to proton irradiation, the most significant damage occurs due to low-energy particles, typically in the range of 50–200 keV [93].

Solar cells are designed to harness solar energy from the terrestrial spectrum, which refers to the sunlight that reaches the Earth's surface. This spectrum encompasses a range of wavelengths in the electromagnetic spectrum, including visible light and some portions of the near-infrared and ultraviolet regions. In the context of broad-band effects, it is important to highlight that aerosol scattering and absorption, primarily governed by Mie theory, followed by molecular scattering and absorption, typically associated with Rayleigh scattering, have the most significant influences. Notably, among molecular constituents, water stands out as the predominant absorber [94]. The optimum bandgap for the terrestrial spectrum lies beyond the absorption range of a traditional dual junction GaInP/GaAs cell, with the bottom GaAs cell having higher bandgap energy than necessary [95].

2.7.3 Key players—examples of companies and R&D institutions

Several companies and research institutions are actively engaged in bandgap engineering for various applications, including solar cells, light emitting diodes (LEDs), and electronic devices. In table 2.4, examples of companies and Research & Development (R&D) labs that are associated with solar cells involving bandgap engineering are summarized. First Solar, a company headquartered in Arizona and

Table 2.4. Examples of R&D labs and manufacturers of III–V and II–VI solar cells.

Company	Product	References
First Solar	Cadmium telluride (CdTe) solar cells Tandem CdTe and perovskites	[97]
Evolar	Tandem thin films Tandem perovskite solar cells	Acquired by first solar
Alta Devices	Gallium Arsenide	[98]
IMEC	Chalcogen based materials ($CuInSe_2$, $AgBiS_2$) Perovskites	N.A.
Fraunhofer ISE	III–V tri-junction GaAs	[99]
NREL	Strain-balanced GaAsP/GaInAs/GaAs	[95]

founded in 1999, is renowned as the world's sixth-largest manufacturer of PV cells and the largest in cadmium telluride (CdTe) solar cells. In May 2023, First Solar agreed to acquire Evolar, a Swedish company that specializes in perovskites. The purchase price of this acquisition is approximately $38 million, which will be initially paid with up to an additional $42 million to be paid subject to certain future technical milestones. Evolar was established in 2019 and has focused on developing solutions and manufacturing equipment for commercializing tandem solar technology utilizing perovskite thin films. The strategic goal of the acquisition is to integrate Evolar's technology into the development of tandem solar cells, which will combine both CdTe and perovskite materials, and with the potential to achieve solar efficiencies exceeding 25% [97]. Alta Devices uses gallium arsenide (GaAs) as the basis for solar technology; it produces and utilizes a very thin layer of GaAs (1–2 μ thick) produced via metal organic chemical vapor deposition (MOCVD) technique [98]. IMEC, Interuniversity Microelectronics Centre, an international R&D Organization, is involved in the development of thin-film organic solar cells and has successfully produced small-sized perovskite solar cells with efficiencies reaching up to 19%. Fraunhofer Institute for Solar Energy Systems, Fraunhofer ISE, has achieved 35.9% conversion efficiency for a III–V monolithic triple-junction solar cells based on silicon [99].

2.8 Conclusion

In conclusion, material selection in solar cell technology is a critical factor that influences efficiency, cost, and environmental sustainability. As the solar cell industry continues to evolve, researchers and engineers must carefully evaluate materials to meet the ever-growing demand for clean and renewable energy sources. Material choices will be central in shaping the future of solar cell technology, as it seeks to meet the world's energy needs while minimizing environmental impact.

References

[1] Kittel C 2005 *Introduction to Solid State Physics* (Wiley)

[2] Mojiri A *et al* 2013 Spectral beam splitting for efficient conversion of solar energy—a review *Renew. Sustain. Energy Rev.* **28** 654–63

[3] Huang G *et al* 2020 Challenges and opportunities for nanomaterials in spectral splitting for high-performance hybrid solar photovoltaic-thermal applications: a review *Nano Mater. Sci.* **2** 183–203

[4] https://en.wikipedia.org/wiki/Copper_indium_gallium_selenide (2023)

[5] Ravindra N M and Srivastava V K 1979 Temperature dependence of the energy gap in semiconductors *J. Phys. Chem. Solids* **40** 791–3

[6] 2022 Grand View Research, I.C.S.P.M.S. Aug, 2020 *Share & Trends Analysis Report by Application (Utility, EOR, Desalination), by Technology, by Region, and Segment Forecasts, 2020–2027* (https://grandviewresearch.com/industry-analysis/concentrated-solar-power-csp-market)

[7] https://ioffe.ru/SVA/NSM/Semicond/SiGe/bandstr.html (2023)

[8] https://nrel.gov/materials-science/photovoltaic-materials.html (accessed 17 December 2024)

[9] Sutherland B R 2020 Solar materials find their band gap *Joule* **4** 984–5

[10] Rühle S 2016 Tabulated values of the Shockley–Queisser limit for single junction solar cells *Solar Energy* **130** 139–47

[11] Gladkikh V *et al* 2020 Machine learning for predicting the band gaps of ABX3 perovskites from elemental properties *J. Phys. Chem.* C **124** 8905–18

[12] Wysocki J J and Rappaport P 1960 Effect of temperature on photovoltaic solar energy conversion *J. Appl. Phys.* **31** 571–8

[13] Zulehner W 2000 Historical overview of silicon crystal pulling development *Mater. Sci. Eng.:* B **73** 7–15

[14] Chittick R, Alexander J and Sterling H 1969 The preparation and properties of amorphous silicon *J. Electrochem. Soc.* **116** 77

[15] Le Comber P and Spear W 1970 Electronic transport in amorphous silicon films *Phys. Rev. Lett.* **25** 509

[16] Deb S K 1996 Thin-film solar cells: an overview *Renew. Energy* **8** 375–9

[17] Catalano A *et al* 1982 Attainment of 10% conversion efficiency in amorphous silicon solar cells *Conf. Rec. IEEE Photovoltaic Spec. Conf.* (Princeton, NJ: RCA Laboratories)

[18] Akerboom E *et al* 2022 Passive radiative cooling of silicon solar modules with photonic silica microcylinders *ACS Photonics* **9** 3831–40

[19] Pankove J I 1984 *Semiconductors and Semimetals: Device Applications* **21** (Academic Press)

[20] Staebler D L and Wronski C 1977 Reversible conductivity changes in discharge—produced amorphous Si *Appl. Phys. Lett.* **31** 292–4

[21] Energy U S D o 2011 *2010 Solar Technologies Market Report (U.S. Department of Energy)*

[22] Kang H 2021 Crystalline silicon vs. amorphous silicon: the significance of structural differences in photovoltaic applications *IOP Conf. Ser.: Earth Environ. Sci.* **726** 012001

[23] Stutzmann M, Jackson W B and Tsai C C 1985 Light-induced metastable defects in hydrogenated amorphous silicon: a systematic study *Phys. Rev. B Condens. Matter.* **32** 23–47

[24] Branz H M 1998 Hydrogen collision model of light-induced metastability in hydrogenated amorphous silicon *Solid State Commun.* **105** 387–91

[25] Meftah A, Meftah M and Merazga A 2004 Modelling of Staebler-Wronski effect in hydrogenated amorphous silicon under moderate and intense illumination *Defect Diffus. Forum* **230–232** 221–32

[26] Kaiser W *et al* 2018 Generalized kinetic monte carlo framework for organic electronics *Algorithms* **11** 37

[27] Almosni S *et al* 2018 Material challenges for solar cells in the twenty-first century: directions in emerging technologies *Sci. Technol. Adv. Mater.* **19** 336–69

[28] 2022 Union of Concerned Scientists https://www.ucsusa.org/

[29] Dominish E, Florin N and Teske S 2019 Responsible minerals sourcing for renewable energy *Report Prepared for Earthworks by the Institute for Sustainable Futures* (University of Technology Sydney) 54

[30] Batra I P, Tekman E and Ciraci S 1991 Theory of Schottky barrier and metallization *Prog. Surf. Sci.* **36** 289–361

[31] Tyagi M S 1984 Physics of Schottky barrier junctions *Metal-Semiconductor Schottky Barrier Junctions and their Applications* ed B L Sharma (Boston, MA.: Springer US) pp 1–60

[32] Rhoderick E 1970 The physics of Schottky barriers ? *Rev. Phys. Technol.* **1** 81

[33] Song Y *et al* 2015 Role of interfacial oxide in high-efficiency graphene–silicon Schottky barrier solar cells *Nano Lett.* **15** 2104–10

[34] https://eepower.com/technical-articles/photovoltaic-cell-fabrication-and-types/#

[35] Bansal I K, Cochran B, Goodrich J, Marcel M and Maniachi J 2002 13th annual IEEE/SEMI advanced semiconductor manufacturing conf *Advancing the Science and Technology of Semiconductor Manufacturing (ASMC)*

[36] Shinde S *et al* 2019 Novel method to address wafer surface condition *30th Annual SEMI Advanced Semiconductor Manufacturing Conf. (ASMC)*

[37] Stark C and Theristis M 2015 The impact of atmospheric parameters on the spectral performance of multiple photovoltaic technologies *2015 IEEE 42nd Photovoltaic Specialist Conference (PVSC) (New Orleans)*

[38] Mohammad Aminul I *et al* 2021 Assessing the impact of spectral irradiance on the performance of different photovoltaic technologies *Solar Radiation* ed A Mohammadreza (Rijeka: IntechOpen) ch 6

[39] Sharma K, Sharma V and Sharma S S 2018 Dye-sensitized solar cells: fundamentals and current status *Nanoscale Res. Lett.* **13** 381

[40] Shariatinia Z 2020 Recent progress in development of diverse kinds of hole transport materials for the perovskite solar cells: a review *Renew. Sustain. Energy Rev.* **119** 109608

[41] Nozik A J 2002 Quantum dot solar cells *Phys.* E **14** 115–20

[42] Su Q *et al* 2011 Green solar electric vehicle changing the future lifestyle of human *World Electr. Veh. J.* **4** 128–32

[43] Singh P and Ravindra N 2012 Temperature dependence of solar cell performance—an analysis *Solar Energ. Mater. Solar Cells* **101** 36–45

[44] Gad R *et al* 2023 Evaluation of thermal management of photovoltaic solar cell via hybrid cooling system of phase change material inclusion hybrid nanoparticles coupled with flat heat pipe *J. Energy Storage* **57** 106185

[45] Zhang Y *et al* 2022 Thermal management of portable photovoltaic systems using novel phase change materials with efficiently enhanced thermal conductivity *Sol. Energy Mater. Sol. Cells* **247** 111936

[46] Saadah M, Hernandez E and Balandin A A 2017 Thermal management of concentrated multi-junction solar cells with graphene-enhanced thermal interface materials *Appl. Sci.* **7** 589

[47] Yang T *et al* 2021 Phase change material heat sink for transient cooling of high-power devices *Int. J. Heat Mass Transfer* **170** 121033

[48] Akbarzadeh A and Wadowski T 1996 Heat pipe-based cooling systems for photovoltaic cells under concentrated solar radiation *Appl. Therm. Eng.* **16** 81–7

[49] Ali A Y M *et al* 2020 Impact of microchannel heat sink configuration on the performance of high concentrator photovoltaic solar module *Energy Rep.* **6** 260–5

[50] Poudel S, Zou A and Maroo S C 2022 Thermal management of photovoltaics using porous nanochannels *Energy Fuels* **36** 4549–56

[51] Zhao B *et al* 2020 Spectrally selective approaches for passive cooling of solar cells: a review *Appl. Energy* **262** 114548

[52] Wang Z *et al* 2020 Lightweight, passive radiative cooling to enhance concentrating photovoltaics *Joule* **4** 2702–17

[53] Meng Y *et al* 2022 Ultra-broadband, polarization-irrelevant near-perfect absorber based on composite structure *Micromachines* **13** 267

[54] PVinsights P 2021 *Solar PV module weekly spot price* http://pvinsights.com/

[55] Oberbeck L *et al* 2020 IPVF's PV technology vision for 2030 *Prog. Photovol.: Res. Appl.* **28** 1207–14

[56] NREL *Solar Manufacturing Cost Analysis* (accessed 28 May 2023)

[57] Dehghanimadvar M, Egan R and Chang N L 2022 Economic assessment of local solar module assembly in a global market *Cell Rep. Phys. Sci.* **3** 100747

[58] Čulík P *et al* 2022 Design and cost analysis of 100 MW perovskite solar panel manufacturing process in different locations *ACS Energy Lett.* **7** 3039–44

[59] Dullweber T and Schmidt J 2016 Industrial silicon solar cells applying the passivated emitter and rear cell (PERC) concept—a review *IEEE J. Photovolt.* **6** 1366–81

[60] Blakers A W *et al* 1989 22.8% efficient silicon solar cell *Appl. Phys. Lett.* **55** 1363–5

[61] Vodapally S N and Ali M H 2023 A comprehensive review of solar photovoltaic (PV) technologies, architecture, and its applications to improved efficiency *Energies* **16** 319

[62] Wang A 1992 *High Efficiency PERC and PERL Silicon Solar Cells* (UNSW) (Sydney)

[63] Taguchi M *et al* 2013 24.7% record efficiency HIT solar cell on thin silicon wafer *IEEE J. Photovolt.* **4** 96–9

[64] De Wolf S and Kondo M 2009 Nature of doped a-Si: H/c-Si interface recombination *J. Appl. Phys.* **105** 103707

[65] Tsunomura Y *et al* 2009 Twenty-two percent efficiency HIT solar cell *Solar Energy Mater. Solar Cells* **93** 670–3

[66] Lee S W *et al* 2020 Historical analysis of high-efficiency, large-area solar cells: toward upscaling of perovskite solar cells *Adv Mater.* **32** e2002202

[67] Green M A *et al* 2022 Solar cell efficiency tables (Version 60) *Prog. Photovolt.: Res. Appl.* **30** 687–701

[68] Zeng Y *et al* 2022 Review on metallization approaches for high-efficiency silicon heterojunction solar cells *Trans. Tianjin Univ.* **28** 358–73

[69] Messmer C *et al* 2020 Efficiency roadmap for evolutionary upgrades of PERC solar cells by topcon: impact of parasitic absorption *IEEE J. Photovolt.* **10** 335–42

[70] TOPCon solar cells: the new PV Module Technology in the Solar Industry *Solar Magzine* (2023) https://solarmagazine.com/solar-panels/topcon-solar-cells/
[71] Wang Y *et al* 2023 Dopant-free passivating contacts for crystalline silicon solar cells: progress and prospects *EcoMat* **5** e12292
[72] Zhou J *et al* 2022 Passivating contacts for high-efficiency silicon-based solar cells: from single-junction to tandem architecture *Nano Energy* **92** 106712
[73] Rhoderick E H and Williams R H 1988 *Metal-semiconductor Contacts* **129** (Oxford: Clarendon Press)
[74] Adari R *et al* 2014 Fermi-level depinning at metal/GaN interface by an insulating barrier *Thin Solid Films* **550** 564–8
[75] Lu H *et al* 2019 Modeling of surface gap state passivation and Fermi level de-pinning in solar cells *Appl. Phys. Lett.* **114** 222106
[76] Dullweber T *et al* 2018 Present status and future perspectives of bifacial PERC + solar cells and modules *Jpn. J. Appl. Phys.* **57** 08RA01
[77] Zhao J, Wang A and Green M A 2001 High-efficiency PERL and PERT silicon solar cells on FZ and MCZ substrates *Solar Energy Mater. Solar Cells* **65** 429–35
[78] Richter A *et al* 2021 Design rules for high-efficiency both-sides-contacted silicon solar cells with balanced charge carrier transport and recombination losses *Nat. Energy* **6** 429–38
[79] Yamaguchi M *et al* 2022 Analysis for efficiency potential of II–VI compound, chalcopyrite, and kesterite-based tandem solar cells *J. Mater. Res.* **37** 445–56
[80] Kurban M, Şimşek Y and Erkoç Ş 2023 II–VI semiconductors bandgap engineering *Handbook of II–VI Semiconductor-Based Sensors and Radiation Detectors: Volume 1, Materials and Technology* ed G Korotcenkov (Cham: Springer International Publishing) pp 109–31
[81] Carmody M *et al* 2010 Single-crystal II–VI on Si single-junction and tandem solar cells *Appl. Phys. Lett.* **96** 153502
[82] Wu X *et al* 2006 13·9%-efficient CdTe polycrystalline thin-film solar cells with an infrared transmission of ~50% *Prog. Photovol.: Res. Appl.* **14** 471–83
[83] Schmid M *et al* 2010 Experimental verification of optically optimized $CuGaSe_2$ top cell for improving chalcopyrite tandems *PV Direct* **1** 10601
[84] Valentini M *et al* 2019 Fabrication of monolithic CZTS/Si tandem cells by development of the intermediate connection *Sol. Energy* **190** 414–9
[85] Essig S *et al* 2017 Raising the one-sun conversion efficiency of III–V/Si solar cells to 32.8% for two junctions and 35.9% for three junctions *Nat. Energy* **2** 17144
[86] Geisz J F *et al* 2020 Six-junction III–V solar cells with 47.1% conversion efficiency under 143-Suns concentration *Nat. Energy* **5** 326–35
[87] Chen Y *et al* 2022 Design and analysis of III–V two-dimensional van der Waals hetero-structures for ultra-thin solar cells *Appl. Surf. Sci.* **586** 152799
[88] Li J *et al* 2021 A brief review of high efficiency III–V solar cells for space application *Front. Phys.* **8** 631925
[89] Torchynska T and Polupan G 2004 High efficiency solar cells for space applications *Superf. vacío* **17** 21–5
[90] Power A S S www.azurspace.com/index.php/en/ (accessed 22 October 2023)
[91] Spectrolab www.spectrolab.com/index.html (accessed 22 October 2023)
[92] SolAero Technologies I (accessed 22 October 2023)

[93] Verduci R *et al* 2022 Solar energy in space applications: review and technology perspectives *Adv. Energy Mater.* **12** 2200125

[94] Matson R J, Emery K A and Bird R E 1984 Terrestrial solar spectra, solar simulation and solar cell short-circuit current calibration: a review *Solar Cells* **11** 105–45

[95] Steiner M A *et al* 2021 High efficiency inverted GaAs and GaInP/GaAs solar cells with strain-balanced GaInAs/GaAsP quantum wells *Adv. Energy Mater.* **11** 2002874

[96] Verduci R *et al* 2022 Solar energy in space applications: review and technology perspectives *Adv. Energy Mater.* **12** 2200125

[97] Clemens K 2023 *First Solar Buys Evolar AB in Quest for Next-Gen Solar Cells*

[98] Admin B 2022 *Why Use Gallium Arsenide Solar Cells?—Alta Devices* https://www.altadevices.com/use-gallium-arsenide-solar-cells/ (accessed 22 October 2023)

[99] Bellini E 2021 *Fraunhofer ISE achieves 35.9% efficiency for III–V triple-junction solar cell based on silicon* https://www.pv-magazine.com/2021/04/23/fraunhofer-ise-achieves-35-9-efficiency-for-iii-v-triple-junction-solar-cell-based-on-silicon/

IOP Publishing

Recent Advances in Solar Cells

N M Ravindra, Leqi Lin and Priyanka Singh

Chapter 3

Copper indium gallium diselenide, cadmium telluride and copper zinc tin sulfide based solar cells

Thin-film solar cells continue to enhance their market share as promising energy conversion devices for indoor and outdoor solar cell applications. In this study, an overview of the recent developments in material specifications, deposition techniques and device architecture of copper indium gallium diselenide (CIGS), cadmium telluride (CdTe) and copper zinc tin sulfide (CZTS) solar cells are presented.

3.1 Introduction

Thin-film solar cells continue to enhance their role as a high-efficiency, reliable and cost-effective alternative to conventional monocrystalline silicon solar cells [1]. Several solar cell technologies including wafer-based, organic and thin films have been studied in the literature. Crystalline silicon has remained the most successful solar cell material candidate from laboratory scale to the commercial sector and makes up ~90% of the global photovoltaic (PV) market [2]. Since the first report of CIGS solar cells by Devaney and Mickelsen in 1987 [3], the technology has grown over the years. As an emerging mature material, CIGS is in high demand because of its better energy conversion efficiency (associated with its high absorption coefficient, tunable bandgap and flexibility) [4] than that of its crystalline counterparts as well as that of amorphous silicon. Normally, the structure of CIGS solar cell consists of a soda-lime glass (SLG) substrate and molybdenum (Mo) back contact at the bottom, CIGS as the interlayer, zinc oxide (ZnO)/aluminum (Al):zinc oxide at the top and n-type cadmium sulfide (CdS) as the window layer, and it is usually fabricated through the co-evaporation and precursor reaction processes (figure 3.1(a)).

CdTe (figure 3.1(b)) is one of the most robust material candidates due to its method of preparation and chemical stability as a semiconductor material for PV cell production. CdTe solar cells were introduced for the first time in 1972, by Bonnet

doi:10.1088/978-0-7503-5994-8ch3

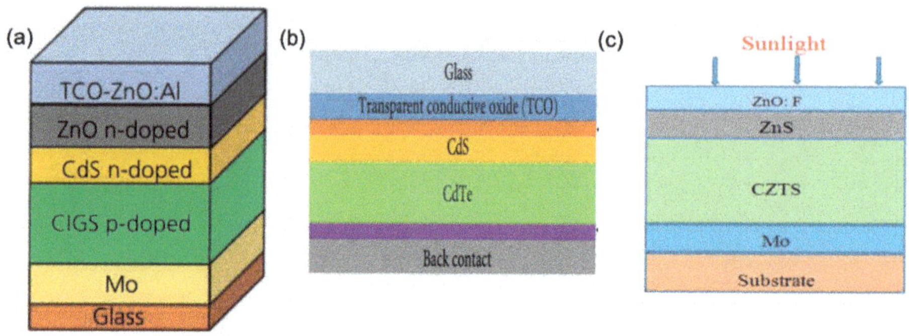

Figure 3.1. The fundamental structure of (a) CIGS, (b) CdTe and (c) CZTS solar cells.

and Rabenhorst with a CdS/CdTe heterojunction delivering a 6% efficiency [5]. The strong improvement in efficiency, during the past few years, was obtained by removing CdS and considering new features such as bandgap grading, copper embedding, and more transparent n-layer [6]. CdTe has an optimum bandgap near 1.5 eV with maximum power conversion efficiency (*PCE*) close to 32% according to the Schockley–Queisser limit with an open circuit voltage of more than 1 V and a short-circuit current density of more than 30 mA cm^{-2} [7]. 22.1% *PCE* is achieved nowadays with an oxygenation process that is introduced during CdSe sputtering to improve its window layer's property [8].

In considering the manufacture of CdTe solar cells, it must be noted that tellurium and indium are relatively scarce and highly expensive. The family of kesterite semiconductor materials such as CZTS (with the chemical formula Cu_2ZnSnS_4) are emerging as a non-toxic and relatively abundant semiconductor (figure 3.1(c)). Kesterites belong to the A_2BCX_4 family, with a tetragonal structure. The certified best *PCE* that has been achieved for these cells is 12.6%, far less than the Schockley–Queisser limit of 32.2%; this can be attributed to the fact that less attention has been paid to CZTS type PV cells than other thin film solar cells in the PV industry [9]. The challenges in the fabrication of kesterite PV cells have been summarized in recent reports; these include device degradation relating dominant factors such as massive deep defects, uncontrolled grain growth, band trailing, unoptimized interfaces. These issues need to be addressed for improved cell efficiency [10, 11]. Thin-film solar cells have significant potential and critical advantages in sunbelt countries with better temperature coefficients and ideal *PCE*s in adverse environments over crystalline silicon solar cells [12].

3.2 Influence of substrate material and substrate temperature

3.2.1 Influence of substrate materials

The substrates in thin film solar cells play a critical role in the development of the entire device. Deposition of the Mo back contact on SLG or flexible substrates will define different selenization conditions [13]. Conventionally, most commercial CIGS solar cells are produced on SLG substrates since SLG provides a sufficient source of

Na (alkali metal) to the absorber naturally and is able to sustain high temperature during deposition and annealing processes [14]. In recent years, thin film solar cells deposited on lightweight and flexible substrates have been studied for versatile applications [15]. For example, flexible CIGS solar cells can be integrated on rooftops or as a bending module with curved surfaces. So far, efficiencies of ~22.8% on SLG substrates, 17.5% on stainless steel (SS) substrates and 20.4% polyimide (PI) substrates for CIGS solar cells have been achieved in the literature [16, 17]. Among various SS substrates, stainless Cr steel is expensive but with lower coefficient of thermal expansion; unalloyed Fe steel is cheaper but with the corrosion issue and higher coefficient of thermal expansion (13 ppm K^{-1} for Cr-free steel compared to 11 ppm K^{-1} for Cr steel at 823 K) [18]. Flexible substrates are more advantageous than conventional SLG substrates for various applications. SLG is well known for its low-costs and can tolerate high temperatures during the fabrication process. PI foils can only tolerate a maximum process temperature below 500 °C, which is a barrier for CIGS solar cells under a high-temperature processing condition [19]. However, due to the significant progress in recent years, flexible PI has become the state-of-the-art option as a substrate compared to its glass counterpart.

Similar substrate materials are being used in the manufacture of CdTe PV cells, and share the same limiting factors of performance, cost and flexibility. Conventional photovoltaic devices have been usually manufactured on glass substrates with high transmittance. Hasani *et al* have investigated the relationships of optical bandgap of conductive [indium tin oxide (ITO) and fluorine-doped tin oxide (FTO)] and non-conductive (glass and quartz) substrates, which reveal that the grain size in the films is smaller on the non-conducting substrates due to their low conductivity than the conductive ones. The nature of the substrate has significant impact on the energy bandgap of CdTe [20]. The extinction coefficient (k) and refractive index (n), as functions of photon energy of CdTe thin films on various substrates, are presented in figure 3.2. Many CdTe back contact technologies utilize small amounts of copper to increase the p-type doping level. However, it is known that, in addition to providing p-type doping, copper introduces deep levels in CdTe, and copper is highly soluble in CdTe [21, 22].

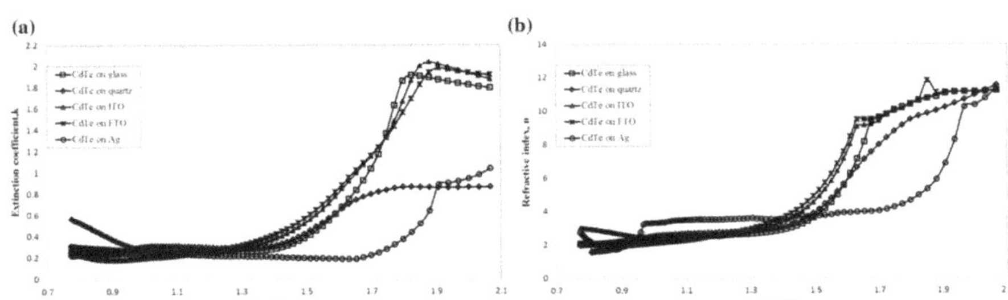

Figure 3.2. (a) Extinction coefficient, (b) refractive index of CdTe thin films on various substrates. Reproduced from [20], with permission from Springer Nature.

Table 3.1. Best efficiencies of CdTe solar cells on flexible substrates [25].

Substrate	Efficiency (%)	V_{oc} (mV)	FF (%)	J_{sc} (mA cm^{-2})
Upilex	10	796	63.8	19.7
Kapton	11.2	810	65.2	21.2
Metal foil	13.6	852	75.3	21.2
Thick glass	14.5	859	72.5	23.3
Ultra-thin glass	16.4	831	76.6	25.5

Metals tend to form Schottky contacts on p-type CdTe, as reported by Ponpon [23]. From the point of view of cost considerations, Ni and carbon are cost-effective and have work function values that are sufficiently high to avoid very large Schottky barriers in comparison to Au and Pt [24]. In table 3.1, it may be noted that the highest *PCE* is obtained when a CdTe PV cell is fabricated on ultra-thin glass, while the cell fabricated on Upilex (a heat-resistant polyimide film) exhibits the lowest *PCE*. This may be due to the expected reduction in the photocurrent density, resulting from the change in substrate transparency [25, 26]. However, the roll-to-roll (R2R) fabrication technique is a low-cost production, high efficiency and high throughout process that is appealing to researchers. Most commonly used flexible substrates in the industry are PI, PolyEthylene Terephthalate (PET), PolyEthylene Naphthalate (PEN) without impurity diffused from metal substrates. As compared to metal foils, flexible glass benefits from low contamination and roughness; as compared to polymer foils, flexible glass is compatible with high temperature processing and possesses high optical transparency which enables more photons to reach the absorber; cells fabricated on these substrates share similar dominant factors to other thin film PV cells [19].

The use of scarce materials such as Ga and In is considered to be a major hindrance for scaling up solar cells that utilize these materials. Cu_2ZnSnS_4 (CZTS), with a similar structure, and low-cost potential is rising as a promising and alternative option in thin film PV cells. Wang *et al* investigated process sequence of CZTS on different substrate materials and built the cost model for the estimation of production cost for each substrate by using Monte Carlo method; this is shown in figure 3.3 [27]. Selenium-containing CZTSSe solar cell has achieved world record efficiency of 12.6%, and pure sulfide, CZTS, has recorded laboratory efficiency of 11% on 0.24 cm^2 area, and 10% on 1 cm^2 [28]. However, CZTS on glass substrates cannot provide alternative advantages to other commercially available PV cells such as its strong competition to monocrystalline silicon which are heavy, non-flexible and rigid. The configuration of being flexible has the advantage of high production volume by roll-to-roll manufacturing. In 2018, UNSW achieved an efficiency of 6.2% on 0.24 cm^2 stainless steel substrates, but the transmittance of metal substrate was very limited [28]. The use of Au, Pt and Ag give good conductivity as the substrate. Nevertheless, the price of these noble metals is too high for wide applications. Molybdenum foil, as one of the most popular metals nowadays, by

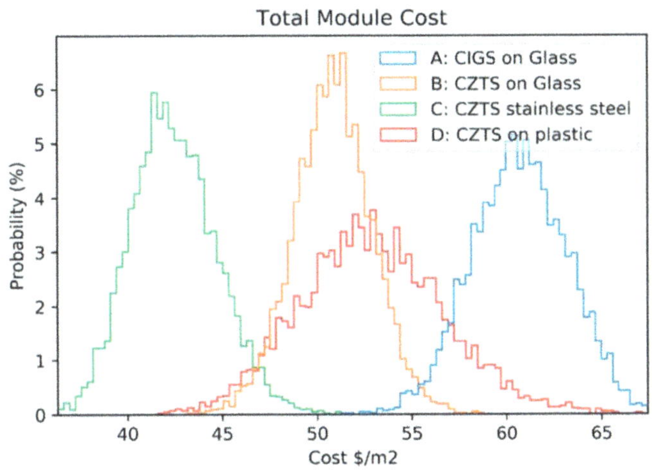

Figure 3.3. Total module cost for different product types based on the assumption of 1 GW per year manufacturing volume. Reproduced from [27] with permission from the Royal Society of Chemistry.

making the foil thin enough as a flexible substrate, has achieved 10% *PCE* [29]. Overall, it has been observed that CZTS on Mo-foil exhibits better properties in terms of crystalline quality, surface microstructure, and optical and electrical properties, which are favorable for PV cells instead of the rigid glass substrates [30]. In 2023, Li *et al* investigated the performance of CZTS PV cell on TCO consisting of ZnO/Cu/ZnO films and obtained *PCE* of 6.01% without bending; this is higher than that of CSCs with ITO [31]. Recently, researchers suggested that ZnO films and metal films formed a dielectric layer/metal/dielectric layer (DMD) structure to improve the films' optoelectronic properties for use in solar cell structures [32, 33].

3.2.2 Influence of substrate temperatures

This subsection has been reproduced with permission from [12].

High substrate temperature will not only enhance the growth of the absorber layer but also favor alkali-diffusion from substrate to absorbers. Most of the high-efficiency thin film modules have been prepared at substrate temperature (T_{sub}) of over 773 K which is very close to the softening temperature of SLG substrates. Low substrate temperature (<723 K) provides the feasibility for reducing the thermally induced stress on the substrate, for example, by depositing Mo on PI [34]. PI substrate is suitable for continuous and rapid roll-to-roll deposition process and hence reduces the manufacturing cost of the thin film solar cell module [35]. The use of PI substrate reduces T_{sub} and can lead to a decrease in manufacturing costs. The use of high substrate temperature with glass substrates promises enough thermal energy to form high-quality thin films.

T_{sub} plays an important role in providing the thermal energy for the interdiffusion between atoms, which is also related to the phase transformations in the growth process of CIGS. In the case CIGS/SLG, with T_{sub} in the temperature range of 623–773 K, by using one-stage co-evaporation process, the enhancement in *PCE*, *FF* and V_{oc} is observed when T_{sub} is increased from 573 K to 773 K [36]. Especially, a significant increase, above a critical temperature of 723 K, is observed due to the drastic increase in carrier concentration. This is because of the diffusion of Na content from SLG into the CIGS absorber at temperatures above 723 K, which also decreases the resistivity. The *FF* and V_{oc} have similar increasing trends as *PCE* which may be ascribed to the improvement in Na diffusion at higher temperature on SLG substrates. Additionally, the decrease in V_{oc} with decreasing T_{sub} can be explained by the recombination occurring in the grain boundaries due to the smaller grain size at low temperatures [37]. Low J_{sc} values at lower T_{sub} (<673 K) can be attributed to the poor quality of CIGS films which reduces the carrier collection. It is worthwhile to note that several papers in the literature have reported slight decrease in J_{sc} after the substrate temperature reaches ~ 823 K, which is in contrast to other parameters such as *FF*, V_{oc} and *PCE* [36]. In the case of CIGS cells/SLG substrates which have the same structure of glass/Mo/CIGS/CdS/i-ZnO/Al–ZnO, it is commonly expected that, with increase in T_{sub} (over 800 K), enhancement in larger grain growth occurs. Eventually, they form big blocks of grains when T_{sub} is in the temperature range of 823 K–873 K, which shows that high T_{sub} can give enough thermal energy to form high quality CIGS films. The high *PCE* at T_{sub} of 823 K is related to Na incorporation and the disappearance of secondary-phase $(In_{0.62}Ga_{0.38})_2Se_3$, which helps to process films with better structural properties. A comparison of *FF–T* and *PCE–T* behavior between Liang's study on SS substrates and other studies in the literature on different substrates (e.g., SLG and PI) is presented in figure 3.4. In addition, the low *FF* values for lower substrate temperature (<723 K) for both SS and SLG substrates can be ascribed to the larger grain boundaries. However, for high T_{sub}, close enough to material softening temperature (in the case of SS substrate, it is ~ 773 K), a high concentration of impurities may diffuse into CIGS modules and act as shunt path [38].

3.3 Recent results

This section has been reproduced with permission from [12].

Currently, most of the high-efficiency thin film solar cells on flexible substrates are made by the incorporation of alkali elements such as sodium and potassium. Besides SLG, PI substrates are preferable due to nonmetallic impurities. In 2020, Kim *et al* proposed CIGS incorporation with both sodium and potassium on PI substrates using low-temperature *in situ* post-deposition treatment (PDT) without affecting the growth and microstructural properties of CIGS films [42]. In recent years, the use of SS substrate has caught researchers' interest due to a better substrate temperature dependence of PV parameters with polished surfaces (low irregularities), which could possibly reduce pin holes and shunts during subsequent processing [43]. It is well known that CIGS absorbers in high-efficiency cells are normally deposited at high T_{sub} (>823 K) on rigid SLG substrates. However, for temperature sensitive flexible PI substrates, its thermal stability limits the process temperatures to about 723 K, which is 100 K less than SLG; thus, a low substrate temperature deposition

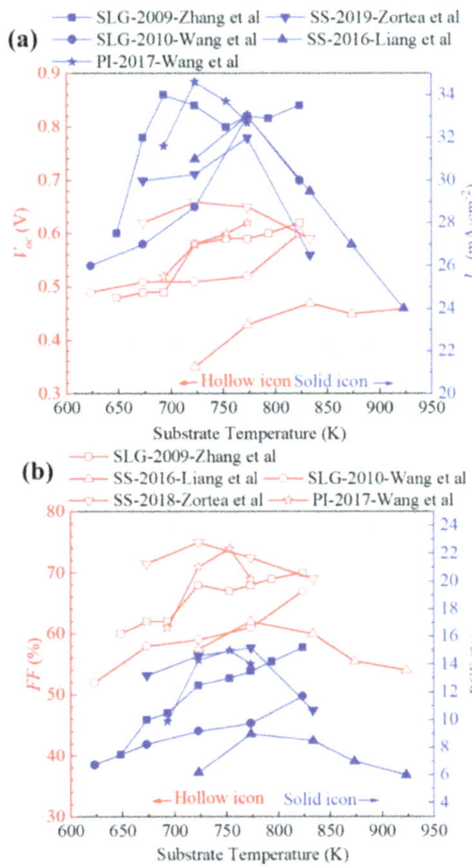

Figure 3.4. J–V parameters of CIGS solar cells on different substrates at different T_{sub}. Data extracted from references [36, 39–41]. Reproduced from [12] with permission from Springer Nature.

technique is necessary while deploying flexible substrates. T_{sub} has a stronger influence on PI with respect to the thermal absorption and emission properties when compared to SLG. Accordingly, the CIGS absorber with relatively low defects, formed on PI, leads to a relatively long carrier lifetime and a reduction in recombination probabilities.

Rahman *et al* designed and numerically analyzed that a *PCE* of 31.86% with V_{oc} of 0.9 V is achieved with a 1 μm-thick-CIGS absorber layer with V_2O_5 BSF and ZnSe buffer layers in this structure [44]. Very recently, Arashti *et al* used SCAPS-1D simulator to simulate the designed cells with CdS/quartz, CdS/ITO, and CdS/FTO layers and obtained a *PCE* of 19.40%, 21.23% and 21.16%, respectively. Furthermore, the optimized cell was obtained for a Au/CdTe/CdS/ITO structure with the efficiency of 22.80% by employing a 3 μm thickness of the CdTe layer at a device temperature of 300 K [45]. In 2023, Li *et al* investigated the performance of CZTS PV cell on TCO, consisting of ZnO/Cu/ZnO, and obtained a *PCE* of 6.01% without bending. In this case, the team obtained *PCE* higher than that of CSCs with

Table 3.2. Recent developments in the performance of CIGS, CdTe and CZTS solar cells.

References	V_{oc} (mV)	J_{sc} (mA/cm^2)	*FF* (%)	*PCE*	Structure
2020 [42]	697	35.9	74.8	18.7	Mo/KF-NaF PI/Mo/CIGS/CdZ/i-ZnO/ITO/(Ni)Al/MgF2
2020 [42]	640	36.1	67.8	16.2	Mo/KF PI/Mo/CIGS/CdZ/i-ZnO/ITO/(Ni)Al/MgF2
2021 [46]	835	35.22	82.29	24.22	Al/ITO/Al–ZnO/i-ZnO/CdS/CIGS/PbS/Mo/Substrate
2022 [47]	840	32.55	85.31	23.42	ZnO/CdS/CIGS/BSF:c-Si:H/Mo
2022 [48]	700	34.8	—	18.5	RF sputtered TCO/Ag/TCO/CIGS
2020 [49]	861	26.9	75.4	17.5	CuCl-RTA treated CdTe
2021 [50]	805	—	—	—	Al/p-CdTe/ZnSe/glass
2021 [51]	1101	35.18	83.86	29.86	FTO/ZnO/CdS/CZTS/CZTSe/Pt
2020 [52]	980	25.31	63.2	15.81	CZTS/CdS(40 nm)/ZnS(40 nm)/ZnO:Al
2020 [52]	1006	26.23	54.4	16.52	CZTS/CdS(40 nm)/In2S3(40 nm)/ZnO:Al

ITO film [31]. Recently, researchers suggested that ZnO films and metal films formed a dielectric layer/metal/dielectric layer (DMD) structure to improve the films' optoelectronic properties [32, 33]. Table 3.2 presents a summary of the recent progress in the performance of CIGS, CdTe and CZTS solar cells.

3.4 Conclusions

An overview of the CIGS, CdTe and CZTS thin film solar cells was presented in this chapter. In this context, the importance and impact of the substrate material and temperature on the solar cell performance was investigated. The recent progress in the device structure and characteristics of thin film solar cells was summarized.

References

[1] Deb S K 1996 Thin-film solar cells: an overview *Renew. Energy* **8** 375–9

[2] Devaney W E and Mickelsen R A 1988 Vacuum deposition processes for $CuInSe_2$ and $CuInGaSe_2$ based solar cells *Sol. Cells* **24** 19–26

[3] Wei S-H, Zhang S B and Zunger A 1998 Effects of Ga addition to $CuInSe_2$ on its electronic, structural, and defect properties *Appl. Phys. Lett.* **72** 3199–201

[4] Sharma S, Jain K and Sharma A 2015 Solar cells: in research and applications—a review *Mater. Sci. Appl.* **06** 1145–55

[5] Bonnet D R H 1972 *Proc. of the 9th Photovoltaic Specialists Conf. (Silver Spring, MD)* 129–31

[6] Romeo A and Artegiani E 2021 CdTe-based thin film solar cells: past, present and future *Energies* **14** 1684

[7] Rühle S 2016 Tabulated values of the Shockley–Queisser limit for single junction solar cells *Sol. Energy* **130** 139–47

[8] Hu A *et al* 2020 Improving CdTe-based thin-film solar cell efficiency with the oxygenated CdSe layer prepared by sputtering process *Phys. Status Solidi (A)* **217** 2000560
[9] Wang W *et al* 2014 Device characteristics of CZTSSe thin-film solar cells with 12.6% efficiency *Adv. Energy Mater.* **4** 1301465
[10] Liu F, Wu S, Zhang Y, Hao X and Ding L 2020 Advances in kesterite $Cu_2ZnSn(S, Se)_4$ solar cells *Sci. Bull.* **65** 698–701
[11] Bade B R *et al* 2021 Investigations of the structural, optoelectronic and band alignment properties of Cu_2ZnSnS_4 prepared by hot-injection method towards low-cost photovoltaic applications *J. Alloys Compd.* **854** 157093
[12] Lin L and Ravindra N 2020 Temperature dependence of CIGS and perovskite solar cell performance: an overview *SN Appl. Sci.* **2** 1361
[13] Ong K H *et al* 2018 Review on substrate and molybdenum back contact in CIGS thin film solar cell *Int. J. Photoenergy* **2018** 9106269
[14] Li W, Yan X, Aberle A G and Venkataraj S 2017 Effect of a TiN alkali diffusion barrier layer on the physical properties of Mo back electrodes for CIGS solar cell applications *Curr. Appl. Phys.* **17** 1747–53
[15] Khelifi S, Marlein J, Burgelman M and Belghachi A 2008 Electrical characteristics of $Cu(In, Ga)Se_2$ thin films solar cells on metallic foil substrates *Proc. of the 23rd European Photovoltaic Solar Energy Conf.* pp 2165–9
[16] Kato T, Wu J L, Hirai Y, Sugimoto H and Bermudez V 2019 Record efficiency for thin-film polycrystalline solar cells up to 22.9% achieved by Cs-treated $Cu(In,Ga)(Se,S)_2$ *IEEE J. Photovolt.* **9** 325–30
[17] Chirilă A *et al* 2013 Potassium-induced surface modification of $Cu(In,Ga)Se_2$ thin films for high-efficiency solar cells *Nat. Mater.* **12** 1107–11
[18] Wuerz R *et al* 2009 CIGS thin-film solar cells on steel substrates *Thin Solid Films* **517** 2415–8
[19] Ramanujam J *et al* 2020 Flexible CIGS, CdTe and a-Si:H based thin film solar cells: a review *Prog. Mater Sci.* **110** 100619
[20] Hasani E, Kamalian M, Gholizadeh Arashti M and Babazadeh Habashi L 2019 Effect of substrate properties on nanostructure and optical properties of CdTe thin films *J. Electron. Mater.* **48** 4283–92
[21] Kharangarh P, Misra D, Georgiou G E and Chin K K 2012 Evaluation of Cu back contact related deep defects in CdTe solar cells *ECS J. Solid State Sci. Technol.* **1** Q110
[22] Capper P 1994 *Properties of Narrow Gap Cadmium-Based Compounds* (IET)
[23] Ponpon J P 1985 A review of ohmic and rectifying contacts on cadmium telluride *Solid-State Electron.* **28** 689–706
[24] Hall R S, Lamb D and Irvine S J C 2021 Back contacts materials used in thin film CdTe solar cells—a review *Energy Sci. Eng.* **9** 606–32
[25] El-Atab N and Hussain M 2020 Flexible and stretchable inorganic solar cells: progress, challenges, and opportunities *MRS Energy Sustain.* **7**
[26] Salavei A *et al* 2012 Flexible CdTe solar cells by a low temperature process on ITO/ZnO coated polymers *27th European Photovoltaic Solar Energy Conf. and Exhibition*
[27] Wang A *et al* 2021 Analysis of manufacturing cost and market niches for Cu_2ZnSnS_4 (CZTS) solar cells *Sustain. Energy Fuels* **5** 1044–58
[28] Yan C *et al* 2018 Cu_2ZnSnS_4 solar cells with over 10% power conversion efficiency enabled by heterojunction heat treatment *Nat. Energy* **3** 764–72

[29] Yang K-J *et al* 2019 Flexible $Cu_2ZnSn(S,Se)_4$ solar cells with over 10% efficiency and methods of enlarging the cell area *Nat. Commun.* **10** 2959

[30] Yagmyrov A, Erkan S, Başol B M, Zan R and Olgar M A 2023 Impact of the ZnS layer position in a stacked precursor film on the properties of CZTS films grown on flexible molybdenum substrates *Opt. Mater.* **136** 113423

[31] Li H *et al* 2023 Investigation on highly flexible CZTS solar cells using transparent conductive ZnO/Cu/ZnO films *Colloids Surf.,* A **663** 131084

[32] Ekmekcioglu M *et al* 2021 High transparent, low surface resistance ZTO/Ag/ZTO multilayer thin film electrodes on glass and polymer substrates *Vacuum* **187** 110100

[33] Ye Z *et al* 2022 Reduction of the water wettability of Cu films deposited on liquid surfaces by thermal evaporation *Colloids Surf.,* A **650** 129569

[34] Zhang L *et al* 2012 Structural, optical and electrical properties of low-temperature deposition $Cu(In_xGa_{1-x})Se_2$ thin films *Sol. Energy Mater. Sol. Cells* **99** 356–61

[35] Reinhard P *et al* 2013 Review of progress toward 20% efficiency flexible CIGS solar cells and manufacturing issues of solar modules *IEEE J. Photovolt.* **3** 572–80

[36] Zhang L *et al* 2009 Effects of substrate temperature on the structural and electrical properties of $Cu(In,Ga)Se_2$ thin films *Sol. Energy Mater. Sol. Cells* **93** 114–8

[37] Lammer M, Klemm U and Powalla M 2001 Sodium co-evaporation for low temperature Cu $(In,Ga)Se_2$ deposition *Thin Solid Films* **387** 33–6

[38] Bae D, Kwon S, Oh J, Kim W K and Park H 2013 Investigation of Al_2O_3 diffusion barrier layer fabricated by atomic layer deposition for flexible $Cu(In,Ga)Se_2$ solar cells *Renew. Energy* **55** 62–8

[39] Liang X *et al* 2016 Substrate temperature optimization for $Cu(In, Ga)Se_2$ solar cells on flexible stainless steels *Appl. Surf. Sci.* **368** 464–9

[40] Zortea L *et al* 2018 $Cu(In,Ga)Se_2$ solar cells on low cost mild steel substrates *Sol. Energy* **175** 25–30

[41] Wang H *et al* 2010 Effect of substrate temperature on the structural and electrical properties of CIGS films based on the one-stage co-evaporation process *Semicond. Sci. Technol.* **25** 055007

[42] Kim K *et al* 2020 Mechanisms of extrinsic alkali incorporation in CIGS solar cells on flexible polyimide elucidated by nanoscale and quantitative analyses *Nano Energy* **67** 104201

[43] Khelifi S *et al* 2010 Characterization of flexible thin film CIGSe solar cells grown on different metallic foil substrates *Energy Procedia* **2** 109–17

[44] Md. Ferdous Rahman N M, Alam I, Md. Hasan A, Moon M M A, Kuddus A, Ishraque Toki G F, Rubel M H K, Abdullah Al Asad M and Khalid Hossain M 2023 Design and numerical analysis of CIGS-based solar cell with V_2O_5 as the BSF layer to enhance photovoltaic performance *AIP Adv.* **13** 045309

[45] Gholizadeh Arashti M, Hasani E, Babazadeh Habashi L and Kamalian M 2023 Effect of substrate type on the physical properties of thermally evaporated CdS Thin films for CdTe/CdS solar cells applications *Phys. Scr.* **98** 065403

[46] Barman B and Kalita P K 2021 Influence of back surface field layer on enhancing the efficiency of CIGS solar cell *Sol. Energy* **216** 329–37

[47] Zouache R, Bouchama I, Saidani O, Djedoui L and Zaidi E 2022 Numerical study of high-efficiency CIGS solar cells by inserting a BSF μc-Si:H layer *J. Comput. Electron.* **21** 1386–95

[48] Kacha K *et al* 2022 Efficiency improvement of CIGS solar cells using RF sputtered TCO/Ag/TCO thin-film as prospective buffer layer *Ceram. Int.* **48** 20194–200

[49] Li D-B *et al* 2020 Maximize CdTe solar cell performance through copper activation engineering *Nano Energy* **73** 104835

[50] Elsaeedy H I, Hassan A A, Yakout H A and Qasem A 2021 The significant role of ZnSe layer thickness in optimizing the performance of ZnSe/CdTe solar cell for optoelectronic applications *Opt. Laser Technol.* **141** 107139

[51] Rana M S, Islam M M and Julkarnain M 2021 Enhancement in efficiency of CZTS solar cell by using CZTSe BSF layer *Sol. Energy* **226** 272–87

[52] Tripathi S, Sadanand, Lohia P and Dwivedi D K 2020 Contribution to sustainable and environmental friendly non-toxic CZTS solar cell with an innovative hybrid buffer layer *Sol. Energy* **204** 748–60

IOP Publishing

Recent Advances in Solar Cells

N M Ravindra, Leqi Lin and Priyanka Singh

Chapter 4

Perovskite solar cells

In recent years, perovskite solar cells (PSCs) have emerged as a highly promising technology in photovoltaics. In this chapter, the key features and advancements in PSCs, with emphasis on their potential in the field of solar energy, are addressed. PSCs offer several advantages over traditional silicon-based photovoltaic (PV) systems, including high absorption coefficient, tunable bandgaps, high efficiency, low production costs, and versatility in design. However, the material and cell degradation as well as their hysteresis related issues continue to pose a challenge from the perspective of their routine industrial manufacture, large scale deployment, reliability and life expectancy.

4.1 Historical perspective

The discovery of calcium titanate ($CaTiO_4$) was made by Gustav Rose in 1839, who identified this perovskite mineral in the Ural mountains, although it was not the correct chemical composition for a perovskite [1, 2]. During the same year, $CaTiO_3$, a calcium titanium oxide, well known as the perovskite, was discovered and named after Gustav Rose's colleague Lev Perovski. In 1926, the crystallographer Victor Goldschmidt determined the crystal structure of the perovskite in his work by proposing a tolerance factor [3]. The crystal structure of $CaTiO_3$ is composed of many three-dimensional arrangements of metal cations within an oxygen octahedral framework. Later on, in 1945, Helen Dick Megaw characterized the crystal structure of perovskites by performing x-ray diffraction (XRD) on barium titanate [4]. In the 20th century, perovskite oxide materials, such as barium titanate ($BaTiO_3$), strontium titanate ($SrTiO_3$), and lead zirconate titanate (PZT-$Pb[Zr_xTi_{1-x}]O_3$), were discovered with many applications in electronic devices, including capacitors and ferroelectric devices for their unique electrical and dielectric properties [5, 6].

In 2009, the first demonstration of the PV effect, using perovskites, was reported by Miyaska and his co-workers, although the *PCE* was relatively low with a value of only 3.8% [7]. Right after that, the groundbreaking discovery of organic–inorganic

doi:10.1088/978-0-7503-5994-8ch4

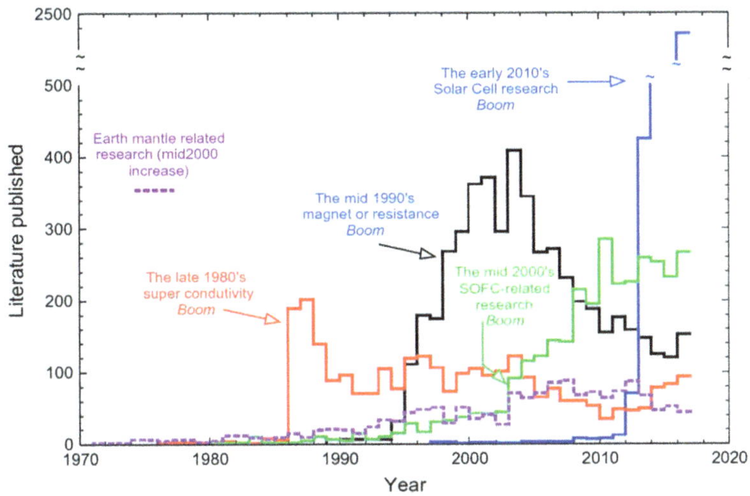

Figure 4.1. The amount of literature published from 1970 to 2017 on perovskite-related research obtained from SCOPUS database. Notes, erratum, editorials, book reviews, and news have been removed from the results. Reproduced from [13] with permission from Springer Nature.

hybrid perovskites as light-absorbing materials in solar cells was reported [8, 9]. Subsequently, PSCs became a hot topic in the renewable energy community, offering a potential low-cost alternative to traditional silicon-based solar cells. Beyond solar cells, perovskites have been explored for a wide range of applications, including catalysis, fuel cells, lasers, light-emitting diodes (LEDs), memory devices and sensors [10–12]. In figure 4.1, the amount of literature published from 1970 to 2017 on perovskite-related research, obtained from SCOPUS database, is presented [13].

4.2 Introduction

PSCs utilize metal-halide perovskite materials as the light absorber. PSCs are usually fabricated with the structure of methylammonium lead halide [$CH_3NH_3PbX_3$ ($MAPbI_3$)], where X = chlorine (Cl), bromine (Br) and/or iodine (I)) semiconductor, which is sandwiched between an electron-transport material (ETM) and a hole-transport material (HTM). Due to the significant improvements and advancements in device fabrication procedures [14], chemical composition [15] and phase stabilization methods [16], PSCs are one of the most efficient and low-cost solution-processing third-generation solar cells that provide unprecedented certified record efficiency (25.7%, certified by NREL) which is almost catching up with the best silicon solar cells (26.7%) [17]. It is worth mentioning that it took four decades for silicon solar cells to achieve the record efficiency, as opposed to only one decade for PSCs to attain similar cell efficiency. The history of the evolution of PSCs-based cell efficiencies is illustrated in figure 4.2(a). PSCs have several advantages over other types of solar cells. The V_{oc} of PSCs is relatively higher, and the perovskite materials utilize solution processing which makes it cost-effective, promising and convenient

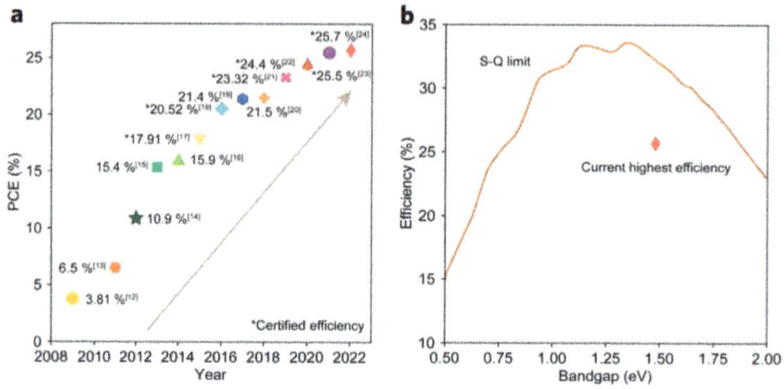

Figure 4.2. (a) The history of evolution of *PCE* of PSCs and (b) S–Q limitation for PSCs [24] John Wiley & Sons. Copyright 2022 Wiley-VCH GmbH.

for the fabrication of thin film solar cells. However, the associated drawbacks with the solution processing are the large number of defects, structural disorder, and chemical inhomogeneity, which lead to a significant reduction in device performance. Nevertheless, PSCs with *PCE* of 25.2% exhibit a V_{oc} of 1.193 V for a material with a bandgap of 1.56 eV, which indicates only 0.367 eV loss from the bandgap and only 0.034 V loss from the radiative V_{oc} limit (S–Q limit) of 1.270 V [18], as shown in figure 4.2(b). These days, several research groups are working on strategies that suppress detrimental effects in reducing energy loss. In 2013, Seok *et al* reported 12.3% *PCE* by using the mixed-halide $CH_3NH_3PbI_{3-x}Br_x$ (10%–15% bromine). A low bromine content (< 10%) yielded better initial efficiency due to the lower bandgap; however, higher bromine content (> 20%) provided better stability under high humidity with a poly-triarylamine HTM [19]. In 2019, Tian *et al* developed a synergic interface design with 16.2% *PCE* by applying an amino-functionalized polymer (PN4N) as cathode interlayer and (PDCBT) as dopant-free anode interlayer. This device exhibited a less than 10% decrease in *PCE* after 400 h of continuous run under 1 sun equivalent illumination [20]. Wen *et al* first proposed carbon quantum dots (CQDs) as an additive in methylammonium lead triiodide for reducing the defects that are caused by grain migration. PSCs achieved 19.17% *PCE* with 0.04 mg·ml^{-1} concentration of CQDs additive and, at the same time, improved the device stability [21].

As a result, perovskite-based single-junction and tandem cells are considered as having a high potential to be commercially available. Several companies such as Oxford PV (UK), Microquanta (China), Utmo-Light (China), Saule Technologies (Poland), and Toshiba (Japan), with the support of huge investments, are already on their way to exploring commercialization by manufacturing and testing large-area PSC single-junction (rigid or flexible) or perovskite/Si tandem PV panels [22]. In 2020, the Helmholtz Association of German Research Centers reported 29.15% *PCE* for the perovskite silicon tandem solar cell [23]. This approach to PSC provides a new electrode contact layer with an improvement in the perovskite compound, which is more stable when illuminated in a tandem solar cell. Initial experiments have already shown these new PSCs to be suitable for large surface areas with the help of vacuum deposition processes.

4.3 Material considerations

Ideal organic–inorganic halide perovskites, with the ABX_3 formula, have a simple cubic structure, where X = chlorine (Cl), bromine (Br) and/or iodine (I) semiconductor, with A = methylammonium ion (MA^+), formamidinium (FA^+), Cs^+; B = Pb^{2+}, Sn^{2+}. The PSCs consist of an absorber layer (for example: $CH_3NH_3PbX_3$), which is inserted between electron-transport layer (ETL) and hole-transport layer (HTL) [25], as shown in figure 4.3. In traditional PSCs, ETL is a colloidal thin-film of SnO_2 [26], TiO_2 [27], ZnO or their mesoporous systems [28], which consist of large grain boundaries and weak recombination at the interface [29]. Specifically, semiconductor ETLs possess natural defects that are due to oxygen vacancies and trap-assisted recombination [30]. The rapid growth of PSCs is mainly attributed to unique properties of metal perovskite materials such as high absorption coefficient and tunable bandgap [31]. However, Pb toxicity poses a barrier for further development; Pb dissolves in liquids and leads to adverse effects on both the environment and the human body. Thus, replacement of Pb with materials with similar properties such as Ge and Sn, without toxicity issues, are proposed as lead-free perovskite materials. Furthermore, researchers have found that replacing the organic components, methylamine (MA) and formamidine (FA), in the perovskite absorber layer with Cs can substantially improve the thermal stability of the material and reduce its sensitivity to humidity [32]. Figure 4.4 shows the band alignment of commonly used metal-oxide ETLs and HTLs with respect to metal-halide perovskite absorbers [33]. In 2023, Zhang *et al* developed an all-inorganic HTL-free $CsGeI_3$ PSC (FTO/ZnOS/$CsGeI_3$/W) with *PCE* up to 26.7%, which provides an alternative metal perovskite material for commercial applications [32]. In recent years, flexible PSCs have attracted the interest of much of the PV research community. Jung *et al* [34], demonstrated Au/PTAA/$MAPbI_3$/MoS_2/Graphene on PET substrate with *PCE* of 11.83%, which exhibited high performance even after 1000 inner and outer bendings at radius of 2 mm.

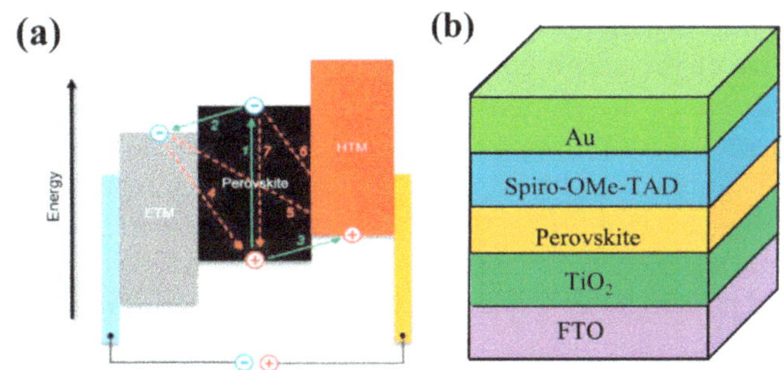

Figure 4.3. (a) The energy levels and charge-transfer process in PSCs and (b) PSC structure [35] CC By 4.0.

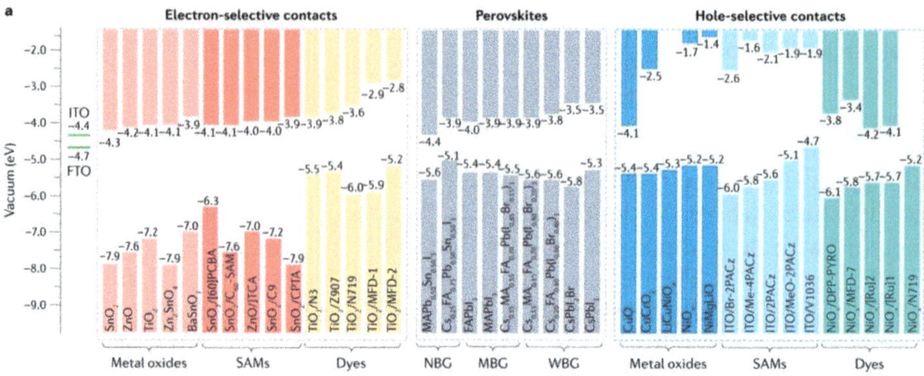

Figure 4.4. Band alignment of ETLs and HTLs with several commonly used perovskites absorbers. Reproduced from [33] with permission from Springer Nature.

In order to improve the efficiency of PSCs, various HTMs have been tested with several types of alternative materials. HTMs play a key role in the process of extraction, collection and transportation of photogenerated holes; they are responsible for the PV performance and stability of PSCs [36]. Although Spiro-OMeTAD (2,2′,7,7′-tetrakis-(N,N-di-4-methoxyphenylamino)-9,9′-spirobifluorene) is the most widely used organic HTM, the verbosely and costly synthetic process hinders the possibility of its commercialization [25]. Thus, it is necessary to develop new HTMs as potentially commercial candidates for PSCs. In 2018, Sun *et al* [37] reported a new fluoranthene-based dopant-free HTMs (dope-free BTF4) with an impressive 18.03% *PCE*. In comparison with conventional Sprio-OmeTAD, this novel dopant-free BTF4 has better *PCE* and enhanced stability. In 2019, Chen *et al* [38] developed two potential alternatives, taking the prototypical spiro-bifluorene core in spiro-OMeTAD as reference, non-spiro (FH-0) and mono-fluorene(FH-3) as the cores to link two methoxyl-triphenylamine terminals. The performance of these two FH-0- and FH-3-based PSCs are assessed under STC with 14.9% and 18.3% *PCE* separately. FH-3, especially, displays moderate energy level, good thermal stability, proper structure packing trends and comparable efficiency with the spiro-OMeTAD type under STC, which can be a promising HTM material for future development. Besides new HTM materials, the interest in dopant-free HTMs are continuing for several reasons, notably that dopants are environmentally unfriendly and can easily cause the stability of the entire device to deteriorate [37, 39]. In 2019, Ye *et al* [39] investigated the effect of dopants on the PV performance with four novel benzotriazole-based donor-acceptor-donor (D-A-D) type HTMs BTA1–2 and DT1–2 in PSCs (figure 4.5). The dopant-free HTMs DT1–2 with DTBT core exhibit better PV performance than BTA1–2 HTMs which achieve 13.22% *PCE*. These four novel HTM materials also show better stability than conventional spiro-OMeTAD.

In the inverted n-i-p-type PSCs, inorganic metal oxides such as TiO_2 and SnO_2 are often used as ETMs, while organic materials such as C60 and phenyl-C61-butyric acid methyl ester (PCBM) are often used as ETMs. In order to overcome some drawbacks of fullerenes such as limited energy level variation, thermal instability

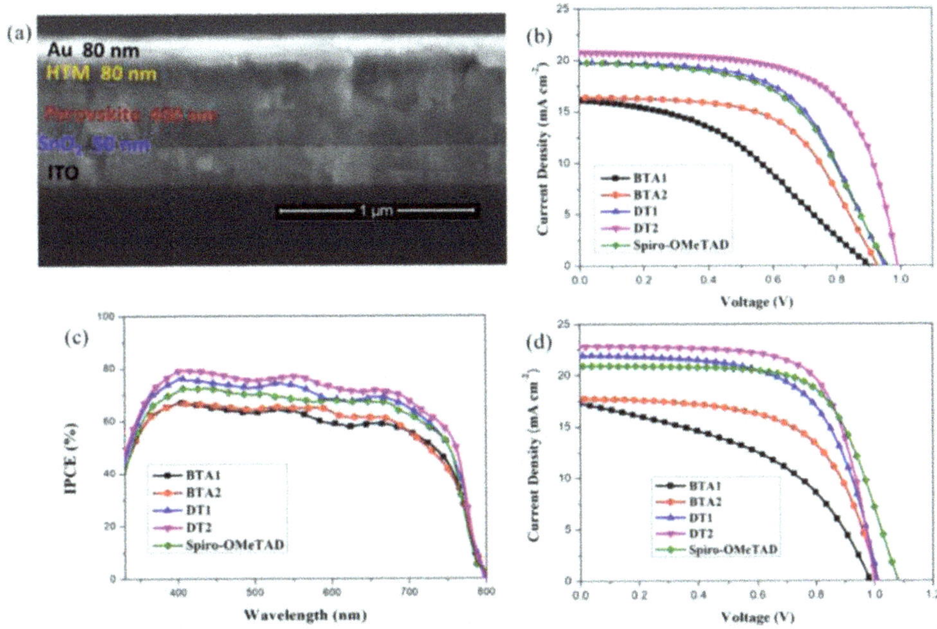

Figure 4.5. (a) PSCs—Cross-sectional SEM image of the DT2 based device. (b) *J–V* curves based on the HTMs BTA1–2, DT1–2 and spiro-OMeTAD without dopants. (c) Incident photo-to-current efficiency (IPCE) performance based on the HTMs BTA1–2, DT1–2 and spiro- OMeTAD without dopants. (d) *J–V* curves of PSCs based on the HTMs BTA1–2, DT1–2 and spiro-OMeTAD with dopants. Reuse with permission from Elsevier [39].

and poor flexibility, non- fullerene organic semiconductors are incorporated into PSCs. The non-fullerene material ETMs can be categorized into electron-transporting materials, fused-ring electron acceptors, interface-modifying materials, additive materials and light-harvesting materials in accordance with their functions. In 2016, Bai *et al* [40] employed a low-temperature solution-derived NiO_x nanoparticle film as the hole contact layer by depositing it on ITO-coated glass substrate and obtained 16.47% *PCE*. The NiO_x nanoparticle hole contact layer can be fabricated at a low temperature of 130 °C without post-treatments which is more suitable for flexible PSC systems. In 2017, Zhang *et al* reported 19.1% *PCE* using a low temperature processing of fused-ring electron acceptors (FREAs) based on indacenodithiophene (IDIC) as new promising non-fullerene ETMs. IDIC ETMs facilitate efficient electron extraction and transportation due to their high mobility and suitable energy levels within perovskites in comparison with the conventional TiO_2 based n-i-p PSCs [41]. In order to enhance the conductivity of ETL layer, Teimouri *et al* demonstrated the effect of lithium (Li) doping on TiO_2, which is prepared by using ultra sonication technique [42]. The resulting *PCE* of 24.23% is achieved which is almost 1.97% higher than that of the undoped composition. This improved conductivity is due to Li doping which provides faster electron transport. In 2023, Zhang *et al* investigated the influence of various ETLs on the performance

Table 4.1. Overview of electrical performance of selected PSCs.

References	V_{oc} (V)	J_{sc} (mA cm^{-2})	*FF* (%)	*PCE* (%)	Structure
2013 [19]	0.91	19.3	70	12.3	FTO/TiO_2/$CH_3NH_3PbI_{3-x}Br_x$ (10–15% Br)/PTAA/Au
2014 [43]	0.84	18.3	50	7.5	FTO/TiO_2/$FAPbI_3$/P3HT
2016 [44]	1.07	22.6	81	19.5	Glass/ITO/Low temperature derived NiO_x nanoparticles
2016 [40]	0.99	19.51	67	12.9	ITO/PTAA/$MAPbI_3$/Doped C60-SAM/TCO
2018 [37]	1.06	22.5	76	18.0	FTO/SnO_2/PCBM/$(FAPbI_3)_{0.85}(MAPbI_3)_{0.15}$/BTF4/$MoO_3$/Au
2019 [38]	0.99	19.5	77	14.9	Glass/FTO/TiO_2/$(FAPbI_3)_{0.85}(MAPbI_3)_{0.15}$ + meso-TiO_2/FH-0 HTM/Au
2019 [38]	1.07	22.1	78	18.4	Glass/ITO/TiO_2/$MAPbI_3$ + meso-TiO_2/FH-3 HTM/Au
2019 [38]	1.09	23.1	78	19.6	Glass/ITO/TiO_2/$MAPbI_3$ + meso-TiO_2/spiro-OMeTAD/Au
2019 [39]	0.93	16.4	56	8.28	ITO/SnO2/CH3NH3PbI$_3$-xClx/BTA2/Au
2019 [39]	1.07	20.6	75	16.5	ITO/SnO2/CH3NH3PbI$_3$-xClx /DT2/Au
2020 [45]	1.1	23.13	75.28	19.17	Cu/Ti/$MAPbI_3$ + CQDs/PTAA/ITO/glass
2021 [34]	0.91	19.82	72.47	12.92	Graphene/MoS2/$MAPbI_3$/PTAA/Au
2023 [46]	1.13	24.7	80.2	22.3	glass/ITO)/MeO-2PACz/perovskite/PC61BM/BCP/Ag
2023 [32]	1.29	23.68	86.75	26.7	Glass/FTO/ETL/$CsGeI_3$/W

of PSCs with HTL-free structure, which reveal that when ZnO is used as ETL, the recess structure is formed on the conduction band, which makes the device exhibit a good *PCE* of 12.68% [32]. Table 4.1 presents a summary of the performance of selected PSCs.

4.4 Device performance considerations

Light absorption, charge separation, charge transport, and charge collection are general solar cell working processes. For instance, a p-i-n junction is suitable as light harvester, while a p-n junction can be either an n- or p-type light harvester. Figure 4.6 shows two typical structures: a mesoscopic nanostructure and a planar structure. It is observed that the efficiency of the mesoscopic structure is lower than that of the planar structure, which is the main factor affecting PV performance [47]. In 2014, Pang *et al* [43] first reported a nearly cubic $NH_2CH = NH\ PbI_3$ (PbI_3) perovskite, poly (3-hexylthiophene) (P3HT) as the HTM for the fabrication of a mesoscopic solar cell with 7.5% *PCE*. PbI_3 has a bandgap of 1.43 eV, and its corresponding absorption edge reaches 870 nm, showing better performance than $MAPbI_3$ (820 nm) as a light harvester. Conventional lead halide PSCs have reached

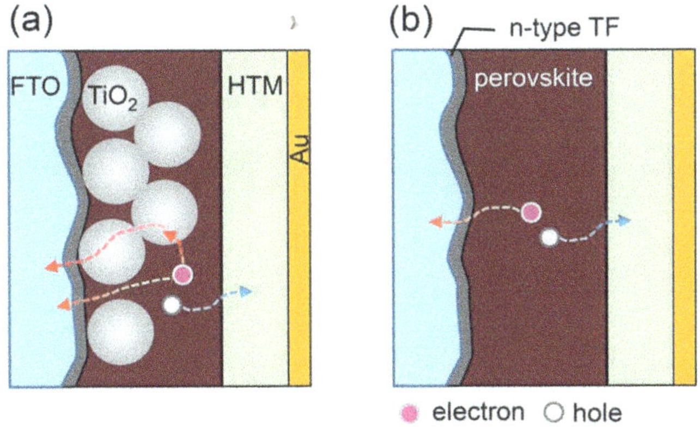

Figure 4.6. (a) Mesoscopic PSC with mesoporous TiO_2 layer and (b) planar structure without a mesoporous TiO_2 layer. Reproduced from [47] John Wiley & Sons. Copyright 2015 WILEY-VCH Verlag GmbH & Co. KGaA, Weinheim

a high 24.2% *PCE*; however, their stability with long running time under STC limit the development [40]. Besides, they suffer from instability under stimuli of heat, oxygen, moisture, light irradiation and electric field due to the weak secondary bonding protection between passivation molecules and the perovskite material surfaces [48]. Yang *et al* reported their studies on a protection layer to reduce the defect density by capping water-insoluble lead (II) oxysalt for PSCs. These PSCs maintained 96.8% of their initial efficiency after operation at maximum power point (MPP) under AM1.5 G irradiation for 1200 h at 65 °C [48]. This step prevents the degradation of PSCs by chemically passivating the defects on the surface of perovskites with this water-insoluble lead (II) oxysalt. A 19.16% *PCE* was achieved under the experimental conditions as described above.

Besides this water-insoluble lead (II) capping technology, there are several alternative ways for protecting the perovskite surface/interface to achieve better stability in PSCs. In 2019, a synergic interface design was developed that yielded 16.2 % *PCE* by applying an amino-functionalized polymer (PN4N) as cathode interlayer and PDCBT (Polybutyloctyl)oxy) carbonyl)(quaterthiophene-diyl) as dopant-free anode interlayer. This device exhibited a less than 10 % drop in *PCE* after 400 h continuous run under 1 sun equivalent illumination [20]. Wen *et al* first proposed carbon quantum dots (CQDs) as additive in $MAPBI_3$ for reducing the defects that are caused by grain migration. PSCs achieved 19.17% *PCE* with 0.04 $mg{\cdot}ml^{-1}$ concentration of CQD additive, and at the same time, improved the stability [45]. As discussed earlier, in 2020, the Helmholtz Association of German Research Centers reported 29.15% *PCE* for the perovskite silicon tandem solar cell [17]. In 2023, Li *et al* [46] reported the fabrication of high-quality perovskite crystalline films by adding a fluorinated polymer, the dipoles of which lowered the formation energy of the perovskite black phase, decreased the defect density, and also tuned the surface work function for charge extraction. Figure 4.7 illustrates the advancements in *PCE* for various PSC devices [33].

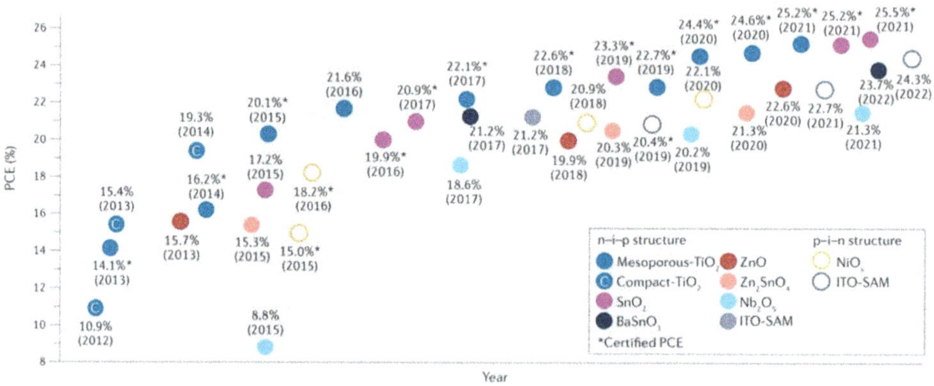

Figure 4.7. Advancements in *PCE* of various PSCs. Reproduced from [33] with permission from Springer Nature.

4.5 Material and cell degradation

The transport of the photogenerated charge carriers (electron–hole pairs) is adversely affected by various recombination processes, where majority carriers control the width of the depletion region and the series resistance. The most common reasons for the instability in PSC devices are thermal decomposition, photo-induced defect formation, asymmetrical chemical composition, phase transition, and electrical bias-induced ion migration. The perovskite compounds with organic molecules are unstable in the presence of moisture, oxygen, light, and heat, which could lead to the degradation in the PSC performance [49]. Goldschmidt tolerance factor (t) is a reliable empirical index to predict which structure is preferred to be formed. It is defined as:

$$t = (R_A + R_X)/[\sqrt{2}(R_B + R_X)]$$

where, R_A is the radius of the A cation, R_B is the radius of the B cation, and R_X is the radius of the anion [50].

Based on the results from first-principles calculations, lead-free iodide perovskites such as $MASnI_3$, $CsSnI_3$, $MASrI_3$, $MABiSI_2$, $MABi_{0.5}Tl_{0.5}I_3$, and $MACaI_3$, have been suggested to be promising candidates for solar cell applications [51–53]. High temperatures during processing and during the device operation lead to serious degradation in the device performance. This is inevitable for solar cells with over 50% in excess of *PCE* resulting as heat. The organic cations such as methylammonium (CH_3NH_3 – MA) and formamidinium [($CH(NH_2)_2$ – FA] are believed to be the main reasons for thermal and environmental instability; they degrade at 120 °C or due to long time exposure to temperatures under 80 °C. Solar cells must remain thermally stable to temperatures of at least up to 85 °C for normal operation. Philippe *et al* observed that $MAPbI_3$ starts to decompose into PbI_2 with an increase in temperature from room temperature to 100 °C and then to 200 °C. In order to avoid the decomposition, researchers have proposed a hybrid of $MAPbI_3$ with Br_3 or I_nBr_{3-n} mixture [54, 55]. Besides, Binek *et al* investigated formamidinium lead iodide ($FAPbI_3$) perovskite, which exhibits an ideal bandgap (≈1.48 eV) and significantly enhanced thermal stability as compared with $MAPbI_3$ [56]. However, the mixed type of perovskite still suffers from phase segregation [57]. Li *et al* designed a high thermal stability perovskite by adding a fluorinated polymer that retained over 90% of cell efficiency after thermal cycle testing between −60 °C and 80 °C for 3000 h [46, 58]. For perovskites, the majority contribution to the hysteresis in *J–V* characteristics is believed to be the ion migration and ion accumulation such as the I^- and MA^+ under external bias [59].

4.6 Hysteresis in *J–V* characteristics

J–V characteristics represents the relationship between voltage applied across the electrical device with respect to the current flowing through it. It is a common way to determine the efficiency and performance of PSCs in a circuit, which is usually carried out under standard AM1.5 illumination. The most important PV parameters can be estimated and obtained from the *J–V* curve accordingly; these include device characteristics such as V_{oc}, J_{sc}, *FF* and *PCE*. However, the forward and reverse scan of the *J–V* curve have hysteresis, which is responsible for the unstable power output of PSCs [60]. Therefore, the hysteresis in *J–V* characteristics restricts the day-to-day use and applications of PSCs and a further understanding into the origin of their hysteresis and methods to minimize this effect are of critical importance. Figure 4.8 shows a typical hysteresis in the *J–V* curve for a PSC. The hysteresis can be ascribed to the internal and external factors of PSCs. The internal factors include ion migration [61], charge capture [62] and carrier accumulation [63] at the interface. The external factors are the characterization conditions such as scan rate, range and the environmental conditions [64]. A flowchart of the potential factors for the hysteresis in PSCs is presented in figure 4.9.

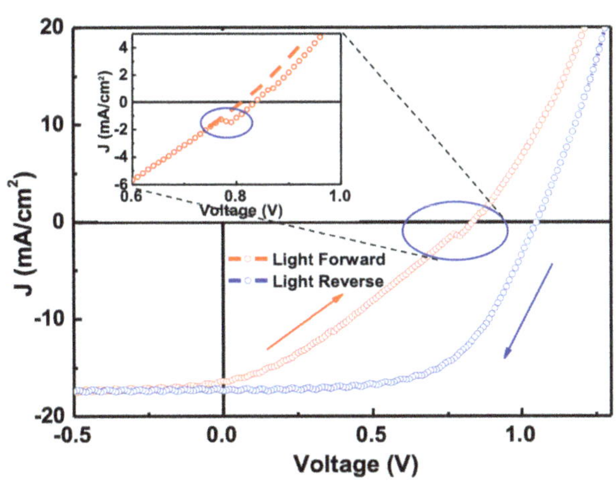

Figure 4.8. A typical *J–V* curve with hysteresis behavior for a PSC at a scan rate of 0.9 V s^{-1}. Reproduced from [65] John Wiley & Sons. Copyright 2016 WILEY-VCH Verlag GmbH & Co. KGaA, Weinheim.

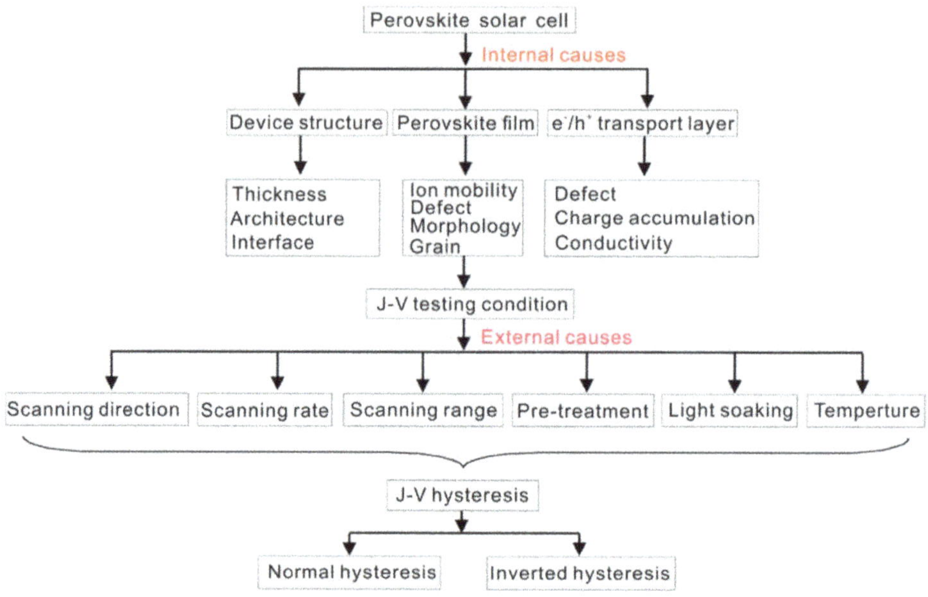

Figure 4.9. Potential factors that contribute to *J–V* hysteresis in PSCs. Reprinted from [59], copyright (2021), with permission from Elsevier.

Several strategies to minimize the hysteresis of *J–V* curves in PSCs, such as the perovskite film surface modification, ETL or HTL engineering methods etc are being investigated. Chen *et al* [66], have proposed the elimination of the hysteresis in PSCs by adding ammonium salts (2-aminoethanesulfonamide hydrochloride (ASCl)) into the perovskite to produce a smoother and high-quality crystalline thin film. Consequently, the PSC shows negligible hysteresis in the *J–V* curves and also achieves higher *PCE*; this approach provides an easy-processing methodology to passivate the surface defects. Neukom *et al* simulated the relationship of hysteresis in *J–V* curves with respect to mobile ions in PSCs, by using a drift-diffusion model and comparing it with the experimental results. The results show that the hysteresis depends on the extent of surface recombination and the diffusion length of the charge carriers [67]. Furthermore, Kari *et al* presented a new strategy of designing two structures with two and three layers of absorbers to mitigate the hysteresis in *J–V* curves for PSCs. The hysteresis in *J–V* curves has been significantly decreased by adjusting precursor ratios and thickness of device layers [68].

Besides the internal factors, the external factors such as scan rate and range have also been studied to realize the hysteresis in *J–V* curves of PSCs. Wu *et al* [65], utilize different sweep rates (fast, medium, and slow) with larger forward bias of 2–3 V to study the impact on the hysteresis in *J–V* curves of PSCs. The results show that the hysteresis is most obvious with the fast sweep mode with a large forward bias and disappears at very slow sweep rate [64]. This provides new insights into the ion movement factor in the *J–V* hysteresis for PSCs. The thickness, material and morphology of ETL and HTL layer improves the charge separation and electron

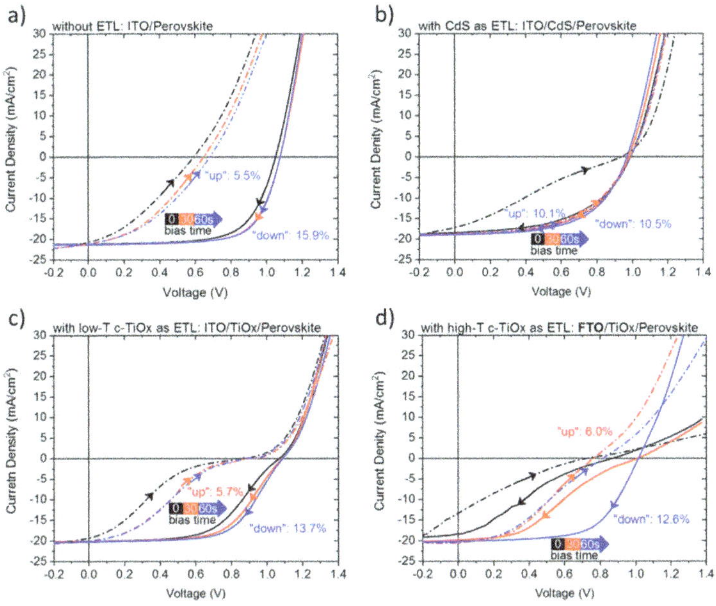

Figure 4.10. The hysteresis in *J–V* characteristics for n–i–p cell stacks of (a) without ETL; (b) with CdS; (c) with low-temperature compact (low-Tc)-TiO_x; (d) with low-Tc-TiO_x/CdS; (e) with CdS/low-Tc-TiO_x; and f) reference with high-Tc-TiO_x . Reproduced from [69] John Wiley & Sons. Copyright 2018 WILEY-VCH Verlag GmbH & Co. KGaA, Weinheim.

collection; these factors also affect the device hysteresis [59]. Researchers have found and compared several conditions for use of CdS as low-cost, easy-processing ETL material in PSCs to achieve low hysteresis in the *J–V* characteristics [69, 70]. It is believed that CdS can suppress hysteresis due to the high electron mobility in comparison with TiO_x. In figure 4.10, the hysteresis of *J–V* characteristics for several combinations of CdS and TiO_x, with and without ETL, are investigated and summarized.

4.7 Conclusion

PSCs represent a transformative innovation in the realm of renewable energy. Their high absorption coefficients, excellent matching with the solar spectrum due to their bandgap tunability and cost-effective processing methods are some of the major advantages of PSCs. The efficiency, scalability, and adaptability of PSCs provide an alternative option in the pursuit of sustainable energy development. Ongoing research and development efforts are necessary to address the lifetime, stability and long-term viability of hybrid organic–inorganic PSCs. PSCs are promising in the global transition towards clean and renewable energy sources. Tandem solar cells, in combination with PSCs, offer significant potential for their commercialization due to their ability to capture and utilize a major fraction of the solar spectrum.

This chapter was reproduced with permission from [70].

References

[1] Rose G 1840 Ueber einige neue Mineralien des Urals *J. Prak. Chem.* **19** 459–68

[2] Rose G 1839 *De novis quibusdam fossilibus quae in montibus Uraliis inveniuntur* (typis AG Schadii)

[3] Goldschmidt V M 1926 Die Gesetze der Krystallochemie *Naturwissensenschaffen* **14** 477–85

[4] Forrester W F and Hinde R M 1945 Crystal structure of barium titanate *Nature* **156** 177–7

[5] Gao W, Zhu Y, Wang Y, Yuan G and Liu J-M 2019 A review of flexible perovskite oxide ferroelectric films and their application *J. Materiom.* **6** 1–16

[6] Tang Q *et al* 2023 Layered $SrTiO_3/BaTiO_3$ composites with significantly enhanced dielectric permittivity and low loss *Ceram. Int.* **49** 23326–33

[7] Kojima A, Teshima K, Shirai Y and Miyasaka T 2009 Organometal halide perovskites as visible-light sensitizers for photovoltaic cells *JACS* **131** 6050–1

[8] Lee M M, Teuscher J, Miyasaka T, Murakami T N and Snaith H J 2012 Efficient hybrid solar cells based on meso-superstructured organometal halide perovskites *Science* **338** 643–7

[9] Stranks S D *et al* 2013 Electron-hole diffusion lengths exceeding 1 micrometer in an organometal trihalide perovskite absorber *Science* **342** 341–4

[10] Ai B, Fan Z and Wong Z J 2022 Plasmonic–perovskite solar cells, light emitters, and sensors *Microsyst. Nanoeng.* **8** 5

[11] Veldhuis S A *et al* 2016 Perovskite materials for light-emitting diodes and lasers *Adv. Mater.* **28** 6804–34

[12] Mathur A, Fan H and Maheshwari V 2021 Organolead halide perovskites beyond solar cells: self-powered devices and the associated progress and challenges *Mater. Adv.* **2** 5274–99

[13] Ortega-San-Martin L 2020 Introduction to perovskites: a historical perspective *Revolution of Perovskite: Synthesis, Properties and Applications* (Springer) pp 1–41

[14] Burschka J *et al* 2013 Sequential deposition as a route to high-performance perovskite-sensitized solar cells *Nature* **499** 316–9

[15] Yang W S *et al* 2017 Iodide management in formamidinium-lead-halide–based perovskite layers for efficient solar cells *Science* **356** 1376–9

[16] Wang Y *et al* 2019 Stabilizing heterostructures of soft perovskite semiconductors *Science* **365** 687–91

[17] *The National Renewable Energy Laboratory (NREL) 2023 is operated for the U.S. Department of Energy (DOE) by Alliance for Sustainable Energy, L. A. Best Research-Cell Efficiencies* https://nrel.gov/pv/module-efficiency.html

[18] Yoo J J *et al* 2021 Efficient perovskite solar cells via improved carrier management *Nature* **590** 587–93

[19] Noh J H, Im S H, Heo J H, Mandal T N and Seok S I 2013 Chemical management for colorful, efficient, and stable inorganic–organic hybrid nanostructured solar cells *Nano Lett.* **13** 1764–9

[20] Tian J *et al* 2019 Dual interfacial design for efficient $CsPbI_2Br$ perovskite solar cells with improved photostability *Adv. Mater.* **31** 1901152

[21] Wang Z *et al* 2020 Lightweight, passive radiative cooling to enhance concentrating photovoltaics *Joule* **4** 2702–17

[22] Guo Z, Jena A K, Kim G M and Miyasaka T 2022 The high open-circuit voltage of perovskite solar cells: a review *Energy Environ. Sci.* **15** 3171–222

[23] Al-Ashouri A *et al* 2020 Monolithic perovskite/silicon tandem solar cell with >29% efficiency by enhanced hole extraction *Science* **370** 1300–9

[24] Gao Z-W, Wang Y and Choy W C H 2022 Buried interface modification in perovskite solar cells: a materials perspective *Adv. Energy Mater.* **12** 2104030

[25] Shariatinia Z 2020 Recent progress in development of diverse kinds of hole transport materials for the perovskite solar cells: a review *Renew. Sustain. Energy Rev.* **119** 109608

[26] Wang L, Si F, Tang F and Xue H 2018 Electronic and optical properties of SnO_2 (110)/ MAPbI3 (100) interface by first-principles calculations *Mater. Res. Express* **6** 026312

[27] Soh M F *et al* 2019 Incorporation of g-C_3N_4/Ag dopant in TiO_2 as electron transport layer for organic solar cells *Mater. Lett.* **253** 117–20

[28] Xiao Y *et al* 2017 W-doped TiO_2 mesoporous electron transport layer for efficient hole transport material free perovskite solar cells employing carbon counter electrodes *J. Power Sources* **342** 489–94

[29] Sherkar T S *et al* 2017 Recombination in perovskite solar cells: significance of grain boundaries, interface traps, and defect ions *ACS Energy Lett.* **2** 1214–22

[30] Stoumpos C C *et al* 2016 Ruddlesden–Popper hybrid lead iodide perovskite 2D homologous semiconductors *Chem. Mater.* **28** 2852–67

[31] Li G *et al* 2020 Impact of perovskite composition on film formation quality and photophysical properties for flexible perovskite solar cells *Molecules* **25** 732

[32] Zhang X *et al* 2023 Investigation of efficient all-inorganic HTL-free $CsGeI_3$ perovskite solar cells by device simulation *Mater. Today Commun.* **34** 105347

[33] Isikgor F H *et al* 2023 Molecular engineering of contact interfaces for high-performance perovskite solar cells *Nat. Rev. Mater.* **8** 89–108

[34] Jung D H, Oh Y J, Nam Y S and Lee H 2021 Effect of layer number on the properties of stable and flexible perovskite solar cells using two dimensional material *J. Alloys Compd.* **850** 156752

[35] Zhang P, Li M and Chen W C 2022 A perspective on perovskite solar cells: emergence, progress, and commercialization *Front Chem.* **10** 802890

[36] Nakar R *et al* 2019 Cyclopentadithiophene and fluorene spiro-core-based hole-transporting materials for perovskite solar cells *J. Phys. Chem.* C **123** 22767–74

[37] Sun X *et al* 2018 Fluoranthene-based dopant-free hole transporting materials for efficient perovskite solar cells *Chem. Sci.* **9** 2698–704

[38] Chen W *et al* 2019 Simply designed nonspiro fluorene-based hole-transporting materials for high performance perovskite solar cells *Synth. Met.* **250** 42–8

[39] Ye X *et al* 2019 Effect of the acceptor and alkyl length in benzotriazole-based donor-acceptor-donor type hole transport materials on the photovoltaic performance of PSCs *Dyes Pigm.* **164** 407–16

[40] Bai Y *et al* 2016 Enhancing stability and efficiency of perovskite solar cells with crosslinkable silane-functionalized and doped fullerene *Nat. Commun.* **7** 12806

[41] Zhang M *et al* 2017 A low temperature processed fused-ring electron transport material for efficient planar perovskite solar cells *J. Mater. Chem.* A **5** 24820–5

[42] Teimouri R *et al* 2020 Synthesizing Li doped TiO_2 electron transport layers for highly efficient planar perovskite solar cell *Superlattices Microstruct.* **145** 106627

[43] Pang S *et al* 2014 $NH_2CH{=}NH_2PbI_3$: an alternative organolead iodide perovskite sensitizer for mesoscopic solar cells *Chem. Mater.* **26** 1485–91

[44] Yin X *et al* 2016 Highly efficient flexible perovskite solar cells using solution-derived NiO_x hole contacts *ACS Nano* **10** 3630–6

[45] Wen Y, Zhu G and Shao Y 2020 Improving the power conversion efficiency of perovskite solar cells by adding carbon quantum dots *J. Mater. Sci.* **55** 2937–46

[46] Li G *et al* 2023 Highly efficient p-i-n perovskite solar cells that endure temperature variations *Science* **379** 399–403

[47] Jung H S and Park N G 2015 Perovskite solar cells: from materials to devices *Small* **11** 10–25

[48] Domanski K *et al* 2017 Migration of cations induces reversible performance losses over day/night cycling in perovskite solar cells *Energy Environ. Sci.* **10** 604–13

[49] Wang R *et al* 2019 A review of perovskites solar cell stability *Adv. Funct. Mater.* **29** 1808843

[50] Jacobsson T J, Pazoki M, Hagfeldt A and Edvinsson T 2015 Goldschmidt's rules and strontium replacement in lead halogen perovskite solar cells: theory and preliminary experiments on $CH_3NH_3SrI_3$ *J. Phys. Chem.* C **119** 25673–83

[51] Bernal C and Yang K 2014 First-principles hybrid functional study of the organic–inorganic perovskites $CH_3NH_3SnBr_3$ and $CH_3NH_3SnI_3$ *J. Phys. Chem.* C **118** 24383–8

[52] Sun Y-Y *et al* 2016 Discovering lead-free perovskite solar materials with a split-anion approach *Nanoscale* **8** 6284–9

[53] Xu P, Chen S, Xiang H-J, Gong X-G and Wei S-H 2014 Influence of defects and synthesis conditions on the photovoltaic performance of perovskite semiconductor $CsSnI_3$ *Chem. Mater.* **26** 6068–72

[54] Aharon S, Cohen B E and Etgar L 2014 Hybrid lead halide iodide and lead halide bromide in efficient hole conductor free perovskite solar cell *J. Phys. Chem.* C **118** 17160–5

[55] Rao H-S, Chen B-X, Wang X-D, Kuang D-B and Su C-Y 2017 A micron-scale laminar $MAPbBr_3$ single crystal for an efficient and stable perovskite solar cell *Chem. Commun.* **53** 5163–6

[56] Binek A, Hanusch F C, Docampo P and Bein T 2015 Stabilization of the trigonal high-temperature phase of formamidinium lead iodide *J. Phys. Chem. Lett.* **6** 1249–53

[57] Tang X *et al* 2018 Local observation of phase segregation in mixed-halide perovskite *Nano Lett.* **18** 2172–8

[58] Cao X *et al* 2016 Modulating hysteresis of perovskite solar cells by a poling voltage *J. Phys. Chem.* C **120** 22784–92

[59] Wu F, Pathak R and Qiao Q 2021 Origin and alleviation of J-V hysteresis in perovskite solar cells: a short review *Catal. Today* **374** 86–101

[60] Wu Y *et al* 2016 On the origin of hysteresis in perovskite solar cells *Adv. Funct. Mater.* **26** 6807–13

[61] Azpiroz J M, Mosconi E, Bisquert J and De Angelis F 2015 Defect migration in methylammonium lead iodide and its role in perovskite solar cell operation *Energy Environ. Sci.* **8** 2118–27

[62] Lee J-W *et al* 2017 The interplay between trap density and hysteresis in planar heterojunction perovskite solar cells *Nano Lett.* **17** 4270–6

[63] Wang Y *et al* 2017 The influence of structural configuration on charge accumulation, transport, recombination, and hysteresis in perovskite solar cells *Energy Technol.* **5** 442–51

[64] Wu F *et al* 2018 Bias-dependent normal and inverted *J–V* hysteresis in perovskite solar cells *ACS Appl. Mater. Interfaces* **10** 25604–13

[65] Li C *et al* 2016 Iodine migration and its effect on hysteresis in perovskite solar cells *Adv. Mater.* **28** 2446–54

[66] Chen W *et al* 2018 Eliminating JV hysteresis in perovskite solar cells via defect controlling *Org. Electron.* **58** 283–9

[67] Neukom M T *et al* 2017 Why perovskite solar cells with high efficiency show small IV-curve hysteresis *Sol. Energy Mater. Sol. Cells* **169** 159–66
[68] Kari M and Saghafi K 2022 Current-voltage hysteresis reduction of $CH_3NH_3PbI_3$ planar perovskite solar cell by multi-layer absorber *Micro Nanostruct.* **165** 207207
[69] Wessendorf C D, Hanisch J, Müller D and Ahlswede E 2018 CdS as electron transport layer for low-hysteresis perovskite solar cells *Sol. RRL* **2** 1800056
[70] Lin L and Ravindra N M 2020 CIGS and perovskite solar cells—an overview *Emerg. Mater. Res.* **9** 812–24

IOP Publishing

N M Ravindra, Leqi Lin and Priyanka Singh

Chapter 5

Polymer solar cells

In this chapter, the evolution of the device architecture, efficiency, and principle of operation of polymer solar cells (PSCs) are discussed. Fabrication and material processing of PSCs for laboratory and various roll-to-roll (R2R) processing techniques such as blade coating, slot-die coating, spray coating, inkjet printing, screen printing and multi-roll transfer are described. The characterization techniques to assess the performance of polymer solar cells are discussed. Photovoltaic (PV) performance parameters such as: open-circuit voltage (V_{oc}), current density (J_{sc}), fill factor (FF) and efficiency (η) of PSCs, based on various polymer donors and acceptors, are reported. Degradation mechanisms observed in PSCs are also discussed briefly.

5.1 Introduction

PV cells are becoming an increasingly important source of renewable energy due to the yearly increase in electricity demand (IRENA) [1]. Over the last two decades, crystalline silicon has dominated the market due to its being cost-effective, with high power conversion efficiency (PCEs) > 26.7% [2–4], and well established technology. However, thin film solar cells, such as amorphous silicon, cadmium telluride/cadmium sulfide, indium phosphide, copper indium gallium diselenide, gallium arsenide and gallium indium phosphide, have been discovered to be less expensive than single-junction solar cells [5–7]. Nonetheless, organic materials [6, 8, 9] appear to have advantages over inorganic PV materials like silicon, and others, such as being lighter and less expensive during solvent processing. The most significant benefit of organic semiconductors is their increased capacity for R2R solvent processing and flexibility in solar cells, both of which will significantly increase production efficiency [10].

PSCs are organic solar cells also referred to as 'plastic solar cells'; they are essentially conjugated organic materials [11]. The discovery by Heeger, MacDiarmid, and Shirakawa (recipients of the 2000 Chemistry Nobel Prize) in

doi:10.1088/978-0-7503-5994-8ch5

1977, that oxidizing conjugated polymer polyacetylene with iodine can increase its conductivity marked the beginning of the field of conjugated materials [12]. In PSCs, the active layer is composed of conjugated polymers, which are long-chain macromolecules with repeating units forming a continuous network. However, in organic solar cells, the active layer typically consists of small organic molecules or oligomers. The polymer chains are functionalized to have alternating electron-donating and electron-accepting units, which promote efficient exciton dissociation and charge transport. Common polymers used in PSCs include poly(3-hexylthiophene)(P3HT), poly(3,4ethylenedioxythiophene):poly(styrenesulfonate) (PEDOT:PSS), and poly (2,7-carbazole) (PCDTBT). Although PSCs are made of organic semiconductors, they function similarly to silicon based solar cells. Despite the intense interest in PSCs in recent years, this technology is still in the research stage and is not yet widely produced commercially. PSCs are generally based on bulk heterojunction (BHJ) concept because of their low cost, flexibility, light weight and ease of fabrication [13–20]. In addition, BHJ based PSCs are promising power sources for flexible electronics [21], as their photoactive layers are composed of n-type acceptor materials (small molecules or polymer acceptors) and p-type polymer donors [22]. PSC efficiencies have improved over the years but are generally less than those of traditional silicon solar cells. Efficient power conversion efficiency (*PCE*) is around 15–18%, while some research prototypes have achieved higher values (> 18%) for organic PV solar cells [4, 23, 24]. Continuous advancements are being made to enhance the efficiency and commercial viability of PSCs. Over the past few decades, the research of PSCs has made significant progress, as seen by the numerous reports of lab-scale efficiencies exceeding 10% [25–31] and even 20% (figure 5.1). The development of novel light absorbing polymer materials is largely responsible for this advancement. The careful design and synthesis of polymer donor and acceptor materials have led to significant increases in PSC performance (with the greatest *PCE* exceeding 18%) [24, 32–35]. The significant advancements made in innovative device structures such as ternary [36] and tandem [37] PSCs have drawn more interest. The ternary blend PSCs, using a wide bandgap polymer donor, have achieved *PCE* of 18.60% which is among the highest values for PSCs to date [38, 39]. Figure 5.1 depicts the efficiency growth of organic solar cells as published by NREL.

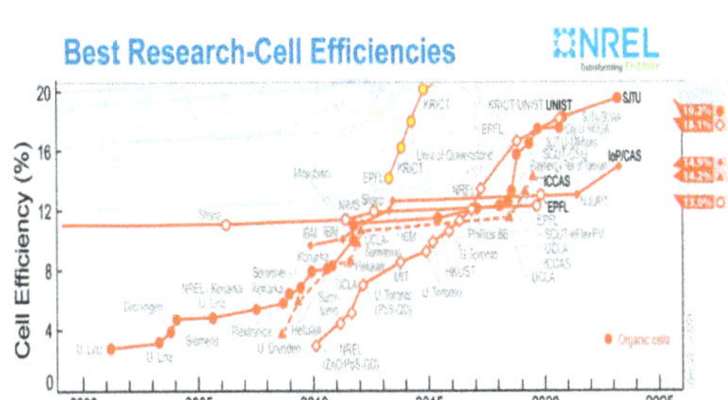

Figure 5.1. Solar efficiency chart provided by National Renewable Energy Laboratory (NREL). This plot is modified for showing the efficiency growth (maximum efficiency—19.2%) for organic solar cells (●). Reproduced from http://www.nrel.gov/pv/cell-efficiency.html. Credit: NREL.

This chapter will provide details of device structure of PSC in accord with its working principle, processing and characterization techniques, role of polymers as donors and acceptors and degradation in PSCs.

5.2 Device structure of polymer solar cells

Initially, PSCs were produced with just one active layer positioned between two electrodes having different work functions (figure 5.2(a)). The two electrodes are generally transparent conducting oxide (TCO) or indium tin oxide (ITO) electrode, and a metal electrode, often aluminum (Al) or silver (Ag). Nonetheless, the single-layer devices had poor *PCE* < 0.1%, due to challenges in producing effective exciton (electron–hole pair) dissociation and therefore causing significant electron and hole recombination [40].

5.2.1 Bilayer heterojunction structure

The development of PSCs has undergone significant evolution since their inception in the late 1980s. Tang [41] reported significant progress in 1986 when he presented a two-layer organic PV device (figure 5.2(b)). This bilayer structure consisted of an electron-donating (donor, D) and electron-accepting (acceptor, A) layer. The device has copper phthalocyanine as donor and perylene tetracarboxylic derivative as acceptor. The structure achieved relatively low *PCE* of about 1%, but demonstrated the potential of organic materials for PV applications. Lower *PCE* of the device was attributed to the small donor/acceptor (D/A) interface area which continued to interfere with the effective exciton diffusion and separation [42].

5.2.2 Bulk heterojunction (BHJ) structure

The BHJ structure of PSC (figure 5.2(c)) was proposed by Yu *et al* [43] in 1995. In a BHJ PSC, the active layer comprises a blend of electron donor and acceptor materials. This structure is also referred as binary PSCs as the active layer consists of a blend of only two materials. This structure facilitates efficient exciton dissociation, charge transport, and an improved D/A interface, and hence leads to enhanced performance of PSC. In BHJ PSCs, conjugated polymers were used as donor and fullerene derivatives or non-fullerene as acceptor. The efficiency of PSCs has soared

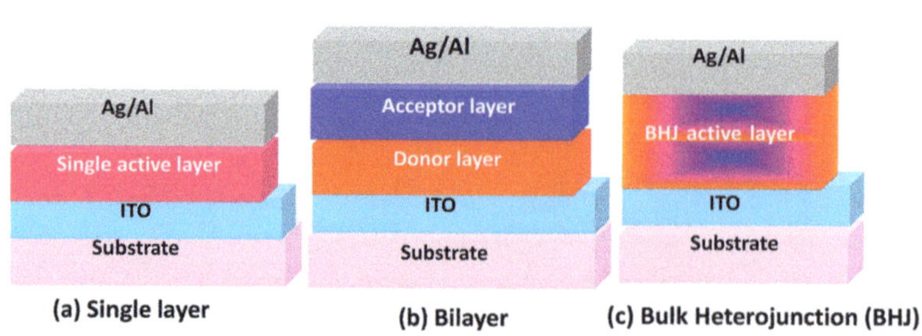

Figure 5.2. Schematic representation of PSCs with three different device structures: (a) single active layer (b) bilayer (c) bulk heterojunction (BHJ).

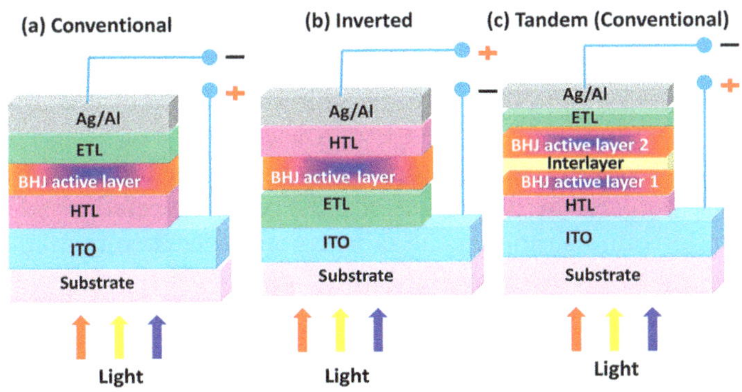

Figure 5.3. Illustration of the structure of (a) conventional PSC (b) inverted PSC and (c) conventional tandem PSC.

to over 18% since the development of the BHJ structure, which was regarded as a breakthrough in PSCs [24, 44, 45], making it a viable PV technology.

5.2.3 Conventional BHJ structure

The conventional or normal structure of a PSC is referred as a BHJ solar cell, as discussed in the previous section. Nonetheless, with continuous research efforts, the structure has improved significantly and it generally consists of several layers arranged in a specific order to facilitate efficient light absorption, charge generation and collection [46, 47], as shown in figure 5.3(a). In a normal structure of PSC, ITO coating on substrate is followed by the deposition of hole transport layer (HTL), BHJ active layer and electron transport layer (ETL). HTL is a conductive polymer layer or small molecule material that facilitates the transport of holes (positive charge carriers), that are generated in the active layer, towards the anode (ITO). PEDOT:PSS is a commonly used material for HTL in PSCs [48, 49]. ETL facilitates the transport of electrons (negative charge carriers) from the active layer to the cathode (metal electrode). Common materials for ETL are zinc oxide (ZnO) or titanium dioxide (TiO_2) [48, 50]. The metal cathode (Al or Ag) is the electron-collecting electrode deposited on top of the ETL [51]. By optimizing the D–A materials, interfaces within each layer and controlling the device structure, researchers aim to enhance the efficiency, stability, and scalability of PSCs for various applications in renewable energy harvesting [52, 53].

5.2.4 Inverted BHJ structure

Inverted structure of PSCs has emerged as a promising alternative architecture since the mid-2000s [11, 54–57]. In this design, the order of layers is reversed compared to conventional PSCs, as shown in figure 5.3(b). In an inverted structure of PSC, ITO coating on substrate is followed by the deposition of ETL, active layer and HTL. Inverted PSCs are generally more stable because the active BHJ layer is shielded from exposure to ambient conditions by the transparent conductive electrode, which acts as a barrier against moisture and oxygen ingress and enhances the device lifetime and performance. Some high-performance PV materials are sensitive to the acidic nature of ITO, which degrades their performance over time. Inverted geometry of PSCs alleviates this issue by eliminating direct contact between the BHJ layer and ITO, thus enabling the use of a wider range of materials. Placing the transparent conductive electrode on top allows for better light trapping within the active layer, as it reduces the likelihood of light being reflected away from the cell. The inverted architecture simplifies the fabrication process compared to traditional PSCs, as it eliminates the need for additional encapsulation layers to protect the active layer. Therefore, the inverted geometry architecture has become increasingly popular in the field of organic photovoltaics due to its potential for enhanced stability, compatibility with high-efficiency materials, and suitability for flexible substrates [58].

5.2.5 Ternary blend

Ternary blend PSCs incorporate three different materials in the active layer including donor and acceptor materials and follow the same structure as conventional or inverted, as shown in figures 5.3(a) and 5.3(b), respectively. These PSCs typically consist of a donor, an acceptor polymer, and a third component, which could be another polymer or a non-fullerene small molecule. Ternary blend structures have garnered significant interest in the field of organic photovoltaics due to their potential to enhance device performance compared to binary blend systems. The third component in ternary blend PSCs can vary widely and is introduced to modify the morphology of the active layer, enhance charge transport, or improve the absorption spectrum [59–61]. Research efforts continue to explore novel ternary blend systems and optimize their composition, morphology, and device architecture to achieve higher efficiencies and long-term stability, bringing organic PV technology closer to commercial viability.

5.2.6 Tandem and multijunction

Tandem and multijunction PSCs have gained attention for their potential to further boost efficiency by combining multiple absorber materials with complementary absorption spectra [62, 63]. Tandem structures stack two or more subcells with different bandgaps to capture a broader range of the solar spectrum and improve overall efficiency, as shown in figure 5.3(c). Multijunction architectures incorporate multiple absorber materials within a single cell, each tuned to absorb specific portions of the solar spectrum. Tandem solar cells will be discussed in detail in chapter 12.

5.2.7 Emerging concepts

Ongoing research efforts focus on exploring novel materials, device architectures, and processing techniques to further enhance the efficiency, stability, and scalability of PSCs. Non-fullerene acceptors [64], perovskite-based materials [65], and advanced interface engineering [53, 66] are among the areas of active investigation. Concepts such as tandem-perovskite/PSC hybrids and integrated flexible devices hold promise for future applications in renewable energy generation. Overall, the evolution of PSC structures has been driven by continuous advancements in materials science, device engineering, and understanding of fundamental PV processes, with the aim of realizing efficient, cost-effective, and environmentally sustainable solar energy conversion technologies.

5.3 Working principle of polymer solar cells

In PSCs, HOMO (highest occupied molecular orbital) and LUMO (lowest unoccupied molecular orbital) energy levels correspond to the highest occupied and lowest unoccupied energy states, respectively, within the polymer donor and fullerene acceptor materials. The HOMO energy level of the donor polymer determines its ability to donate electrons, while the LUMO energy level of the acceptor material determines its ability to accept electrons. In PSCs, an appropriate energy level offset, between the HOMO of the donor and the LUMO of the acceptor, facilitates exciton dissociation and charge separation at the D/A interface. In PSCs, excitons are generated when photons are absorbed by the donor polymer, leading to the excitation of electrons from the HOMO to the LUMO, as shown in figure 5.4. Excitons diffuse within the active layer and reach the D/A interface where excitons dissociate into free electrons and holes due to the energy level offset between the donor and acceptor. The efficiency of exciton dissociation depends on factors such as the morphology of the donor–acceptor blend, the energetic disorder within the

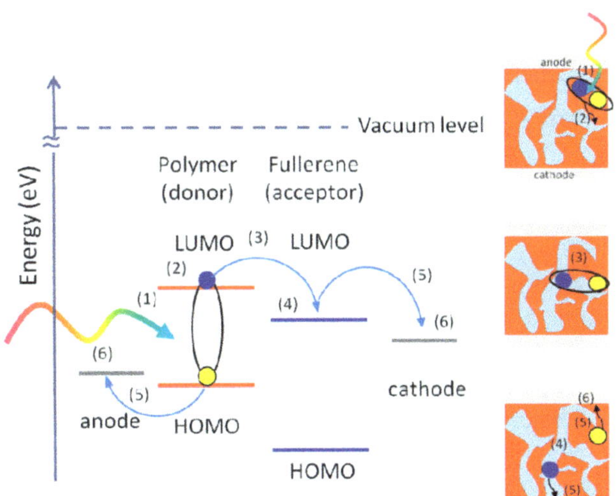

Figure 5.4. Schematic representation of the working mechanism of BHJ PSC having polymer-based donor and fullerene as acceptor. A simplified energy diagram is shown on the left, whereas electron and hole behavior at various stages (1–6) occurring is represented on the right. (1) photon absorption leading to the creation of the exciton (electron–hole pair), (2) diffusion of exciton to the D/A interface, (3) dissociation of exciton at the interface through charge transfer, (4) the electron–hole pair that is still Coulomb bound is separated, (5) the hole in the donor and the electron in the acceptor migrate towards their respective electrodes, and (6) charge collection occurs. Figure is adapted with permission from [128], copyright Lietuvos mokslų akademija, 2018.

material, and the interfacial properties. Upon dissociation, the free electrons and holes move through their respective pathways within the active layer towards the electrodes. The electrons are collected at the ETL, while holes are collected at the HTL, and then transported to cathode and anode, respectively. Efficient charge transport and collection are essential for minimizing losses due to recombination and maximizing the photocurrent output of the PSC. Understanding the interplay between HOMO and LUMO energy levels, exciton generation, dissociation, and charge transport mechanisms is critical for designing high-performance PSCs with improved efficiency and stability. Researchers continue to explore novel materials, device architectures, and processing techniques to optimize these photophysical processes and advance the development of PSC technology for renewable energy applications.

The charge transfer process is depicted in figure 5.4, which presents the operating principle of a PSC and highlights all significant steps. Since charge transfer is mediated by the energy gain of transferred charges, the donor material needs to have greater HOMO and LUMO energy levels than the corresponding levels of the acceptor material in order for charge transfer to occur. In conventional and inverted PSCs, ETLs need to have a low work function in order to facilitate the transfer of electrons from LUMO levels in acceptor material from the active layer. Additionally, in order to prevent hole transport, ETLs must provide sufficient electron mobility and band alignment. Furthermore, high ETL transmittance is required in the case of inverted PSCs in order to permit light transmission to the active layer [67, 68].

5.4 Fabrication and material processing of polymer solar cells

This section will provide fabrication and material processing for laboratory and R2R schemes of PSCs.

5.4.1 Spin coating

Spin coating is a common technique used in the fabrication of PSCs. McCullogh [69] reported the synthesis of regioregular poly(3-hexylthiophene, or P3HT). Numerous researches have been devoted to optimizing the synthesis and processing parameters for P3HT and other conjugated polymers for PSCs [53, 70]. High-efficiency polymer devices have been fabricated by spin coating a polymer film on a small area in a N_2-glovebox. Thin polymer films are created when a polymer solution is deposited onto a revolving substrate, where centrifugal forces encourage the solvent to evaporate, as shown in figure 5.5 [8, 71, 72]. While spin coating is a popular coating technique used at the laboratory level, it is not suitable for R2R and commercial processes due to its non-continuous nature, inability to work with a flexible substrate, over 98% material waste during the spinning process (figure 5.5) [73–75]. The parameters, such as spin speed, spin time, and concentration of the polymer solution, need to be carefully optimized to achieve the desired film thickness, uniformity, and morphology, which can significantly impact the performance of the solar cell.

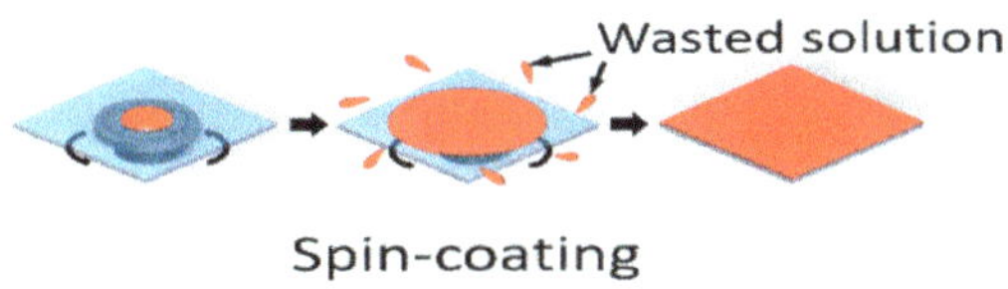

Figure 5.5. Schematic illustration of spin coating process via the four coating methods under investigation. Reproduced from [129], copyright (2022) with permission from Elsevier.

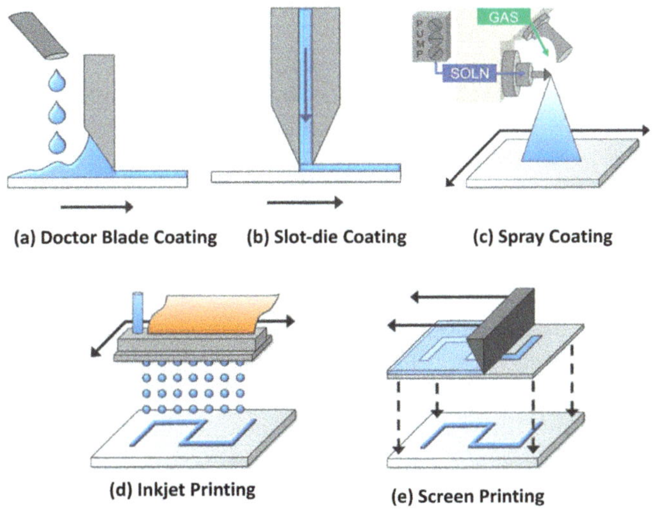

Figure 5.6. Schematic illustration of common deposition techniques for R2R PSCs production. Reproduced from [130] with permission from Royal Society of Chemistry.

5.4.2 Roll-to-roll process

R2R processing is a continuous or discrete manufacturing technique used to fabricate polymer solar cells, on flexible substrates such as plastic films or metal foils [73, 76, 77]. It is better to use continuous R2R processing since it can reduce fabrication time and processing energy [73, 76, 77]. The process begins with feeding a continuous roll of flexible substrate such as polyethylene terephthalate (PET) or polyethylene naphthalate (PEN)) onto a series of rollers within the R2R system. R2R processing has significant advantages for the large-scale production of PSCs; it enables efficient and cost-effective manufacturing while maintaining high quality and performance standards. Nevertheless, developing efficient R2R processing for mass production is a challenging issue since it necessitates a well-defined multilayer architecture and a number of processing steps, including annealing, ultraviolet curing, drying, and encapsulation etc [74, 75, 77, 78]. As shown in the figure 5.6, doctor blade coating, slot-die coating, spray coating, inkjet printing and screen printing are the most often utilized R2R fabrication processes for PSCs. The R2R processing can be categorized as one or two-dimensional based on how the substrate or coating apparatus move. Doctor blade, slot-die coating and screen printing techniques only permit one-dimensional coating along the substrate's unroll direction, whereas spray coating and inkjet printing provide two-dimensional movement of the coating apparatus. A certain R2R approach may be more advantageous than another, depending on the necessary feature. Screen printing [79], for instance, can offer more practical patterning [73, 76, 79–82]. The following section will explain the various R2R processing techniques such as blade coating, slot-die coating, spray coating, inkjet printing, screen printing and multi-roll transfer used for PSCs fabrication.

5.4.2.1 Doctor blade coating

Doctor blade coating, also known as knife-over-roll coating, is a method commonly used in the fabrication of PSCs for depositing thin and uniform layers of polymer onto a substrate [83–85]. In this process, concentration of the polymer solution is optimized for the desired film thickness and quality. The solution is dispensed onto the substrate (glass or flexible plastic) using a coating applicator, commonly referred to as a doctor blade, as shown in figure 5.6(a). Doctor blade is a flat, rigid blade typically made of metal or plastic, which is mounted above the substrate. The gap between the doctor blade and the substrate is carefully controlled and can be adjusted manually or automatically to achieve the desired film thickness. As the substrate moves beneath the doctor blade, the blade spreads the coating solution across the substrate surface, resulting in a thin, uniform layer of solution. Any excess coating solution that exceeds the desired film thickness is scraped off by the doctor blade, ensuring that only the desired amount of material remains on the substrate.

Doctor blade coating offers several advantages for the production of PSCs, including precise control over film thickness, uniformity, and scalability for large-scale manufacturing. However, it also requires careful optimization of coating parameters such as blade gap, coating speed, and viscosity of the coating solution to achieve the desired film properties.

5.4.2.2 Slot-die coating

Slot die coating is another common technique for the fabrication of PSCs. It is a precision coating method that allows for the deposition of thin and uniform layers of polymers onto a glass or flexible substrate [83, 86–88]. Slot die consists of a narrow slot through which the polymer solution is dispensed onto the substrate, as shown in figure 5.6(b). The slot die is mounted above the substrate and positioned at a precise distance to control the thickness of the deposited film. The cleaned substrate is placed on a moving platform beneath the slot die. The polymer solution is pumped into the slot die, where it forms a thin liquid film across the width of the slot. As the substrate moves beneath the slot die at a controlled speed, the polymer solution is continuously dispensed onto its surface through the slot. The gap between the slot die and the substrate is carefully adjusted to control the thickness of the deposited film. By optimizing parameters such as the solution flow rate, slot die geometry, substrate speed, and gap distance, a uniform coating of the polymer solution is achieved across the entire surface of the substrate. The photograph of slot-die coating of active layer on a PSC is shown in the figure 5.7(a), which consists of several tightly spaced stripes. Figure 5.7(b) depicts coating of 48 stripes, each 3 mm wide and 1 mm apart, which are coated at the same time [84]. The viscosity of the

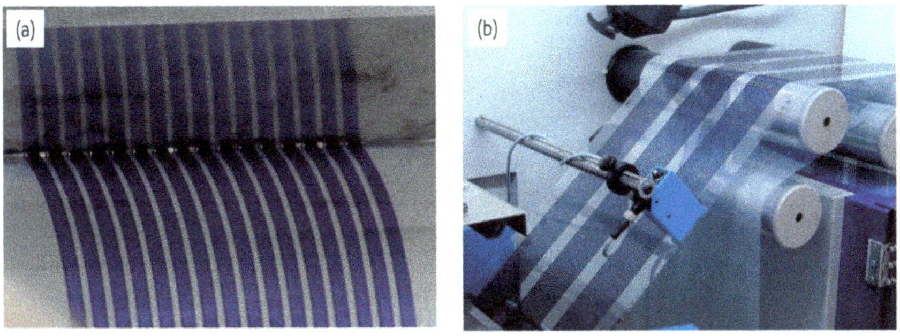

Figure 5.7. (a) Image of the standing meniscus during slot-die coating of active layer of PSCs consisting of several tightly spaced stripes. (b) Total 48 stripes, each 3 mm wide and 1 mm apart, are coated at the same time. Reprinted from [84], copyright (2012) with permission from Elsevier.

polymer solution determines the continuity of the ink column and has a significant impact on the wet and dry thickness of the device that is manufactured. However, pump pressure, rotation speed, and meniscus thickness can all be adjusted very precisely to optimize the wet thickness [83, 86–88]. Slot die coating has numerous benefits for the production of PSCs, including high coating uniformity, precise control over film thickness, and compatibility with R2R manufacturing processes, making it suitable for large-scale production.

5.4.2.3 Spray coating

Spray coating is a common technique for producing large-area PSCs [83, 86–89]. It involves the deposition of a polymer solution or nanoparticle dispersion onto a substrate (glass or flexible) using a spray nozzle or atomizer as shown in figure 5.6(c). The polymer solution should be well-mixed and homogenous to ensure uniform coating. The solution or nanoparticle dispersion is loaded into a spray gun or nozzle. The substrate is placed on a flat surface or within a controlled environment chamber. The spray gun or nozzle is positioned above the substrate, and the solution or dispersion is atomized into fine droplets. These droplets are then directed towards the substrate surface in a controlled manner. As the droplets land on the substrate, they spread out and coalesce to form a thin film. The motion of the spray gun or substrate, as well as the spray parameters such as spray rate, spray distance, and angle, are carefully controlled to achieve the desired film thickness and uniformity.

Spray coating offers several advantages for the production of PSCs, including simplicity, versatility, and the ability to coat large-area substrates with high throughput. However, it also requires careful optimization of spray parameters to achieve uniform coating and minimize defects [83, 86–89].

5.4.2.4 Inkjet printing

Inkjet printing is an additive manufacturing technique that has gained significant interest in the fabrication of polymer solar cells due to its ability to precisely deposit materials in a controlled manner [83, 84, 90]. Inkjet printing requires the preparation of an ink solution containing the polymer in a suitable solvent and may also include other additives to achieve the desired rheological properties and compatibility for printing process. The ink is loaded into the inkjet printer cartridge, and precise droplets of the ink are ejected onto the substrate from the print head as depicted in figure 5.6(d). The print head moves across the substrate in a controlled manner, depositing droplets of ink at specific locations according to the desired pattern or design. Inkjet printing allows for precise control over droplet placement and spacing, enabling the deposition of complex patterns with high resolution. Multiple layers of materials, including the active layer, charge transport layers, and electrodes can be deposited onto the substrate using inkjet printing. Each layer is printed sequentially, allowing for the fabrication of multilayered solar cell devices.

Inkjet printing offers several advantages for the production of PSCs, such as high resolution, precise control of deposition, and the ability to print complex patterns or designs. Additionally, inkjet printing is a non-contact printing technique, which minimizes substrate damage and enables the printing on flexible substrates. However, inkjet printing also presents challenges such as ink formulation optimization, nozzle clogging, and compatibility with large-scale manufacturing processes [83, 84, 90].

5.4.2.5 Screen printing

Screen printing is a widely used technique in the fabrication of PSCs, particularly for depositing the active layer and other functional layers [83, 86, 87, 91, 92]. In screen printing, the desired polymer is dissolved in a suitable solvent, along with other additives to prepare the ink. A mesh screen made of fine material (such as polyester or stainless steel) is prepared. The screen is coated with a stencil, which defines the pattern or shape of the material to be deposited as illustrated in figure 5.6(e). The stencil can be created using various methods, such as photochemical etching or direct film deposition. The prepared ink is applied onto the screen, and a squeegee is used to spread the ink evenly over the stencil. The screen is then placed in contact with the substrate, and pressure is applied to force the ink through the stencil and onto the substrate, as shown in figure 5.6(e). This transfers the ink onto the substrate in the desired pattern or shape. Multiple layers can also be deposited onto the substrate using screen printing to deposit other functional layers of the solar cell, such as the active layer, charge transport layers, and electrodes. Each layer is deposited using a separate stencil and ink formulation. A schematic illustration of a flexible screen, squeegee, ink, printing procedure, and printed pattern is shown in figure 5.8.

Figure 5.8. Screen printing representation of (a) flat-bed screen printing of silver paste, (b) the printed pattern is seen in the lower part, (c) photograph of rotary screen printing of conducting graphite ink onto a clear polyester foil. Reprinted from [84], copyright (2012) with permission from Elsevier.

Screen printing is very useful for the production of polymer solar cells, including high throughput, cost-effectiveness, and the ability to print large-area devices with relatively simple equipment [83, 86, 87, 91, 92]. However, achieving precise control over layer thickness and morphology may require careful optimization of ink formulation, printing parameters, and substrate surface preparation.

5.4.2.6 Multi-roll transfer

Multi-roll transfer (MRT) is a R2R manufacturing technique used in the fabrication of thin-film solar cells, including polymer solar cells [93]. The procedure consists of several consecutive fabrication phases, as shown in figure 5.9.

It allows for the continuous deposition and transfer of functional layers onto a flexible substrate using multiple rolls. MRT begins with feeding a continuous roll of flexible substrate such as PET or metal foil onto a series of rollers within the MRT system. Various functional layers of the PSC, including the active layer, charge transport layers, and electrodes, are deposited onto separate carrier rolls. As the substrate moves through the MRT system, the functional layers are sequentially transferred from the carrier rolls onto the substrate. Each layer is brought into contact with the substrate using pressure rollers or other mechanisms to ensure adhesion. The MRT system may operate within a controlled environment chamber to maintain optimal conditions for layer deposition and transfer, such as temperature, humidity, and the ambiance. Precision alignment and registration mechanisms ensure that the deposited layers are accurately positioned and aligned with respect to each other and the substrate. This is crucial for the performance and efficiency of the resulting solar cell device. After all the functional layers are transferred onto the substrate, the multilayered structure may undergo R2R lamination to ensure proper bonding and adhesion between the layers. Once the

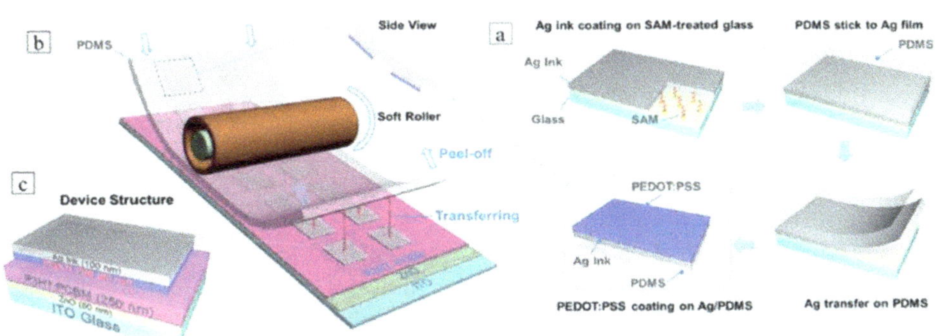

Figure 5.9. The MRT process, as proposed by Youn *et al* [93] (a) it involves silver (Ag) ink coating on self-assembled monolayer(SAM) treated glass followed by sticking it to the polydimethylsiloxane (PDMS) surface and adding PEDOT:PSS; (b) A soft roller is used to press the PDMS film against the active layer, PDMS film contains the back electrode and the hole transport layer. This process encourages the transfer of two more layers, completing the inverted architectural device. (c) The structure of the device such as the electron transport layer ZnO nanoparticles, the photoactive layer P3HT:PCBM, and the multilayer PEDOT:PSS/Ag generated using the MRT process. Reprinted from [93] with permission from Royal Society of Chemistry.

layers are laminated, the continuous roll of solar cell material may be patterned and cut into individual cell modules using techniques such as laser scribing or mechanical cutting. Throughout the multi-roll transfer process, various quality control measures and testing procedures are implemented to ensure the performance and reliability of the fabricated solar cells. This may include in-line monitoring of key parameters such as layer thickness, composition, and electrical properties.

MRT enables large-scale manufacturing, and the ability to deposit complex multilayer structures with high precision and uniformity [93]. However, it also requires careful optimization of process parameters and equipment design to achieve optimal performance and efficiency.

5.5 Electrical & optical characterization of polymer solar cells

Generally, the primary methods of characterizing PSCs are current–voltage (*I–V*) and spectral response (*SR*) measurements. *I–V* characteristics are utilized to determine the diode parameters such as junction ideality factor (n), saturation current (I_o), series resistance (R_s) and shunt resistance (R_{sh}). This section will provide a brief overview of the techniques that are used to assess the performance of PSCs. Nonetheless, chapter 11 will cover advanced computing methods and characterization approaches in detail.

5.5.1 Illuminated *I–V* characteristics

I–V characteristics of a conventional p-n junction solar cell, under steady state illumination, can most simply be described using a single exponential model as,

$$I = -I_{ph} + I_o(e^{qV_j/nkT} - 1) + V_j/R_{sh} \tag{5.1}$$

where,

$$V_j = V - IR_s \tag{5.2}$$

I_{ph} represents the photogenerated current, V_j is the voltage developed or dropped across the junction, k is the Boltzmann constant, T is the temperature of the cell, V is the terminal voltage, n is the ideality factor and I_o is the reverse saturation current. Figure 5.10(a) represents the schematic diagram of a single exponential model of a p-n junction solar cell.

In equations (5.1) and (5.2), R_s, R_{sh} represent the values of series and shunt resistance, respectively. The resistivity of the material and ohmic contact resistance between the metal and the polymer contacts contribute to the total series resistance of the solar cell. The shunt resistance can arise from imperfections on the device surface as well as leakage currents across the edges of the cell. For ideal case, values of $R_s = 0$ and $R_{sh} = \infty$ are desired. But in practice, R_s and R_{sh} have some finite values.

5.5.2 Dark *I–V* characteristics

For an ideal solar cell, the expression for the *I–V* characteristics is given by,

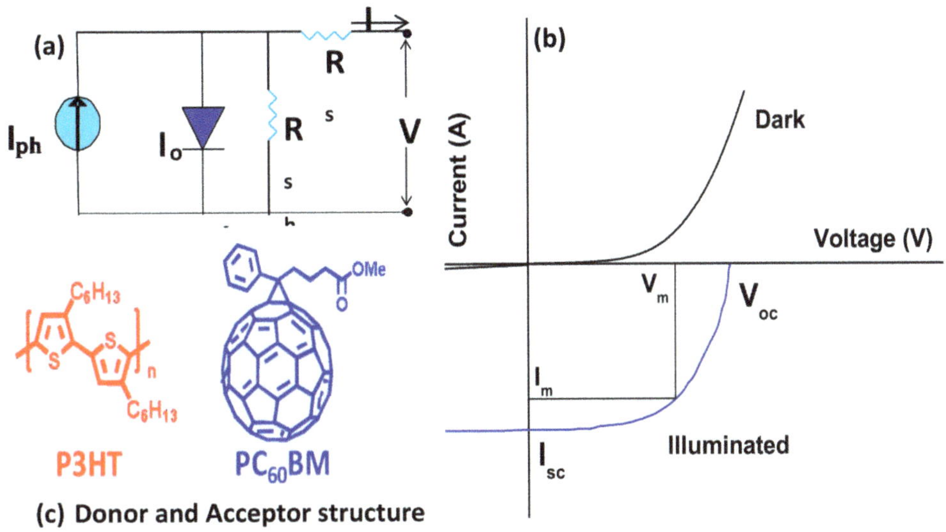

Figure 5.10. Schematic illustration of (a) solar cell single-diode model having series (R_s) and shunt resistances (R_{sh}), and (b) dark and illuminated *I–V* curves of solar cell, (c) chemical structure of polymer donor (P3HT) and acceptor ($PC_{60}BM$) commonly used in BHJ active layer structure.

$$I = -I_{ph} + I_d \tag{5.3}$$

where, I_d is the diode current. The current–voltage relation for a solar cell having a finite series and shunt resistance values can be expressed using a single-diode model as,

$$I_d = I_o\left[\exp\frac{q\,V_j}{n\,k\,T} - 1\right] \tag{5.4}$$

For practical analysis of solar cell performance, the dark and illuminated *I–V* characteristics are shown in figure 5.10(b). The dark *I–V* curve is shifted downward by a light generated current I_{ph} resulting in the illuminated *I–V* characteristics.

5.5.3 Solar cell performance parameters

Four parameters of solar cells are used to characterize illuminated solar cells: the short-circuit current (I_{sc}), open-circuit voltage (V_{oc}), fill factor (*FF*), and the efficiency (η).

The open-circuit voltage, V_{oc}, is defined as the voltage when terminals are open in the solar cell. By setting the total current *I* to zero in equation (5.1), the relationship between V_{oc}, I_o and I_{ph} for a single exponential model, assuming $R_{sh}\rightarrow\infty$, is,

$$V_{oc} = \frac{nkT}{q}\ln\left\{\frac{I_{ph}}{I_O} + 1\right\} \tag{5.5}$$

V_{oc} is related to I_{ph} and I_o. For a high V_{oc}, a low I_o is absolutely necessary.

For an optimum value of load resistance, maximum output power is obtained. The condition for maximum power point and expression for power output is given by,

$$\frac{d(IV)}{dV} = 0 \tag{5.6}$$

$$P_{\max} = I_m V_m \tag{5.7}$$

Here, I_m and V_m represent the values of the current and voltage of the cell at the maximum power point (as mentioned in figure 5.10(b)). The fill factor (*FF*) is defined as the ratio of the maximum power output ($P_{\max}$) at the maximum power point to the product of the open-circuit voltage (V_{oc}) and short-circuit current ($I_{sc}{\approx}I_{ph}$) and can be expressed as,

$$FF = [P_{\max}/V_{oc}I_{sc}] \tag{5.8}$$

FF gives a measure of the maximum *IV* product (i.e. V_m, I_m) and its value is always less than unity. The output power per unit area of the cell can be determined from the *I–V* characteristics by evaluating the area of the rectangle formed between the operating point (maximum power point) and the two axes. The operating point depends on the load resistance. The increase in series resistance and decrease in shunt resistance severely degrades the *FF* values.

The efficiency of a solar cell is defined as the ratio of power output corresponding to the maximum power point to the power input and is represented as,

$$\eta = \frac{P_{\max}}{P_{in}.\,\text{Area}} \tag{5.9}$$

where, P_{in} is the intensity of the incident radiation.

In terms of the V_{oc}, J_{sc} and *FF*, the efficiency of the cell can be represented as,

$$\eta = \frac{V_{oc}.\,J_{SC}.\,FF}{P_{in}} \tag{5.10}$$

where, J_{sc} is the current density (current per unit area) and P_{in} is the intensity of incident light. The short-circuit current density of the cell depends on the solar spectral irradiance and is given by,

$$J_{sc} = q\int_{h\nu=E_g}^{\infty}\frac{dN_{ph}}{dh\nu}d(h\nu) \tag{5.11}$$

Here, N_{ph} is the incident photon flux. In a practical solar cell, the value of J_{sc} may be limited by reflection losses, ohmic losses (series and shunt resistance) and recombination losses.

5.5.3.1 Spectral response

The spectral response (*SR*) is defined as the ratio of the photocurrent generated by a solar cell under monochromatic illumination of a given wavelength, to the value of the incident power radiation at the same wavelength. Mathematically, spectral response is given by,

$$SR = \frac{J_{sc}}{P_{in}} \tag{5.12}$$

where, J_{sc} and P_{in} correspond to radiation of photon energy $h\nu \geqslant E_g$. The short-circuit current density of the cell depends on the incident wavelength of solar radiation.

5.6 Donors and acceptors in polymer solar cells

In PSCs, the donor and acceptor materials are crucial components that directly influence their performance [53]. Furthermore, the bandgap of polymer donors and acceptors plays an important role in determining the PV performance. Generally, the bandgap of polymer donors used in PSCs from about 1.2 to 2.0 electron volts (eV) for absorption of a significant portion of the solar spectrum. Polymer donors with narrower bandgaps absorb light in the longer wavelength region (red and near-infrared), while those with wider bandgaps absorb light in the shorter wavelength region (blue and ultraviolet). Researchers continually explore new polymer materials and modify existing ones to achieve optimal bandgaps and improve the overall efficiency of PSCs. The choice of donor and acceptor materials depends on factors such as their optical and electronic properties, energy levels, compatibility, processability, stability, and overall device performance requirements. This section will provide a brief overview of polymer donors and acceptors that are utilized in active layer formation of BHJ PSCs. These are categorized on the basis of the use of polymers as: wide bandgap polymers, medium bandgap polymers, and narrow bandgap polymers. Table 5.1 summarizes the PV parameters: V_{oc}, J_{sc}, *FF* and *PCE* (η) of BHJ PSCs based on various polymer donors and acceptors that have recently been reported in literature.

5.6.1 Donor polymers

Poly (3-hexylthiophene) (P3HT) is one of the most widely used conjugated polymers as a donor material in PSCs. It has good processability and relatively high charge carrier mobility. Poly(3,3′-didodecylquarterthiophene) (PQT-12) is another conjugated polymer that has been used as a donor material in PSCs. In addition, Poly(2,6-(4,8-bis(5-(2-ethylhexyl)thiophen-2-yl)-benzo[1,2-b:4,5-b′]dithiophene))-alt-(5,5-(1′,3′-di-2-thienyl-5′,7′-bis(2-ethylhexyl)benzo[1′,2′-c:4′,5′-c′]dithiophene-4,8-dione)) polymer has been utilized as a donor material in high-efficiency PSCs [53, 70]. Figure 5.10(c) depicts the chemical structure of widely used polymer donor (P3HT). Donor polymers can be categorized as wide bandgap polymers, medium bandgap polymers and narrow bandgap polymers.

5.6.1.1 Wide bandgap polymers

Table 5.1 summarizes the PV parameters, V_{oc}, J_{sc}, *FF* and *PCE* (η) using wide bandgap polymers as donors for BHJ PSC. *PCE* of 8.4 % was reported for an imine-

Table 5.1. PV parameters, open-circuit voltage (*Voc*), current density (*Jsc*), fill factor (*FF*) and power conversion efficiency (*PCE-η*) for BHJ PSCs. BHJ PSCs are categorized on the basis of the use of polymers as: wide bandgap polymers as donors, medium ban gap polymers as donors, narrow bandgap polymers as donors and polymers as acceptors. [131] is the source of the data, which has been modified accordingly.

Polymer	Structure (BHJ)	V_{oc} (mV)	J_{sc} (mAcm^{-2})	*FF*	*PCE* (η)	References
Wide bandgap polymers as donors	ITO/PEDOT:PSS/PBDT-TTZ:N2200/PFN-Br/Ag	870	14.4	0.67	8.40	[94]
	ITO/PEDOT:PSS/PTzTz:IT-4F/PFN-Br/Al	820	18.81	0.69	10.63	[95]
	ITO/PEDOT:PSS/PBiTPD:Y6/PFN-Br/Ag	830	25.6	0.66	14.20	[96]
	ITO/PEDOT:PSS/P106:DBTBT-IC:Y18-DMO/PFN/Al	910	24.82	0.73	16.49	[97]
	ITO/PEDOT:PSS/W1:Y6/PDIN/Ag	890	25.92	0.69	16.16	[98]
	ITO/PEDOT:PSS/D18:Y6/PDIN/Ag	865	27.31	0.75	18.22	[24]
	ITO/PEDOT:PSS/D18:Y6:PC61BM /PDIN/Ag	859	27.70	0.76	17.89	[99]
	ITO/ZnO/[PTB7-Th:Si-BDT:DCNBT-TPIC/MoO_3/Ag	860	22.32	0.68	13.45	[100]
Medium bandgap polymers as donors	ITO/PEDOT:PSS/POBDFB:ITIC: PCBM/PFN/Al	720	17.65	0.62	7.91	[101]
	ITO/PEDOT:PSS/PM6:Y6/PDINO/Al	830	25.3	0.75	15.7	[35]
	ITO/PEDOT:PSS/P:ITIC-m:Y6/PFN/Al	990	20.65	0.74	15.13	[102]
	ITO/PEDOT:PSS/P130:Y6/PFN/Al	890	23.84	0.72	15.28	[103]
	ITO/PEDOT:PSS/PM6:Y6: $PC_{71}BM$/PDINO/Al	850	25.7	0.76	16.67	[104]
	ITO/PEDOT:PSS/PM6:MF1:Y6/PDIN/Al	853	25.68	0.77	17.22	[105]
	ITO/PEDOT:PSS/PM6:PM7-Si:C9/PFN-Br/Ag	864	26.35	0.77	17.7	[106]
Narrow bandgap polymers as donors	ITO/PEDOT:PSS/P3:$PC_{71}BM$(4)/LiF/Al	770	7.59	0.41	2.45	[107]
	ITO/PEDOT:PSS/P(T2BDY−TBDT)/PNDIT-F3N −Br/Ag	780	12.07	0.47	4.40	[108]
	ITO/PEDOT:PSS/PTT-EFQX:PCBM/PFN-Br/Ag	690	11.19	0.68	5.37	[109]
	ITO/PEDOT:PSS/PffBT-DPP/PCBM/MeIC/ZrAcAc/Al	760	16.1	0.73	9.0	[110]
	ITO/PEDOT:PSS/PTQ10:Y6/PFN-Br/Al	820 ± 1	23.9 ± 0.1	0.73	14.5 ± 0.1	[111]
polymers as an acceptor	ITO/ZnO/PBDB-T:PIID(CO) 2FT/MoO_3/Ag	640	8.30	0.50	2.65	[112]
	ITO/ZnO/PTB7-Th:NDP-V/V_2O_5/Al	740	17.07	0.67	8.59	[113]
	ITO/ZnO/PEI/BSS10:PBDB-T/MoO_3/Ag	860	18.55	0.64	10.10	[114]
	ITO/PEDOT:PSS/PTzBISi:N2200/C60N/Ag MTHF-TA + SVA	880	17.62	0.76	11.25	[27]
	ITO/PEDOT:PSS/PBDB-T:PYT/PDINN50/Ag	883	22.70	0.72	14.57	[115]

substituted wide bandgap polymer donor for the structure ITO/PEDOT:PSS/PBDT-TTZ:N2200/PFN-Br/Ag [94]. Furthermore, in a related study, the BHJ structure ITO/PEDOT:PSS/PTzTz:IT-4F/PFN-Br/Al, which has donor polymer PTzTz in conjunction with a non-fullerene acceptor (IT-4F), produced *PCE* of 10.63% [95]. In addition, bithieno[3,4-c]pyrrole-4,6-dione (PBiTPD), a donor based on the thieno [3,4-c]pyrrole-4,6-dione (TPD), was reported by Zhao *et al* and *PCE* of 14.20% was achieved using the solar cell configuration ITO/PEDOT:PSS/PBiTPD:Y6/PFN-Br/Ag [96]. In another study, binary and ternary BHJ solar cells were fabricated by Keshtov *et al* [97] using a D–A polymer P106 as a donor and two non- fullerene acceptors, Y18-DMO and DBTBT-IC. P106 has dithieno [2,3-e;3′2′-g]isoindole-7,9 (8H) (DTID) as an acceptor and 2-dodecylbenzo[1,2-b:3,4-b′:6,5-b″]trithiophene (3TB) as a donor. *PCE* of 16.49 % was reported for a ternary structure ITO/PEDOT:PSS/P106:DBTBT-IC:Y18-DMO/PFN/Al [97]. Wide-bandgap (2.16 eV) polymer W1, 1,2-difluoro-4,5-bis(octyloxy)benzene was synthesized by Wang *et al* for fabrication of a BHJ solar cell having structure ITO/PEDOT:PSS/W1:Y6/PDIN/Ag, where Y6 is a non-fullerene electron acceptor (BTP-4F). They were able to achieve a *PCE* of 16.16% [98]. Liu *et al* [24] developed wide-bandgap (1.98 eV) donor copolymer for fabricating BHJ solar cell and reported *PCE* of 18.22% for ITO/PEDOT:PSS/D18:Y6/PDIN/Ag device structure [24]. The *J–V* characteristics for this high efficiency cell are shown in figure 5.11(a). Qin *et al* [99] have achieved *PCE* of 17.89% for the device architecture ITO/PEDOT:PSS/D18:Y6:PC61BM/PDIN/Ag by making thick active layer via incorporating PCBM into a D18-Y6 blend for BHJ solar cells [99]. Recently Gokulnath *et al* [100] reported wide bandgap donor polymers to fabricate a ternary solar cell using siloxane-functionalized polymer Si-BDT. *PCE* of 13.45% was obtained for device structure ITO/ZnO/[PTB7-Th(0.6):Si-BDT(0.4):DCNBT-TPIC(0.6)/MoO_3/Ag [100].

5.6.1.2 Medium bandgap polymers

An enhanced PV performance was reported by Chen *et al* [101] using medium bandgap polymer as donor for BHJ PSCs. Thiophene, a copolymer based on diflurobenzothiadiazole (FBT) and benzodithiophene (BDT) was synthesized to fabricate ternary and binary BHJ solar cells. Non fullerene3,9-bis(2-methylene-(3-(1,1-dicyanomethylene)-indanone))-5,5,11,11-tetrakis(4-hexylphenyl) dithienol[2,3-d:2′,3′-d′]-s-indaceno[1,2-b:5,6-b′]dithiophene (ITIC) and [6,6]-phenyl-C71-butyric acid methyl ester (PC71BM) were used as acceptors. *PCE* of 7.91% has been achieved for the structure ITO/PEDOT:PSS/POBDFBT:PCBM:ITIC/PFN/Al [101]. Furthermore, Yuan *et al* were able to achieve a *PCE* of 15.7% by using PM6 as a donor and a ladder-type Y6 as acceptor for the device architecture ITO/PEDOT:PSS/PM6:Y6/PDINO/Al [35]. The *J–V* characteristics of this high efficiency solar cell are shown in figure 5.11(b). Sharma *et al* [102] used a conjugated polymer based on BODIPY-thiophene to create a ternary solar cell by combining two polymers with two distinct acceptors, Y6 and ITIC-m. A *PCE* of 15.13% was achieved for the ternary solar cell with architecture of ITO/PEDOT:PSS/P:ITIC-m:Y6/PFN/Al [102]. A medium bandgap copolymer donor, D-A1-D-A2, was synthesized in a more recent work, where A1 is fluorinated benzothiadiazole, A2 is a new

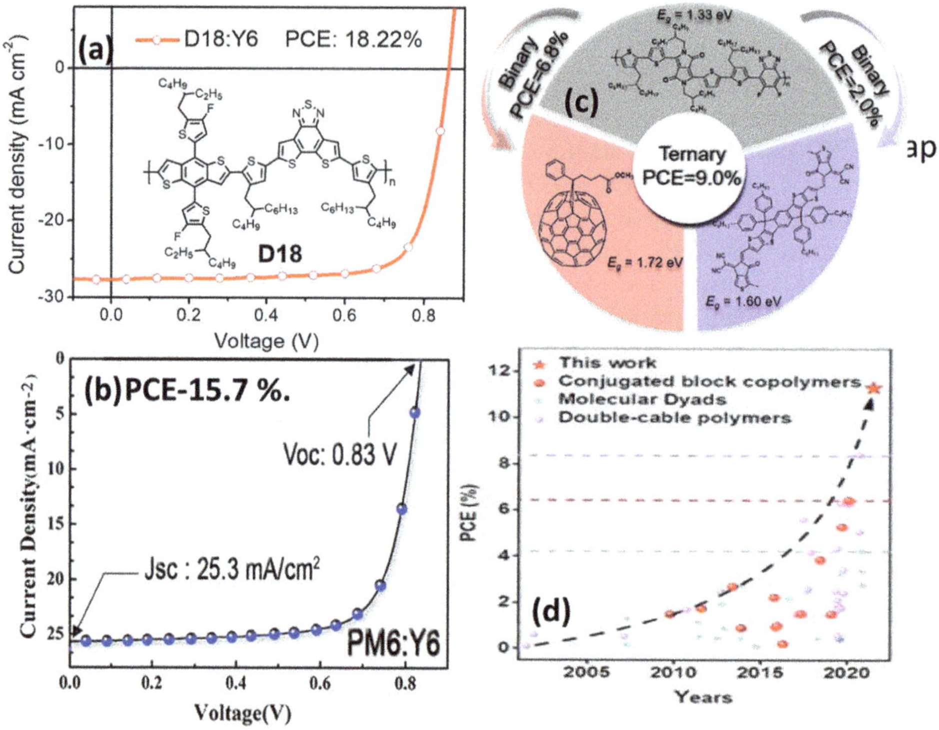

Figure 5.11. (a) The *J–V* characteristics of a BHJ solar cell having *PCE* of 18.22% based on a wide-bandgap (1.98 eV) donor copolymer (D18) device structure ITO/PEDOT:PSS/D18:Y6/PDIN/Ag. Reprinted from [24], copyright (2020) with permission from Elsevier. (b) The *J–V* characteristics for a medium bandgap-based polymer donor (PM6) solar cell having device architecture ITO/PEDOT:PSS/PM6:Y6/PDINO/Al achieved *PCE* of 15.7% by using a ladder-type Y6 acceptor. Reprinted from [35], copyright (2019) with permission from Elsevier.(c) *PCE* of 9.0% was demonstrated for ternary solar cells (binary structure-*PCE*-2.0% and 6.8%) with the architecture ITO/PEDOT:PSS/PffBT-DPP:PCBM:MeIC/ZrAcAc/Al for a narrow bandgap polymer donor (1.33 eV) PffBT-DPP, based on diketopyrrolopyrrole (DPP). Reprinted with permission from [110], copyright (2020) American Chemical Society. (d) *PCE* of 14.57% was reported for structure ITO/PEDOT:PSS/PBDB-T:PYT/PDINN50/Ag with block copolymer PBDB-Tb-PYT, acceptor to create active layer. Reprinted from [115], copyright (2021) with permission from Elsevier.

anthra[1,2-b:4,3,b′:6,7-c″]trithiophene-8.12-dione (A3T), and D is thiophene. *PCE* of 15.28% has been achieved for the structure ITO/PEDOT:PSS/P130:Y6/PFN/Al [103]. Ternary solar cells were fabricated by adding PCBM as a third component to the PM6-Y6 binary mixture by Yan *et al* A *PCE* of 16.67% with the architecture of ITO/PEDOT:PSS/PM6:Y6:$PC_{71}BM$/PDINO/Al was obtained for rigid solar cell [104]. An *et al* [105] have achieved *PCE* of 17.22% for a ternary solar cell having structure ITO/PEDOT:PSS/PM6:MF1:Y6/PDIN/Al using Y6 and MF1 as acceptors [105]. Peng *et al* [106] used a medium bandgap polymer as donor in BHJ devices. D–A type polymer, PM7-Si, was synthesized by modifying PM6 with the substitution of chorine for fluorine and ethylhexyl group with alkylsilyl chains. Ternary

BHJ solar cells exhibited *PCE* of 17.7% for the structure ITO/PEDOT:PSS/PM6: PM7-Si:C9/PFN-Br/Ag [106].

5.6.1.3 Narrow bandgap polymers

A donor material P3, based on benzodithiophene by attaching a 2-(2-octyldodecyl) selenophene ring to the fourth and eighth positions of the benzene ring in BDT fabricated by Caliskan *et al* [107] and *PCE* of 2.45% was reported for device architecture ITO/PEDOT:PSS/P3:PC71BM/LiF/Al [107]. Low bandgap (1.30–1.35 eV) D–A copolymers 4,4-Difluoro-4-bora-3a,4a-diaza-s-indacene (BODIPY) as donor with benzo[1,2-b:4,5-b′]dithiophene (BDT) functioning as acceptor were synthesized by Can *et al* [108] for a structure ITO/PEDOT:PSS/P(T2BDY-TBDT)/PNDIT-F3N−Br/Ag, demonstrating the best *PCE* of 4.40% [108]. Furthermore, *PCE* of 5.37% was achieved using the D–A (PTT-EFQX) material in the architecture ITO/PEDOT:PSS/PTT-EFQX:PCBM/PFN-Br [109]. The polymer PffBT-DPP, based on diketopyrrolopyrrole (DPP), exhibits a small bandgap of 1.33 eV, as reported by Pan *et al. PCE* of 9.0% was demonstrated for ternary solar cells with the architecture ITO/PEDOT:PSS/ PffBT-DPP:PCBM:MeIC/ ZrAcAc/Al [110]. Figure 5.11(c) depicts the efficiency of a ternary solar cell and binary structures (*PCE*-2.0% and 6.8%). Rech *et al* reported a *PCE* of ~15% for a device structure (ITO)/PEDOT:PSS/PTQ10:Y6/PFN-Br/Al using a donor polymeric material based on PT10 [111].

5.6.2 Acceptor polymers

The most commonly used acceptor polymer is [6, 6]-Phenyl-C61-butyric acid methyl ester (PCBM) in PSCs. It is a fullerene derivative and has been extensively studied due to its good electron mobility and compatibility with various donor materials. Non-fullerene acceptors (NFAs) have gained significant attention in recent years due to their potential to overcome the limitations of fullerene-based acceptors. Various NFAs have been developed, including small molecules and polymers such as ITIC, ITIC-4F, Y6, Y6-C, and many others [53, 70]. Figure 5.10(c) depicts the chemical structure of widely used polymer donor–acceptor ($PC_{60}BM$). Nevertheless, few studies are reported for other than fullerene and non-fullerene- based polymer in BHJ PSC fabrication (table 5.1). In a related work, N-acyl-substituted isoindigo (IID) motifs (IID(CO)) was used as acceptor for fabrication of BHJ solar cells and produced *PCE* of 2.65% [112]. Naphthalene-diimide (NDP-V) used as a polymeric acceptors inverted structure device achieved *PCE* of 8.59% for inverted structure ITO/ZnO/PTB7-Th:NDP-V/V_2O_5/Al [113]. Kolhe *et al* reported a series of inverted photovoltaic cells with NDI-biselenophene/NDI-selenophene copolymer as an acceptor with an equivalent proportion of 90:10 (BSS10) that yielded *PCE* of 10.10% for the structure ITO/ZnO/PEI/BSS10:PBDB-T/MoO_3/Ag. [114]. With N2200 poly as an acceptor, Zhu *et al* [27] fabricated solar cells with the structure ITO/PEDOT:PSS/PTzBISi:N2200/C60N/Ag, which resulted in a maximum *PCE* of 11.25%. Block copolymer PBDB-Tb-PYT, with donor and acceptor when mixed to create active layer, a *PCE* of 14.57% was achieved for structure ITO/PEDOT:PSS/

PBDB-T:PYT/PDINN50/Ag [115]. A block copolymer containing donor and acceptor moieties (PBDB-Tb-PYT) was used. A *PCE* of 14.57% was recorded for the blends obtained by mixing the polymers in the active layer.

5.7 Degradation of polymer solar cells

In spite of the high efficiency and low cost processing, PSCs still struggle to maintain long-term stability [116–120]. This is due to the degradation of PSCs which is a complex process influenced by various factors, including materials, device architecture, environmental conditions, and operational parameters. Some common degradation mechanisms observed in PSCs are due to the following reasons:

5.7.1 Photochemical degradation

Exposure to light, especially UV radiation, induces photochemical degradation in PSCs [121, 122]. This degradation leads to changes in the molecular structure of organic materials, such as polymer chains and acceptor molecules, and affects their optical and electrical properties. Photochemical degradation may also cause the formation of reactive species, such as free radicals, which can further accelerate degradation processes.

5.7.2 Thermal degradation

High temperatures accelerate chemical reactions in PSCs, leading to thermal degradation of organic materials [123, 124]. Heat promotes bond cleavage, cross-linking, and rearrangement of molecular chains, resulting in reduced charge transport properties, increased trap states, and changes in the morphology of the active layer. Thermal degradation may also lead to the loss of volatile components, such as additives and solvents, which can affect device stability and performance.

5.7.3 Oxidative degradation

Exposure to oxygen and other oxidizing agents lead to oxidative degradation of PSCs [118, 125]. Oxygen can react with organic materials, causing oxidation of functional groups and the formation of carbonyl and hydroperoxide species. Oxidative degradation may lead to changes in the chemical structure, morphology, and electronic properties of PSCs, ultimately resulting in a decrease in device efficiency and lifetime.

5.7.4 Hydrolytic degradation

Moisture ingress into PSCs may lead to hydrolytic degradation of organic materials, particularly at interfaces and defects within the device [117, 126]. Hydrolysis reactions can break chemical bonds in polymers and other organic molecules, leading to chain scission, depolymerization, and loss of mechanical integrity. Hydrolytic degradation may also promote the formation of undesirable species, such as acids and salts, which can further accelerate degradation processes.

5.7.5 Electrochemical degradation

Electrochemical reactions at interfaces and within the device may contribute to degradation in PSCs [117, 126]. For example, redox reactions involving metal electrodes or impurities may lead to corrosion, degradation of charge transport layers, and formation of charge traps. Electrochemical reactions may also cause ion migration, polarization effects, and changes in the electronic properties of PSCs, impacting device performance and stability.

5.7.6 Environmental degradation

Other environmental factors, such as humidity, pollutants, and mechanical stress can also contribute to the degradation of PSCs [117, 126, 127]. Humidity promotes hydrolysis and mold growth, while pollutants cause chemical contamination and surface degradation. Mechanical stress from handling, bending, or thermal cycling can lead to mechanical damage, delamination, and cracking of device components.

Addressing degradation mechanisms in PSCs requires a multidisciplinary approach, involving materials science, device engineering, encapsulation technologies, and reliability testing. Strategies for improving the stability and lifetime of PSCs include the development of more stable organic materials, optimization of device architectures, implementation of effective encapsulation and packaging methods, and rigorous testing under accelerated aging conditions.

5.8 Conclusions

An overview of the status of PSCs has been presented in this chapter. Various device structures and configurations, the working principle, methods of processing and electrical and optical methods of characterization have been briefly discussed. Donors and acceptors and the degradation mechanisms are described. PSCs continue to evolve as technology transfer from lab scale to research and development to manufacturing finds direction from the perspective of device performance stability, repeatability, reproducibility and subsequently, yield.

References

[1] Hazarika G 2023 *Renewables Accounted for 83% of the Total Generation Capacity Added in 2022* https://mercomindia.com/renewables-generation-capacity-added-in-2022

[2] Zhao J, Wang A and Green M A 1999 24·5% Efficiency silicon PERT cells on MCZ substrates and 24·7% efficiency PERL cells on FZ substrates *Prog. Photovolt. Res. Appl.* **7** 471–4

[3] Green M A 2009 The path to 25% silicon solar cell efficiency: history of silicon cell evolution *Prog. Photovolt. Res. Appl.* **17** 183–9

[4] Green M A *et al* 2023 Solar cell efficiency tables (Version 61) *Prog. Photovolt. Res. Appl.* **31** 3–16

[5] Cho A 2010 *Energy's Tricky Tradeoffs* (American Association for the Advancement of Science)

[6] Graetzel M *et al* 2012 Materials interface engineering for solution-processed photovoltaics *Nature* **488** 304–12

[7] Deng X *et al* 2003 *Handbook of Photovoltaic Science and Engineering* ed A Luque (Chichester: Wiley)

[8] Service R F 2011 Outlook brightens for plastic solar cells *Science* **332** 293–3

[9] You J *et al* 2013 A polymer tandem solar cell with 10.6% power conversion efficiency *Nat. Commun.* **4** 1446

[10] Huang Y-C *et al* 2023 Highly efficient flexible roll-to-roll organic photovoltaics based on non-fullerene acceptors *Polymers* **15** 4005

[11] Chen L M *et al* 2009 Recent progress in polymer solar cells: manipulation of polymer: fullerene morphology and the formation of efficient inverted polymer solar cells *Adv. Mater.* **21** 1434–49

[12] Shirakawa H *et al* 1977 Synthesis of electrically conducting organic polymers: halogen derivatives of polyacetylene, (CH) *J. Chem. Soc., Chem. Commun.* 578–80

[13] Li G, Zhu R and Yang Y 2012 Polymer solar cells *Nat. Photonics* **6** 153–61

[14] Inganäs O 2018 Organic photovoltaics over three decades *Adv. Mater.* **30** 1800388

[15] Matsuhisa N *et al* 2019 Materials and structural designs of stretchable conductors *Chem. Soc. Rev.* **48** 2946–66

[16] Cui C and Li Y 2019 High-performance conjugated polymer donor materials for polymer solar cells with narrow-bandgap nonfullerene acceptors *Energy Environ. Sci.* **12** 3225–46

[17] Pagliaro M, Ciriminna R and Palmisano G 2008 Flexible solar cells *ChemSusChem.* **1** 880–91

[18] Angmo D and Krebs F C 2013 Flexible ITO-free polymer solar cells *J. Appl. Polym. Sci.* **129** 1–14

[19] Po R *et al* 2012 Polymer-and carbon-based electrodes for polymer solar cells: Toward low-cost, continuous fabrication over large area *Sol. Energy Mater. Sol. Cells* **100** 97–114

[20] Lu S *et al* 2017 Recent development in ITO-free flexible polymer solar cells *Polymers* **10** 5

[21] Roncali J 2009 Molecular bulk heterojunctions: an emerging approach to organic solar cells *Acc. Chem. Res.* **42** 1719–30

[22] Cooper A I 2009 Conjugated microporous polymers *Adv. Mater.* **21** 1291–5

[23] Zhang M *et al* 2021 Single-layered organic photovoltaics with double cascading charge transport pathways: 18% efficiencies *Nat. Commun.* **12** 309

[24] Liu Q *et al* 2020 18% Efficiency organic solar cells *Sci. Bull.* **65** 272–5

[25] Zhao W *et al* 2016 Fullerene-free polymer solar cells with over 11% efficiency and excellent thermal stability *Adv. Mater.* **28** 4734–9

[26] Lin Y *et al* 2016 High-performance electron acceptor with thienyl side chains for organic photovoltaics *JACS* **138** 4955–61

[27] Zhu L *et al* 2019 Aggregation-induced multilength scaled morphology enabling 11.76% efficiency in all-polymer solar cells using printing fabrication *Adv. Mater.* **31** 1902899

[28] Zhao R *et al* 2020 Organoboron polymer for 10% efficiency all-polymer solar cells *Chem. Mater.* **32** 1308–14

[29] Chen J D *et al* 2015 Single-junction polymer solar cells exceeding 10% power conversion efficiency *Adv. Mater.* **27** 1035–41

[30] Zhao J *et al* 2016 Efficient organic solar cells processed from hydrocarbon solvents *Nat. Energy* **1** 1–7

[31] Zhang S *et al* 2018 Over 14% efficiency in polymer solar cells enabled by a chlorinated polymer donor *Adv. Mater.* **30** 1800868

[32] Tong Y *et al* 2020 Progress of the key materials for organic solar cells *Sci. China Chem.* **63** 758–65

[33] Liang Y *et al* 2010 For the bright future—bulk heterojunction polymer solar cells with power conversion efficiency of 7.4% *Adv. Mater.* **22** E135–8
[34] Lin Y *et al* 2015 An electron acceptor challenging fullerenes for efficient polymer solar cells *Adv. Mater.* **27** 1170–4
[35] Yuan J *et al* 2019 Single-junction organic solar cell with over 15% efficiency using fused-ring acceptor with electron-deficient core *Joule* **3** 1140–51
[36] Lee J-W *et al* 2024 High efficiency, thermally stable, and mechanically robust ternary all-polymer solar cells achieved by alloyed vinyl-linked polymerized small-molecule acceptors *Nano Energy* **122** 109338
[37] Shi Z *et al* 2019 Tandem structure: a breakthrough in power conversion efficiency for highly efficient polymer solar cells *Sustain. Energy Fuels* **3** 910–34
[38] Wu P *et al* 2024 18.6% efficiency all-polymer solar cells enabled by a wide bandgap polymer donor based on Benzo[1,2-d:4,5-d′]bisthiazole *Adv. Mater.* **36** 2306990
[39] Hummelen J C *et al* 1995 Preparation and characterization of fulleroid and methanofullerene derivatives *J. Organ. Chem.* **60** 532–8
[40] Chamberlain G A 1983 Organic solar cells: a review *Sol. Cells* **8** 47–83
[41] Tang C W 1986 Two-layer organic photovoltaic cell *Appl. Phys. Lett.* **48** 183–5
[42] Knupfer M 2003 Exciton binding energies in organic semiconductors *Appl. Phys.* A **77** 623–6
[43] Yu G *et al* 1995 Polymer photovoltaic cells: enhanced efficiencies via a network of internal donor-acceptor heterojunctions *Science* **270** 1789–91
[44] Lin Y *et al* 2020 Self-assembled monolayer enables hole transport layer-free organic solar cells with 18% efficiency and improved operational stability *ACS Energy Lett.* **5** 2935–44
[45] Lin Y *et al* 2020 A simple n-dopant derived from diquat boosts the efficiency of organic solar cells to 18.3 *ACS Energy Lett.* **5** 3663–71
[46] Mayer A C *et al* 2007 Polymer-based solar cells *Mater. Today* **10** 28–33
[47] Singh R P and Kushwaha O S 2013 Polymer solar cells: an overview *Macromolecular Symposia* (Wiley Online Library)
[48] Lattante S 2014 Electron and hole transport layers: their use in inverted bulk heterojunction polymer solar cells *Electronics* **3** 132–64
[49] Xu H *et al* 2020 Hole transport layers for organic solar cells: recent progress and prospects *J. Mater. Chem.* A **8** 11478–92
[50] Moiz S A, Alzahrani M S and Alahmadi A N 2022 Electron transport layer optimization for efficient PTB7: PC70BM bulk-heterojunction solar cells *Polymers* **14** 3610
[51] Sachs-Quintana I *et al* 2014 Electron barrier formation at the organic-back contact interface is the first step in thermal degradation of polymer solar cells *Adv. Funct. Mater.* **24** 3978–85
[52] Cai W, Gong X and Cao Y 2010 Polymer solar cells: recent development and possible routes for improvement in the performance *Sol. Energy Mater. Sol. Cells* **94** 114–27
[53] Li Y *et al* 2022 Recent progress in organic solar cells: a review on materials from acceptor to donor *Molecules* **27** 1800
[54] Li G *et al* 2006 Efficient inverted polymer solar cells *Appl. Phys. Lett.* **88** 253503
[55] He Z *et al* 2012 Enhanced power-conversion efficiency in polymer solar cells using an inverted device structure *Nat. Photonics* **6** 591–5
[56] Hau S K, Yip H-L and Jen A K-Y 2010 A review on the development of the inverted polymer solar cell architecture *Polym. Rev.* **50** 474–510
[57] Gong X 2012 Toward high performance inverted polymer solar cells *Polymer* **53** 5437–48

[58] Hau S K *et al* 2008 Interfacial modification to improve inverted polymer solar cells *J. Mater. Chem.* **18** 5113–9

[59] Xu X, Li Y and Peng Q 2022 Ternary blend organic solar cells: understanding the morphology from recent progress *Adv. Mater.* **34** 2107476

[60] Doumon N Y, Yang L and Rosei F 2022 Ternary organic solar cells: A review of the role of the third element *Nano Energy* **94** 106915

[61] Lu L *et al* 2015 Status and prospects for ternary organic photovoltaics *Nat. Photonics* **9** 491–500

[62] Wang J *et al* 2021 A tandem organic photovoltaic cell with 19.6% efficiency enabled by light distribution control *Adv. Mater.* **33** 2102787

[63] Xue J *et al* 2004 Asymmetric tandem organic photovoltaic cells with hybrid planar-mixed molecular heterojunctions *Appl. Phys. Lett.* **85** 5757–9

[64] Yan C *et al* 2018 Non-fullerene acceptors for organic solar cells *Nat. Rev. Mater.* **3** 1–19

[65] Roghabadi F A *et al* 2018 Bulk heterojunction polymer solar cell and perovskite solar cell: concepts, materials, current status, and opto-electronic properties *Sol. Energy* **173** 407–24

[66] Hu J *et al* 2022 Self-assembled monolayers for interface engineering in polymer solar cells *J. Polym. Sci.* **60** 2175–90

[67] Yin Z, Wei J and Zheng Q 2016 Interfacial materials for organic solar cells: recent advances and perspectives *Adv. Sci.* **3** 1500362

[68] Lai T-H *et al* 2013 Properties of interlayer for organic photovoltaics *Mater. Today* **16** 424–32

[69] McCullough R D *et al* 1993 Design, synthesis and control of conducting polymer architectures: structurally homogeneous poly(3-alkylthiophenes) *J. Org. Chem.* **58** 904

[70] Facchetti A 2013 Polymer donor–polymer acceptor (all-polymer) solar cells *Mater. Today* **16** 123–32

[71] Hashmi G *et al* 2011 Review of materials and manufacturing options for large area flexible dye solar cells *Renew. Sustain. Energy Rev.* **15** 3717–32

[72] Jung E H *et al* 2019 Efficient, stable and scalable perovskite solar cells using poly (3-hexylthiophene) *Nature* **567** 511–5

[73] Krebs F C 2009 Fabrication and processing of polymer solar cells: a review of printing and coating techniques *Sol. Energy Mater. Sol. Cells* **93** 394–412

[74] Shen X, Hu W and Russell T P 2016 Measuring the degree of crystallinity in semicrystalline regioregular poly (3-hexylthiophene) *Macromolecules* **49** 4501–9

[75] Sahu N, Parija B and Panigrahi S 2009 Fundamental understanding and modeling of spin coating process: a review *Indian J. Phys.* **83** 493–502

[76] Rubio Arias J J *et al* 2021 Solution processing of polymer solar cells: towards continuous vacuum-free production *J. Mater. Sci., Mater. Electron.* **32** 11367–92

[77] Krebs F C 2009 All solution roll-to-roll processed polymer solar cells free from indium-tin-oxide and vacuum coating steps *Org. Electron.* **10** 761–8

[78] Zuo C *et al* 2018 One-step roll-to-roll air processed high efficiency perovskite solar cells *Nano Energy* **46** 185–92

[79] Krebs F C *et al* 2009 A complete process for production of flexible large area polymer solar cells entirely using screen printing—first public demonstration *Sol. Energy Mater. Sol. Cells* **93** 422–41

[80] Hall D B, Underhill P and Torkelson J M 1998 Spin coating of thin and ultrathin polymer films *Polym. Eng. Sci.* **38** 2039–45

[81] Du X *et al* 2019 Efficient polymer solar cells based on non-fullerene acceptors with potential device lifetime approaching 10 years *Joule* **3** 215–26
[82] Zhang Y *et al* 2018 Thermally stable all-polymer solar cells with high tolerance on blend ratios *Adv. Energy Mater.* **8** 1800029
[83] Søndergaard R R, Hösel M and Krebs F C 2013 Roll-to-Roll fabrication of large area functional organic materials *J. Polym. Sci., Part B: Polym. Phys.* **51** 16–34
[84] Søndergaard R *et al* 2012 Roll-to-roll fabrication of polymer solar cells *Mater. Today* **15** 36–49
[85] Ji G *et al* 2019 12.88% efficiency in doctor-blade coated organic solar cells through optimizing the surface morphology of a ZnO cathode buffer layer *J. Mater. Chem.* A **7** 212–20
[86] Gilot J, Wienk M M and Janssen R A 2007 Double and triple junction polymer solar cells processed from solution *Appl. Phys. Lett.* **90** 143512
[87] Tanenbaum D M *et al* 2012 Edge sealing for low cost stability enhancement of roll-to-roll processed flexible polymer solar cell modules *Sol. Energy Mater. Sol. Cells* **97** 157–63
[88] Zhong W *et al* 2019 In situ structure characterization in slot-die-printed all-polymer solar cells with efficiency over 9% *Sol. RRL* **3** 1900032
[89] La Notte L *et al* 2018 Fully-sprayed flexible polymer solar cells with a cellulose-graphene electrode *Mater. Today Energy* **7** 105–12
[90] Corzo D *et al* 2019 Digital inkjet printing of high-efficiency large-area nonfullerene organic solar cells *Adv. Mater. Technol.* **4** 1900040
[91] Larsen-Olsen T T *et al* 2012 Roll-to-roll processed polymer tandem solar cells partially processed from water *Sol. Energy Mater. Sol. Cells* **97** 43–9
[92] Angmo D *et al* 2013 Roll-to-roll inkjet printing and photonic sintering of electrodes for ITO free polymer solar cell modules and facile product integration *Adv. Energy Mater.* **3** 172–5
[93] Youn H, Lee T and Guo L J 2014 Multi-film roll transferring (MRT) process using highly conductive and solution-processed silver solution for fully solution-processed polymer solar cells *Energy Environ. Sci.* **7** 2764–70
[94] Cao Z *et al* 2019 Understanding of imine substitution in wide-bandgap polymer donor-induced efficiency enhancement in all-polymer solar cells *Chem. Mater.* **31** 8533–42
[95] Xue C *et al* 2020 Achieving efficient polymer solar cells based on benzodithiophene–thiazole-containing wide band gap polymer donors by changing the linkage patterns of two thiazoles *New J. Chem.* **44** 13100–7
[96] Zhao J *et al* 2020 Bithieno[3,4-c]pyrrole-4,6-dione-mediated crystallinity in large-bandgap polymer donors directs charge transportation and recombination in efficient nonfullerene polymer solar cells *ACS Energy Lett.* **5** 367–75
[97] Keshtov M L *et al* 2021 Binary and ternary polymer solar cells based on a wide bandgap D-a copolymer donor and two nonfullerene acceptors with complementary absorption spectral *ChemSusChem.* **14** 4731–40
[98] Wang T *et al* 2020 A 2.16 eV bandgap polymer donor gives 16% power conversion efficiency *Sci. Bull.* **65** 179–81
[99] Qin J *et al* 2020 Over 16% efficiency from thick-film organic solar cells *Sci. Bull.* **65** 1979–82
[100] Gokulnath T *et al* 2022 Facile strategy for third component optimization in wide-band-gap π-conjugated polymer donor-based efficient ternary all-polymer solar cells *ACS Appl. Mater. Interfaces* **14** 11211–21

[101] Chen Y *et al* 2019 Enhanced photovoltaic performances via ternary blend strategy employing a medium-bandgap DA type alternating copolymer as the single donor *Sol. Energy* **183** 350–5

[102] Sharma G D *et al* 2020 Polymer solar cell based on ternary active layer consists of medium bandgap polymer and two non-fullerene acceptors *Sol. Energy* **207** 1427–33

[103] Keshtov M L *et al* 2022 New medium bandgap donor D-A1-D-A2 type copolymers based on Anthra[1,2-b: 4,3-b':6,7-c''] trithiophene-8,12-dione groups for high-efficient non-fullerene polymer solar cells *Macromol. Rapid Commun.* **43** 2100839

[104] Yan T *et al* 2019 16.67% rigid and 14.06% flexible organic solar cells enabled by ternary heterojunction strategy *Adv. Mater.* **31** 1902210

[105] An Q *et al* 2020 Alloy-like ternary polymer solar cells with over 17.2% efficiency *Sci. Bull.* **65** 538–45

[106] Peng W *et al* 2021 Using two compatible donor polymers boosts the efficiency of ternary organic solar cells to 17.7% *Chem. Mater.* **33** 7254–62

[107] Caliskan M *et al* 2020 Narrow band gap benzodithiophene and quinoxaline bearing conjugated polymers for organic photovoltaic applications *Dyes Pigm.* **180** 108479

[108] Can A *et al* 2022 Meso-π-extended/deficient BODIPYs and low-band-gap donor–acceptor copolymers for organic optoelectronics *ACS Appl. Polym. Mater.* **4** 1991–2005

[109] Guo L *et al* 2021 Novel narrow bandgap polymer donors based on ester-substituted quinoxaline unit for organic photovoltaic application *Sol. Energy* **220** 425–31

[110] Pan L *et al* 2020 Efficient organic ternary solar cells employing narrow band gap diketopyrrolopyrrole polymers and nonfullerene acceptors *Chem. Mater.* **32** 7309–17

[111] Rech J J *et al* 2021 Designing simple conjugated polymers for scalable and efficient organic solar cells *ChemSusChem.* **14** 3561–8

[112] Cruciani F *et al* 2019 N-acylisoindigo derivatives as polymer acceptors for 'all-polymer' bulk-heterojunction solar cells *Macromol. Chem. Phys.* **220** 1900029

[113] Guo Y *et al* 2017 Improved performance of all-polymer solar cells enabled by naphthodiperylenetetraimide-based polymer acceptor *Adv. Mater.* **29** 1700309

[114] Kolhe N B *et al* 2019 New random copolymer acceptors enable additive-free processing of 10.1% efficient all-polymer solar cells with near-unity internal quantum efficiency *ACS Energy Lett.* **4** 1162–70

[115] Wu Y *et al* 2021 A conjugated donor-acceptor block copolymer enables over 11% efficiency for single-component polymer solar cells *Joule* **5** 1800–15

[116] Jørgensen M, Norrman K and Krebs F C 2008 Stability/degradation of polymer solar cells *Sol. Energy Mater. Sol. Cells* **92** 686–714

[117] Krebs F C 2012 *Stability and Degradation of Organic and Polymer Solar Cells* (John Wiley & Sons)

[118] Norrman K *et al* 2010 Degradation patterns in water and oxygen of an inverted polymer solar cell *JACS* **132** 16883–92

[119] Yan L and Ma C-Q 2021 Degradation of polymer solar cells: knowledge learned from the polymer:fullerene solar cells *Energy Technol.* **9** 2000920

[120] Li N *et al* 2017 Abnormal strong burn-in degradation of highly efficient polymer solar cells caused by spinodal donor-acceptor demixing *Nat. Commun.* **8** 14541

[121] Liu T *et al* 2023 Photochemical decomposition of Y-series non-fullerene acceptors is responsible for degradation of high-efficiency organic solar cells *Adv. Energy Mater.* **13** 2300046

[122] Yousif E and Haddad R 2013 Photodegradation and photostabilization of polymers, especially polystyrene: review *Springerplus* **2** 398
[123] Mohammed Y A *et al* 2023 The roles of acceptors in the thermal-degradation of P3HT based organic solar cells *Phys. B: Condens. Matter* **653** 414666
[124] Ning Y *et al* 2014 Investigation on thermal degradation process of polymer solar cells based on blend of PBDTTT-C and PC70BM *Int. J. Photoenergy* **2014** 1–9
[125] Seemann A *et al* 2011 Reversible and irreversible degradation of organic solar cell performance by oxygen *Solar Energy* **85** 1238–49
[126] Lee J U *et al* 2012 Degradation and stability of polymer-based solar cells *J. Mater. Chem.* **22** 24265–83
[127] Speller E M *et al* 2019 Toward improved environmental stability of polymer:fullerene and polymer:nonfullerene organic solar cells: a common energetic origin of light- and oxygen-induced degradation *ACS Energy Letter* **4** 846–52
[128] Wincukiewicz A *et al* 2018 Radiative recombination and other processes related to excess charge carriers, decisive for efficient performance of electronic devices *Lith. J. Phys.* **58** 49–61
[129] Vohra V *et al* 2022 A comparative study of low-cost coating processes for green & sustainable organic solar cell active layer manufacturing *Opt. Mater.:* X **13** 100127
[130] Pasquarelli R M, Ginley D S and O'Hayre R 2011 Solution processing of transparent conductors: from flask to film *Chem. Soc. Rev.* **40** 5406–41
[131] Gnida P *et al* 2022 Polymers in high-efficiency solar cells: the latest reports *Polymers* **14** 1946

IOP Publishing

N M Ravindra, Leqi Lin and Priyanka Singh

Chapter 6

Concentrator photovoltaics systems

The use of concentrators to convert the incident solar energy into useful electricity occurs via two approaches: (a) concentrator solar (thermal) power (CSP) systems, and (b) concentrator photovoltaics (CPV) systems. They represent innovative approaches to enhancing the power output from solar energy conversion. These systems work by utilizing optical elements, such as lenses or mirrors, to concentrate sunlight onto: (a) drive traditional steam engines via the use of the thermal component of the energy and (b) solar cells by utilizing the photovoltaic effect, converting, thereby, the incident solar energy to useful power. CSP technologies provide continuous electricity irrespective of day or night, and have been considered as the promising alternative to traditional fuel sources for sustainable development. CPV systems could be combined with agriculture as agri-photovoltaics which provides the benefits of dual-land use, low-carbon emissions and low maintenance. In particular, these integrated multifunctional approaches can be more land-efficient as they generate more power for a given area due to the higher solar flux as well as the higher efficiency. In this chapter, the impact of CPVs on the environment is also discussed.

6.1 Introduction

6.1.1 Concentrator solar power systems

CSP systems convert solar energy into electricity by using mirrors or lenses to concentrate the incident sunlight onto a receiver. Unlike CPV, CSP does not directly convert sunlight into electricity. Instead, it focuses sunlight, using mirrors to reflect and concentrate sunlight onto a receiver via a high temperature fluid to generate heat that is used to produce steam, which then drives a turbine to generate electricity. There are several CSP technologies including linear Fresnel reflector, central receiver (power tower), parabolic dish/engine systems and parabolic trough system [1], as shown in figure 6.1 [2].

doi:10.1088/978-0-7503-5994-8ch6

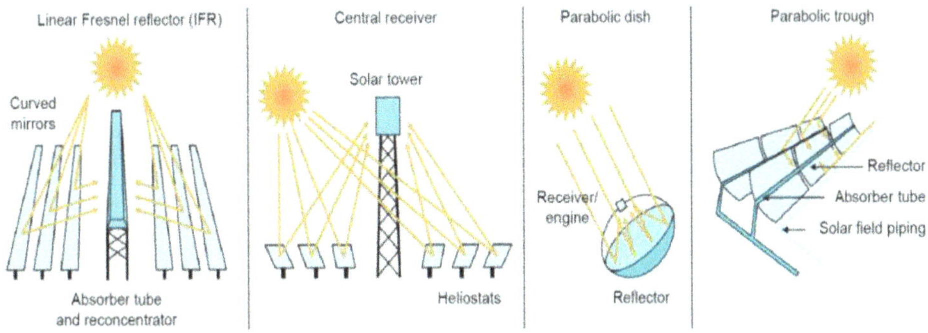

Figure 6.1. Concentrated solar power systems. Reprinted from [2], copyright (2021) with permission from Elsevier.

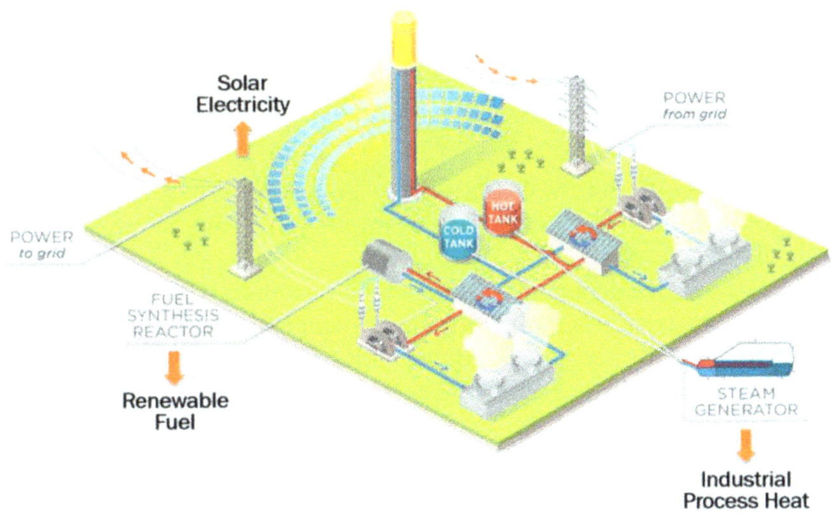

Figure 6.2. Illustration of a CSP system. Reproduced from [4] Credit: USA Department of Energy.

Most of the CSP systems operate with heat transfer fluids (HTFs) and a thermal energy storage system (TES) as in power plants to provide electricity continuously irrespective of day or night. CSP systems are one of the best candidates as replacement for combustion of fuel sources for sustainable development. It has been reported that the global CSP market size was valued at USD 4.5 billion in 2019 and is expected to grow at a compound annual growth rate (CAGR) of 9.7% from 2020 to 2027 [3].

An illustration of a CSP system is presented in figure 6.2 [4]. These systems are generally used for utility-scale projects.

In power tower systems, mirrors are arranged around a central tower that functions as the receiver. In utilizing linear systems, rows of mirrors concentrate the sunlight onto parallel tube receivers that are positioned above them [4]. The main contributors to CSP plants are Spain and the United States, sharing a power

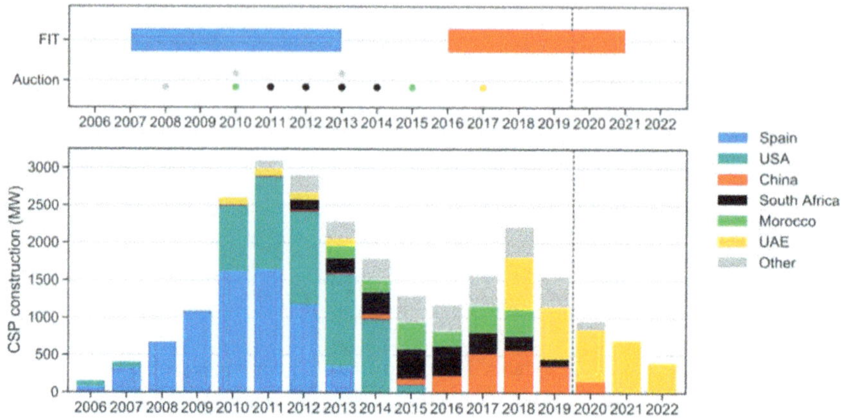

Figure. 6.3. Policies triggering new CSP projects (upper panel) and CSP capacity under construction per year (lower panel), 2006–22. Upper panel: only support that triggered construction is displayed; hence, Morocco has no entry for 2019 as Midelt 1[1] has not broken ground (in April 2020). The US PPA deals are not included. Lower panel: includes stations under construction in January 2020 and scheduled for completion before the end of 2022. Source: [5] Taylor & Francis Online http://tandfonline.com.

generation capacity of 2300 MW (48%) and 1738 MW (36%), respectively, as of 2019 [4]. Figure 6.3 illustrates some of the examples of recent CSP projects in various parts of the world [5].

With a global commitment to decreasing greehouse gases (GHGs) in the developed and developing countries and faced with the challenge of fluctuation in fuel oil price, coupled with increase in energy demand, the CSP systems offer great promise of green and low-cost energy compared with conventional fuel combustion plants, especially for tropical countries with vast area. Some of the concerns in the use of CSP systems are being addressed throughout the world. These include issues such as site selections, technological evaluations, performance analyses, economic investigations, and developments in processing, materials and the associated supply-chain [6].

With enhanced use of CSP systems, their prices are decreasing at a significant rate, down 69% from 2012 to 2022. This is illustrated in table 6.1 [7].

An example of a test facility for CSP systems is the High-Flux Solar Furnace (HFSF), designed, developed and constructed by the National Renewable Energy Laboratory. HFSF has the capability 'to quickly concentrate solar radiation to 10 kW over a 10-cm diameter, equivalent to 2,500 Suns', attaining temperatures of 1800 °C. By utilizing secondary optics, it can reach 3000 °C at 'peak solar fluxes of 20 000 Suns' [8].

6.1.2 Concentrating photovoltaics systems

CPV systems utilize optical techniques, such as lenses or curved mirrors, to focus sunlight onto small, high-efficiency solar cells. By concentrating sunlight, the

[1] Noor Midelt Solar PV Park 1, Morocco is a 210 MW Solar PV Power Plant; the project is to be commissioned in 2024.

Table 6.1. Cost summary of the use of various sources of renewable energy and their associated percentage change from 2010 to 2022 [7].

	Total installed costs (2022 USD/kW)			Capacity factor (%)			Levelised cost of electricity (2022 USD/kWh)		
	2010	2022	Percent change	2010	2022	Percent change	2010	2022	Percent change
Bioenergy	2 904	2 162	-26%	72	72	1%	0.082	0.061	-25%
Geothermal	2 904	3 478	20%	87	85	-2%	0.053	0.056	6%
Hydropower	1 407	2 881	105%	44	46	4%	0.042	0.061	47%
Solar PV	5 124	876	-83%	14	17	23%	0.445	0.049	-89%
CSP	10 082	4 274	-58%	30	36	19%	0.380	0.118	-69%
Onshore wind	2 179	1 274	-42%	27	37	35%	0.107	0.033	-69%
Offshore wind	5 217	3 461	-34%	38	42	10%	0.197	0.081	-59%

systems can achieve higher solar flux, increasing the amount of sunlight that reaches the solar cells, thus improving their overall throughput. CPVs can achieve higher solar conversion efficiencies compared to traditional flat-plate PV systems.

The concentration ratio in CPVs can range from tens to hundreds of times the normal sunlight intensity. The cells, utilized in CPVs, are often made of III–V materials such as gallium arsenide. In recent years, CPV multi-junction solar cell efficiencies of 46% have been obtained compared to conventional solar power tower steam engine efficiency of 14% and average PV panel efficiency of 15% [9]. The main idea of CPV was introduced for the first time by Shockley and Queisser in 1961. A maximum conversion efficiency of about 30% for mono-junction solar cells, under an illumination of 1000 W m^{-2}, has been reported in the literature [10]. CPV systems require: (a) direct sunlight rather than diffuse light and (b) dual-axis tracking systems to follow the sun's movement throughout the day, ensuring that sunlight is always focused on the solar cells. CPV systems can be categorized into low-to-high concentration PV based on the amount of their solar concentration. In particular, CPVs operate most efficiently in high concentrated sunlight and can be more land-efficient as they generate more power for a given area.

6.2 Concentrator photovoltaics

An illustration of the details of CPV is presented in figure 6.4 [11]. This figure is adapted from a recent paper—'The emergence of concentrator photovoltaics for perovskite solar cells' by Sadhukhan *et al* [11].

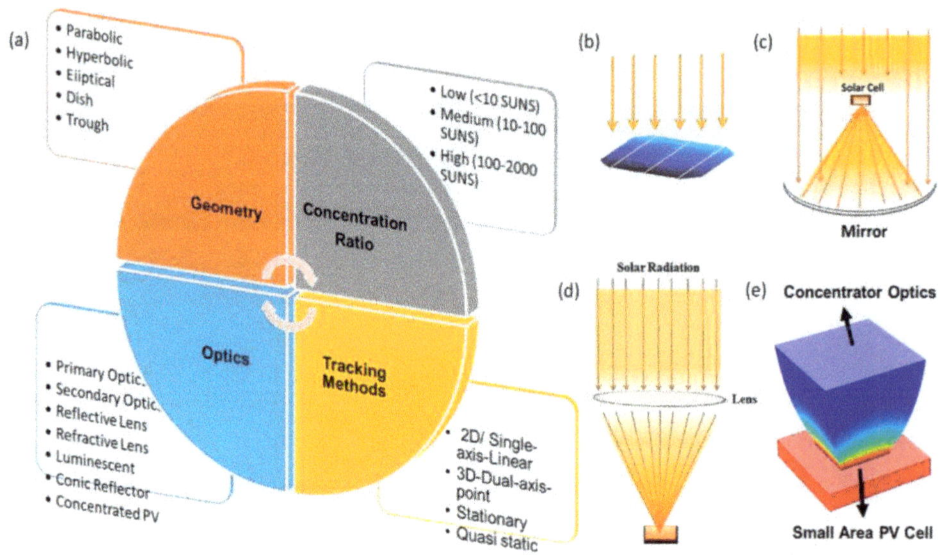

Figure 6.4. Fundamental features of concentrator optics. (a) Different classes of concentrator optics. (b) A solar cell without concentrator. Some common types of concentrators: (c) Reflector type. (d) Lens-based refractor type. (e) Reflector–refractor combined. Reprinted with permission from [11], copyright 2021 The Author(s). Published under an exclusive license by AIP Publishing.

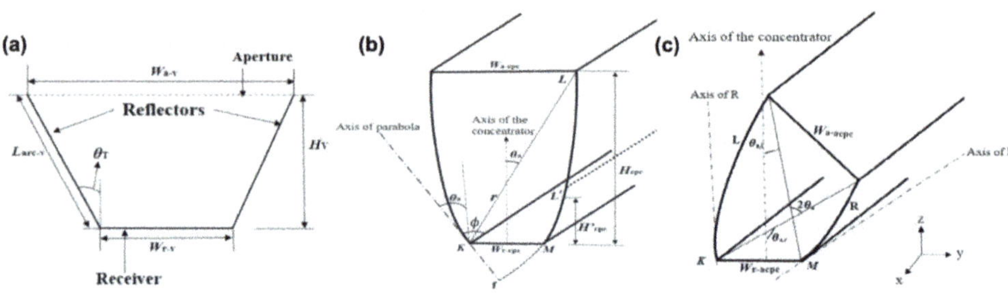

Figure 6.5. (a) V-trough concentrator, (b) compound parabolic concentrator (CPC) and (c) asymmetric compound parabolic concentrator (ACPC) [13–16]. Reprinted from [13], copyright (2020), with permission from Elsevier.

6.2.1 Low concentration PV (LCPV)

A low-concentration photovoltaic (LCPV) system involves the use of PV modules with optics to concentrate sunlight onto solar cells, at a lower concentration in comparison to high-concentration photovoltaics (HCPV). According to a new survey, the LCPV market is projected to reach US$ 2125 million in 2029, increasing from US$ 1360 million in 2022, with the CAGR of 6.7% during the period of 2023 to 2029 worldwide [12].

The geometries of CPV can be categorized into V-trough concentrator, compound parabolic concentrator (CPC) and asymmetric compound parabolic concentrator (ACPC). As shown in figure 6.5(a), a V-trough concentrator consists of two flat reflectors inclined at an angle (θ_T) to the axis normal to the receiver, which helps direct and concentrate sunlight onto a specific target. Figure 6.5(b) shows the symmetric compound parabolic concentrator which consists of two parabolic reflectors with equal half acceptance angle (θ_a) located on the both sides of the receiver. Figure 6.5(c) presents the cross-sectional view of an asymmetric CPC with acceptance angles [13–16].

6.2.2 Medium concentration PV (MCPV)

The medium concentration PV (MCPV) system is typically made with single and dual-axis tracker. Generally, the optical concentration ratio ranges from 3 to 100X. The higher concentrating capability implies that MCPVs track sunlight more frequently. The dual-axis tracker provides higher efficiencies than a single-axis one. However, the need for accurate tracking systems, potential increased maintenance requirements and the management of the system are challenges in MCPV [17].

6.2.3 High concentration PV (HCPV)

HCPV typically perform at concentration ratios beyond hundred to thousand times and the solar cells that perform well under these conditions are made of a series of III–V semiconductor multi-junctions. HCPV systems are based on a technology which in a short–medium term will reach a higher trajectory in terms of efficiency and energy production [18]. HCPV systems are promising power supply systems that have caught the attention of a significant number of users in the solar energy industry. In particular, substantial focus has been given to the thermal management of HCPV systems concerning the heat removal during the last decade [10, 19].

Ali *et al* proposed a three-dimensional thermal model, coupled with a conjugate heat transfer model, for DP-HCPVM (dense packed—high concentrator photovoltaic module/thermal construct to deal with different active cooling conditions [20]. This technology facilitates the reduction in cell temperature non-uniformity by 11.4% and 37.8% without sacrificing the overall performance and efficiency of the cell, as shown in figure 6.6 [20].

6.3 Concentrated solar power

6.3.1 Parabolic trough systems

Parabolic trough has easier construction and maintenance among all types of CSP in which countries such as Spain has the highest number of parabolic troughs followed by the USA and China. Linear Fresnel reflector and parabolic trough are both linear focusing systems while the other two are point focusing, which can produce solar induced heat at higher temperature due to the higher sunlight concentration ratio [2].

Table 6.2 presents a detailed comparison of each type of CSP systems. The structure of the parabolic trough is constructed with hollow carbon-steel profiles with rectangular cross-section, which is basically made of aluminum to reduce the weight burden of the structure. A vacuum tube with total length of 1800 mm is used as a heat receiver and is positioned in the focal line of the parabolic trough [21]. CSP

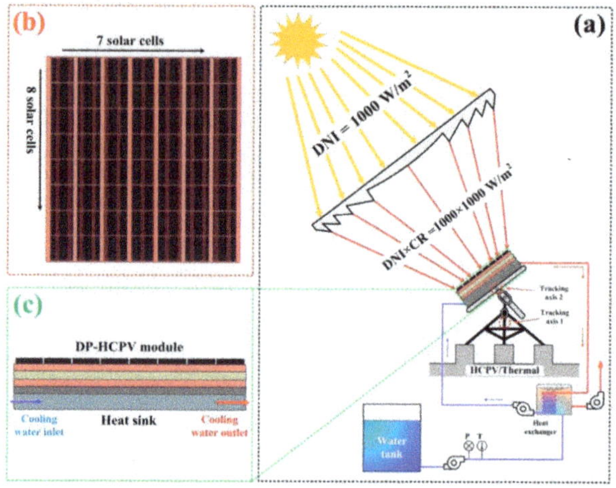

Figure 6.6. Detailed schematic of HCPVM/T system assembly (a) the dish concentration system (b) the geometry of the DP-HCPVM 56-cell assembly (c) the geometry of the DP-HCPVM integrated with the heat sink (DP-HCPVM/T). Reprinted from [20], copyright (2021), with permission from Elsevier.

Table 6.2. Characteristic data for various CSP technologies [31–37].

Parameters	Parabolic trough	Fresnel reflector	Parabolic dish systems	Solar power tower
Classification	Linear focus		Point focus	
Capacity (MW_{el})	10–280	10–100	0.01–0.025	10–150
Operating temperature (°C)	125–400	125–400	120–1500	300–1000
Concentration ratio	70–100	25–100	1000–3000	300–1000
Water requirement (m^3 MWh^{-1})	3	3	0.05–0.1	2–3
Thermal cycle efficiency (%)	35–42	30–42	30–40	30–45
Total installed cost ($US kW^{-1})	3900–4100 (without TES); 6300–8300 (with TES)	—	—	5700–6400 (6–7.5 h TES); 8100–9000 (12–15 h TES)
O&M cost	0.012–0.02	—	0.21	0.034
Technology risk	Low	Moderate	Moderate	Moderate
Land use (km^2 MW^{-1})	0.025	0.008	0.011	0.036

systems have been associated with the possibility of integrating them with large-scale TES systems to adapt to the electricity production based on daily energy demand [22]. The materials selection for TES systems in parabolic troughs is the most important step to achieve high performance and efficiency.

Normally, phase change materials (PCMs) such as the commercial molten salt for TES is a non-eutectic salt mixture of $NaNO_3/KNO_3$, with a limitation of the material thermal decomposition temperature of ~565 °C. $MgCl_2/KCl/NaCl$ is still considered as one of the most promising chloride salt mixtures. This thermal decomposition temperature further limits the maximum heat storage capacity. Thus, advanced improvement in the efficiency can be achieved by operating TES systems with higher thermal stability. NREL proposed one of the novel molten salts, TES/HTF system, with the operating temperature of 520 °C–720 °C combining with supercritical carbon dioxide (sCO_2) Brayton power cycle (Operating temperature 500 °C–700 °C), which can have energy efficiency higher than 50% [23]. Four main kinds of promising technical concepts have been researched intensively including solid particles as TES/HTF, molten salts (especially chlorides) as TES/HTF, gases as HTF (e.g., helium) with indirect TES (e.g., solid media, PCMs), and liquid metals as HTF with different TES options (e.g., PCMs) [2]. Future work will target the mitigation of the corrosion problem and the thermal durability of the molten salts and subsequently, on the scale-up issues from the laboratory to commercial scale.

6.3.2 Solar power tower systems

Solar power tower systems are considered to be one of the most useful methods for concentration of solar power due to their high efficiency of ~18% [24]. They are expected to be the first choice for implementation by many countries for the construction of CSP. These systems use a large field of mirrors called heliostats to concentrate sunlight onto a receiver at the top of a tall tower. The concentrated sunlight heats a working fluid, typically a liquid or gas, which is then used to generate steam to drive a turbine and produce electricity. Tower systems allow higher concentration ratios to be achieved, which in turn means higher fluid operating temperatures and thus higher power cycle efficiencies [25]. Besides, the systems can be easily scaled up to adapt to the power usage for various conditions. However, these systems require a very large area for the installation of the heliostat field, which leads to high cost of production.

6.3.3 Parabolic dish systems

A typical parabolic dish system (figure 6.7) is composed of a parabolic concentrator connected with a receiver and an alternator. Currently, with the overall efficiencies of 13%–32%, 1.0–38.8 kW power is generated from the parabolic dish type CSP depending on the differences in size, design and radiation utility. The highest efficiency of 29.4 % with 25.3 kW was reported in Boeing, California while the reported maximum theoretical efficiency was 32% in a lab study [26]. It also highlighted the hybridization mode of CSP with simultaneous use in agriculture, water desalination, thermal storage and solar cooking which led to the widespread applications of CSP.

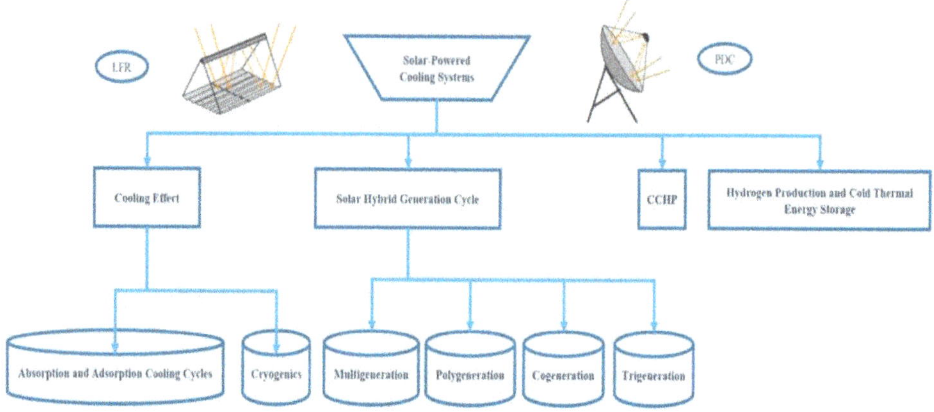

Figure 6.7. Various solar-powered systems equipped with PDC (parabolic dish collector) and LFR (linear Fresnel reflector). Reprinted from [30], with permission from Springer Nature.

The parabolic dish systems concentrate sunlight onto a small focal point with very high temperatures, contributing to high thermal efficiency. Besides, the system can respond quickly to changes in sunlight intensity, making them suitable for applications that require rapid adjustments.

6.3.4 Fresnel reflector systems

Fresnel reflector systems utilize a specific type of optical lens called a Fresnel lens to concentrate sunlight onto a receiver. The Fresnel lens is a flat optical lens that can focus light like a traditional curved lens but is much thinner, making it more cost-effective to manufacture. The Fresnel collector can be considered as a broken parabolic reflector so that each reflector can independently track and reflect sunlight to the receiver. As the CSP systems usually accumulate and absorb a high amount of heat during the generation of electricity, there is a necessity for strong and efficient cooling systems for CSP. Various fluids such as water, air, nanofluids and thermal oils can be used to transfer heat that is absorbed by the collectors [27–29]. In recent years, with progress in nanotechnology, nanofluids have become promising materials for their good heat transfer capability [30].

In table 6.2, a summary of the characteristic data for various CSP technologies is presented [31–37].

6.3.5 Concentrator photovoltaic (CPV) technology—examples

CPV technology enables high efficiency cells at low system cost. Additionally, rapid scale-up of these systems requires low capital investment [38]. However, the high efficiency multi-junction cells that are utilized in CPV systems, while small in size, are generally expensive. The multi-junction III–V semiconductor cells with high efficiency >45% are made by spectral matching of the device structure with specific absorber layers having specific bandgaps. Bandgap engineering and lattice matching

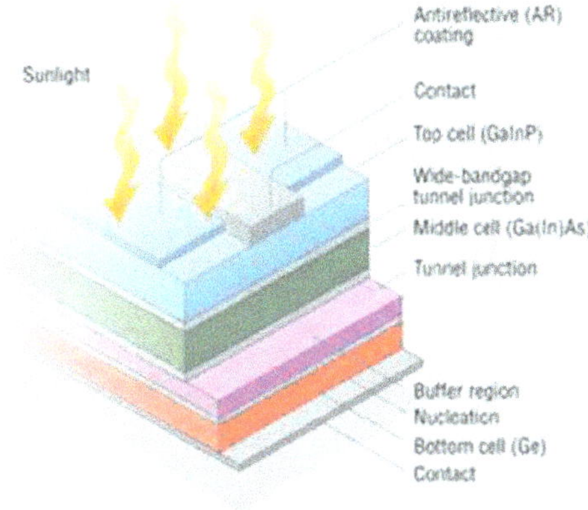

Figure 6.8. An illustration of a high efficiency multi-junction solar cell. Reproduced with permission from [39] credit: US Department of Energy.

play critical roles in the development of multi-junction solar cells for CPV technology.

An illustration of a multi-junction solar cell is presented in figure 6.8 [39].

A summary of the record efficiencies of III–V multi-junction cells and CPV modules, from the results of a collaborative research program on CPV technology between Fraunhofer Institute for Solar Energy Systems (Fraunhofer ISE) and the National Renewable Energy Laboratory (NREL) [40], is presented in figure 6.9.

In recent years, Fraunhofer ISE has been working on the 'development of a highly concentrating CPV module based on modern micro-production technology' [41]. An example of a measured current–voltage characteristics of a μ-CPV test module that consists of III–V multi-junction micro solar cell is presented in figure 6.10. For the fabrication of these μ-CPV test modules, Fraunhofer ISE has been evaluating low-cost technologies from traditional areas of semiconductor manufacturing such as microelectronics, optoelectronics and display manufacturing to significantly reduce the costs of high efficiency CPV modules.

Another recent study reports the fabrication of a 'Stretchable micro-scale concentrator photovoltaic module with 15.4% efficiency for three-dimensional curved surfaces' [42]. In this study, the authors investigate the design, fabrication, testing and performance of a solar module that consists of commercial GaInP/GaInAs/Ge triple junction cells that are integrated into screen-printed silicone lenses. The objective of this study is to enable potential solutions for power sources for curved surfaces in biomedical devices, buildings, flexible/collapsible electronics and vehicles.

6.3.6 Concentrator photovoltaic (CPV) technology—device fundamentals

The device characteristics of CPV systems can be summarized by the equations that traditionally govern the solar cells [43–48]:

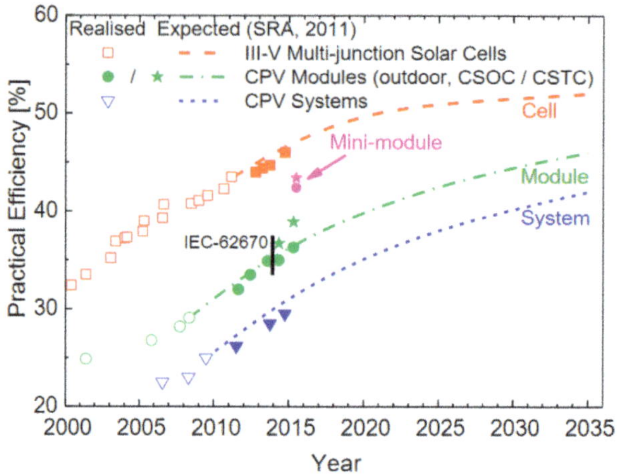

Figure 6.9. Development of record efficiencies of III–V multi-junction solar cells and CPV modules (cells: x*AM1.5 d; modules: outdoor measurements). Progress in top-of-the-line CPV system efficiencies is also indicated. (AM1.5 d lab records according to Green *et al* Solar cell efficiency tables from 1993 to 2016; CPV module and system efficiencies collected from various publications). The trend lines show expected efficiencies from the Strategic Research Agenda (SRA) developed by the European Photovoltaics Technology Platform in 2011. Recent efficiency values (full symbols) follow the trend very well. Reprinted with permission from [40] credit: NREL.

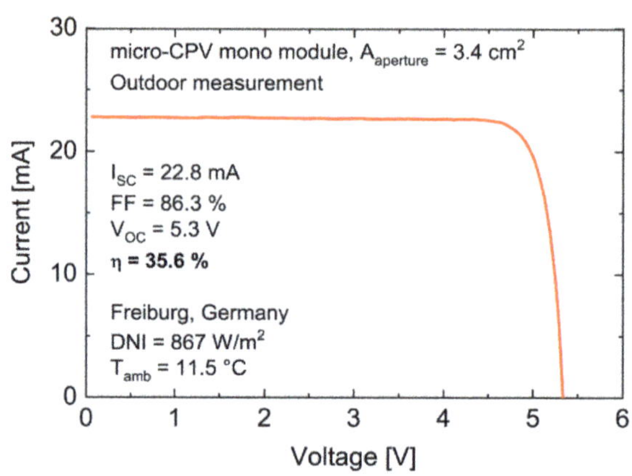

Figure 6.10. Current–voltage curve of a μ-CPV test module that consists of a micro solar cell, a secondary ball lens optics, and a primary lens. The measurement was conducted at the tracking unit of the CPV outdoor test setup at Fraunhofer ISE. Reproduced with permission from [41], copyright Fraunhofer ISE.

$$P = I_{sc} \times V_{oc} \times FF \tag{6.1}$$

Under a cell illumination equivalent to x—Suns,

$$P_x = I_{xsc} \times V_{xoc} \times FF_x \tag{6.2}$$

where,

$$I_{xsc} = x \times I_{sc} \tag{6.3}$$

and

$$V_{xoc} = V_{oc} + \left(\frac{kT}{q}\right)\ln(x) \tag{6.4}$$

The corresponding cell efficiencies, η, are given by:

$$\eta = (P/QA) \tag{6.5}$$

and

$$\eta_x = (P_x/xQA) \tag{6.6}$$

Thus, the cell efficiency, as a function of solar concentration, can be summarized as:

$$\eta_x = \eta \times (P_x/xP) = \eta[1 + \{(kT/q)(\ln(x)/V_{oc})\}] \times (FF_x/FF) \tag{6.7}$$

The symbols used in the above equations represent the following cell characteristics:

P = Power generated by a cell under One Sun illumination on the surface of the earth; this corresponds to a peak solar irradiance of $Q = 1000$ Watts m^{-2};
V_{oc} = Open-circuit voltage;
I_{sc} = Short-circuit current;
FF = Fill factor;
A = Area;
x = Solar concentration;
kT/q = Thermal voltage = ~26 mV at 300 K;
η_x = Cell efficiency at concentration x.

In general, the increase in cell temperature, with increased solar concentration, needs to be controlled to prevent degradation and potential irreversible damage to the cells. This requires the use of heat sinks. While the heating of the cells is one of the parameters for decreased performance, there are other additional mechanisms that lead to losses in efficiency in CPV Systems. This is illustrated in figure 6.11 [49].

The cell efficiency, η_{cell}, is a function of its operating temperature. It is given by the following equation:

$$\eta_{\text{cell}} = \eta_{\text{ref}}\,[1 - \beta_{\text{ref}}\,(T_{\text{cell}} - T_{\text{ref}})] \tag{6.8}$$

where, $T_{\text{ref}} = 25$ °C, η_{ref} and β_{ref} are constants with values of 0.18 and 0.0045 K^{-1}, respectively for silicon. In this case, η_{cell} refers to a silicon solar cell [50]. For a multi-junction high-concentrator PV cell, $\eta_{\text{ref}} = 42\%$ at a reference temperature $T_{\text{ref}} = 25$ °C and β_{ref} represents the relative temperature coefficient of the cell and is = −0.106%/°C for TJ cell AZURE SPACE product [51].

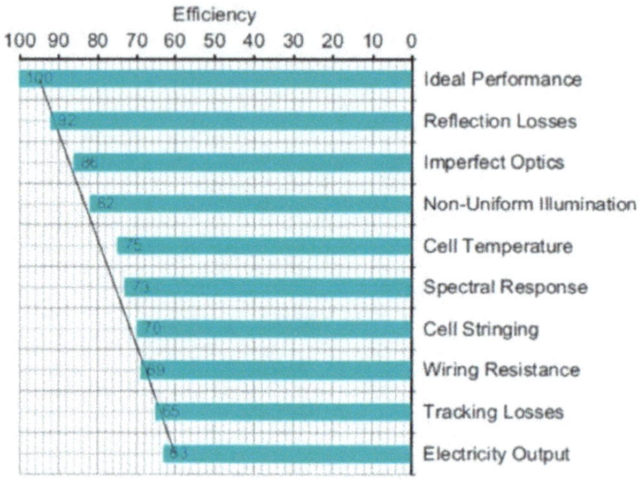

Figure 6.11. Losses in CPV systems. Reproduced from [49], copyright (2012), with permission from Elsevier.

6.4 Agri-photovoltaics

With advancements in a variety of technologies, rapid expansion of the manufacturing sector and improved living quality per capita around world, the demands on energy have increased significantly. The decreasing mortality and the increasing population have also led to the increasing requirement of foods worldwide. Thus, a water-energy-food-land ecosystem needs to be considered and a balance needs to be maintained in allocating resources to large-scale solar farms. Solar energy, as one of the most environmentally friendly and feasible green energy generation sources, copes with climate change and global warming, and is well accepted and utilized by many countries. The market share of CSP systems, spread over a large area, is also increasing, leading to land scarcity in some parts of the world. Thus, it is critical that the land is shared for a variety of uses and applications such as the installation and commissioning of renewable energy resources such as windmills, CSP and CPV systems, animal farming and agriculture. Consequently, agri-photovoltaics facilitates the benefits of dual-land use, low-carbon emissions, low maintenance, and the interventions and incentives-related policies from the governments. Additionally, agri-photovoltaics provides various operational incentives that can potentially improve crop production by reducing the reliance on non-renewable resources [52, 53].

Agri-photovoltaic systems can be installed on open-field farms or integrated with protected crop cultivation environments of greenhouses in which the interspace photovoltaics is mainly applied in pasture and arable farming, while overhead photovoltaics is mainly suitable for horticulture, as shown in figure 6.12 [54]. The plants under the PV modules benefit from effective water/rain redeployment [55], wind mitigation and temperature deviation protection [56], reduction in evapotranspiration, perfection in soil moisture, security in contrast to climatic uncertainty and risky natural events such as hailstones [57].

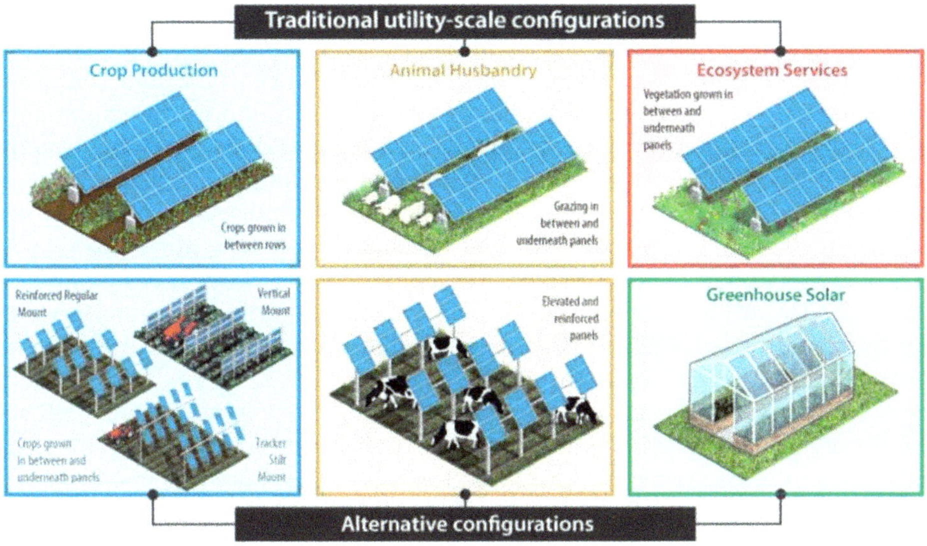

Figure 6.12. Agrivoltaics includes many different uses. Agrivoltaics systems can be installed in the same basic row layout as a traditional large-scale solar plant—or they can be modified to provide extra space for light, animals, or farm equipment to move under and between them. Courtesy: NREL [61]

The first concept of agri-photovoltaics was proposed in 1982 by Goetzberger and Zastrow. They reported that almost the same amount of radiation can be achieved if the collectors are elevated by about 2 m above the ground with the periodic distance between collectors [58]. Recently, an agri-photovoltaic case study in Niger, located in West Africa in the Sahel Zone, has been reported. Niger is exposed to mostly hot and dry desert weather with most of the terrain in desert and sand dunes. With the estimation of 23 million population, only 17% live in urban areas and have higher opportunity to access safe drinking water [59]. Niger urgently needs sustainable development in energy and food generation. The results of this study showed that combining the use of solar energy with food production and water supply by using agri-photovoltaics is promising in the villages in Niger. The largest agri-photovoltaic research facility has been installed by Fraunhofer ISE near Lake Constance with a capacity of 194.4 kW in Germany [60]. Today, more and more countries in the world have begun to encourage agri-photovoltaics from the perspective of policy and financial subsidy. However, agri-photovoltaics is still in its infancy with insufficient details on the performance of crops and energy generation as function of the module types, tilt angle, orientation and different types of crops. Moreover, additional investigations on the practical business relationships between local farmers, potential investors and traders, as well as case studies in different parts of the world are also needed. Some of the related applications are illustrated in figure 6.12 [61].

6.5 Impact on the environment

Solar energy is naturally abundant and more environmentally sustainable than combustion of fossil fuel energy sources. The use of solar energy leads to a reduction in the amount of GHGs and air pollution emissions. However, there are some drawbacks that are associated with solar energy developments during construction and operation of solar power plants such as the disturbance to the land and eco-life, and the impact on the soil, water, and air. In addition to requiring large amounts of water, it must be noted that the production of solar cells is an energy intensive process and like many semiconductor manufacturing industries, it uses materials that are highly toxic. Furthermore, it is important to manage solar panels safely once they reach the end of their lives and become electronic waste.

The construction of concentrated solar facilities requires large areas of land (~5–10 acres for 1 MW) which may lead to interference with grazing, agriculture use or other business utilities. Thus, it is necessary to have proper policy coordinates with appropriate decisions on the position of concentrated solar facilities to largely prevent the disturbance of land use. In recent years, several state-of-the-art-technologies have been proposed to prevent land disturbance such as floating of photovoltaics on reservoirs [62], partial-shading on agriculture [63], mounting solar panel on buildings and siting on degraded lands [64]. In 2021, McKuin *et al* reported the use of cadmium telluride (CdTe) solar PV panels over canals which can effectively generate electricity and reduce a significant amount of water from evaporation at a reasonable cost [65]. This approach will address the concern that conventional over-ground solar may disturb agricultural lands. At the same time, problems of bird mortality and loss of wildlife habitat due to deforestation are minimized. Although the use of water per unit of generation of solar power is less than many other energy technologies (0.02–0.07 m^3 per MW h^{-1}), due to their size, they can use substantial amounts of water for construction and operation, mostly for cleaning PV solar panels and for dust suppression from disturbed soils [66–68]. Parabolic trough and central tower systems typically use conventional steam plants to generate electricity, which commonly consume water for cooling [68].

In addition to the above-mentioned issues, the impact on the surrounding animals and biodiversity with the use of solar energy are the predominant problems with renewable energy developments. Large areas of land are currently being evaluated for utility-scale solar energy development (USSED) including areas with high biodiversity and protected species [69]. It has been reported that PV panels with a 10% conversion efficiency would need to cover an area of about 32 000 square kilometers, or approximately an area equal to the state of Maryland [70]. Such commitments involve significant ground disturbance and direct (e.g., mortality) and indirect (e.g., habitat loss, degradation, modification) impacts on wildlife and their habitat [71]. For example, the potential impacts on deer, sheep and even the tortoises in the desert include impediments to their free movement, the creation of migration bottlenecks and a reduction in effective winter range size [69]. Moreover, the fans and pumps that are associated with the cooling systems in solar power developments generate noise which may cause hearing loss in animals and interfere with the ability to hunt. Last but not least, the alteration in landscape creates microclimate effects which change the characteristics of the environment that affects wildlife. For example, the unused heat surrounding the CSP facilities may damage wildlife by creating localized drought or high temperature conditions. In order to minimize risks to biodiversity, solar project developers should avoid areas of high environmental significance such as protected areas and conserved areas in accordance with the guidelines [72]. Barron-Gafford *et al* report the 'Photovoltaic heat island effect' in which larger solar power plants exhibit increased local temperatures of 3 °C–4 °C warmer than the surroundings [73]. On an unrelated note, the US Federal Aviation Administration (FAA) has a clear policy on the construction of solar PV projects near airports. The policy stipulates that airports must measure the visual impact of PV projects on pilots and air traffic control personnel. The objective of this policy is to ensure that solar PV projects near airports do not create hazardous glare [74].

6.6 Conclusion

Concentrator PV systems offer advantages in terms of efficiency. However, they are not as widely deployed as traditional flat-plate PV systems, primarily due to their higher cost, complexity, and specific environmental requirements. Ongoing research and development continues to improve the technology and address these challenges, making concentrator systems an interesting area for the future of solar power generation. Policy-based support from the various levels of local and state administration and the associated incentives can contribute to the foundation for a more sustainable development of this technology, wherein economic development is strongly coupled with the development of the environment.

CSP systems continue to evolve by taking advantage of significant developments in new phase change materials as well as the associated technology. By their inherent ability to provide energy, day and night, they circumvent the need to be concerned too much with energy storage.

This chapter is reproduced with permission from [75].

References

[1] Agency I E 2014 Technology roadmap—solar thermal electricity *Technology Report* (Paris: International Energy Agency)

[2] Ding W and Bauer T 2021 Progress in research and development of molten chloride salt technology for next generation concentrated solar power plants *Engineering* **7** 334–47

[3] Grand View Research I 2022 *Concentrated Solar Power Market Size, Share & Trends Analysis Report By Application (Utility, EOR, Desalination), by Technology, by Region, and Segment Forecasts, 2020—2027* https://grandviewresearch.com/industry-analysis/concentrated-solar-power-csp-market

[4] https://energy.gov/eere/solar/concentrating-solar-thermal-power-basics (accessed 25 March 2024)

[5] Lilliestam J *et al* 2021 The near- to mid-term outlook for concentrating solar power; mostly cloudy, chance of sun *Energy Sources, Part* B **16** 23–41

[6] REN21 2017 *Renewable Global Status Report-REN21 Secretariat; REN21* (Paris, France)

[7] Renewable Power Generation Costs in 2022 https://mc-cd8320d4–36a1–40ac-83cc-3389-cdn-endpoint.azureedge.net/-/media/Files/IRENA/Agency/Publication/2023/Aug/IRENA_Renewable_power_generation_costs_in_2022.pdf?rev=3b8966ac0f0544e89d7110d90c9656a0 (accessed 25 March 2024)

[8] High-Flux Solar Furnace https://nrel.gov/csp/facility-hfsf.html (accessed 30 March 2024)

[9] SOLARTRON *HCPV Solar Parabolic Solar Concentrator Technology* https://solartronenergy.com/hcpv-solar/ (Accessed 25 March 2024)

[10] El Himer S, El Ayane S, El Yahyaoui S, Salvestrini J P and Ahaitouf A 2020 Photovoltaic concentration: research and development *Energies* **13** 5721

[11] Sadhukhan P *et al* 2021 The emergence of concentrator photovoltaics for perovskite solar cells *Appl. Phys. Rev.* **8** 041324

[12] Global Low-Concentration Photovoltaic(LCPV) Market Research Report 2023 (https://reports.valuates.com/market-reports/QYRE-Auto-17I15560/global-low-concentration-photovoltaic-lcpv, 2023)

[13] Parupudi R V, Singh H and Kolokotroni M 2020 Low concentrating photovoltaics (LCPV) for buildings and their performance analyses *Appl. Energy* **279** 115839

[14] Fraidenraich N and Almeida G 1991 Optical properties of V-trough concentrators *Sol. Energy* **47** 147–55

[15] Carvalho M, Collares-Pereira M, Gordon J and Rabl A 1985 Truncation of CPC solar collectors and its effect on energy collection *Sol. Energy* **35** 393–9

[16] Rabl A 1976 Comparison of solar concentrators *Sol. Energy* **18** 93–111

[17] Kroupa M *et al* 2015 A semiconductor radiation imaging pixel detector for space radiation dosimetry *Life Sci. Space Res.* **6** 69–78

[18] Pérez-Higueras P, Muñoz E, Almonacid G and Vidal P G 2011 High Concentrator photovoltaics efficiencies: present status and forecast *Renew. Sustain. Energy Rev.* **15** 1810–5

[19] Alzahrani M, Baig H, Shanks K and Mallick T 2020 Estimation of the performance limits of a concentrator solar cell coupled with a micro heat sink based on a finite element simulation *Appl. Therm. Eng.* **176** 115315

[20] Ali Y M, A *et al* 2021 Thermal analysis of high concentrator photovoltaic module using convergent-divergent microchannel heat sink design *Appl. Therm. Eng.* **183** 116201

[21] Resende M d O *et al* 2023 Experimental study of alternative receiver tube for parabolic trough system for typical summer and winter months in a tropical region *Appl. Therm. Eng.* **232** 121023

[22] Ruiz-Cabañas F J, Prieto C, Madina V, Fernández A I and Cabeza L F 2017 Materials selection for thermal energy storage systems in parabolic trough collector solar facilities using high chloride content nitrate salts *Sol. Energy Mater. Sol. Cells* **163** 134–47

[23] Mehos M, Turchi C, Vidal J, Wagner M, Ma Z, Ho C, Kolb W, Andraka C and Kruizenga A 2017 Concentrating solar power Gen3 demonstration roadmap *Technical Report* NREL/TP-5500-67464 NREL

[24] Mutaz B and Elbeh A K S 2021 Analysis and optimization of concentrated solar power plant for application in arid climate *Energy Sci. Eng.* **9** 784–97

[25] Guédez R, Topel M, Spelling J and Laumert B 2015 Enhancing the profitability of solar tower power plants through thermoeconomic analysis based on multi-objective optimization *Energy Procedia* **69** 1277–86

[26] Zayed M E *et al* 2021 A comprehensive review on dish/stirling concentrated solar power systems: design, optical and geometrical analyses, thermal performance assessment, and applications *J. Clean. Prod.* **283** 124664

[27] Saffarian M R, Moravej M and Doranehgard M H 2020 Heat transfer enhancement in a flat plate solar collector with different flow path shapes using nanofluid *Renew. Energy* **146** 2316–29

[28] Moravej M *et al* 2021 Experimental study of a hemispherical three-dimensional solar collector operating with silver-water nanofluid *Sustain. Energy Technol. Assess.* **44** 101043

[29] Gholamalipour P, Siavashi M and Doranehgard M H 2019 Eccentricity effects of heat source inside a porous annulus on the natural convection heat transfer and entropy generation of Cu-water nanofluid *Int. Commun. Heat Mass Transfer* **109** 104367

[30] Esfanjani P, Jahangiri S, Heidarian A, Valipour M S and Rashidi S 2022 A review on solar-powered cooling systems coupled with parabolic dish collector and linear Fresnel reflector *Environ. Sci. Pollut. Res.* **29** 42616–46

[31] Zarza-Moya E 2018 *A Comprehensive Guide to Solar Energy Systems* (Elsevier) pp 127–48

[32] Pitz-Paal R 2020 *Future Energy* (Elsevier) pp 413–30

[33] Ummadisingu A and Soni M 2011 Concentrating solar power—technology, potential and policy in India *Renew. Sustain. Energy Rev.* **15** 5169–75

[34] Belgasim B, Aldali Y, Abdunnabi M J, Hashem G and Hossin K 2018 The potential of concentrating solar power (CSP) for electricity generation in Libya *Renew. Sustain. Energy Rev.* **90** 1–15

[35] Stein W and Buck R 2017 Advanced power cycles for concentrated solar power *Sol. Energy* **152** 91–105

[36] Parrado C, Marzo A, Fuentealba E and Fernández A 2016 2050 LCOE improvement using new molten salts for thermal energy storage in CSP plants *Renew. Sustain. Energy Rev.* **57** 505–14

[37] Fernández A G and Cabeza L F 2020 Corrosion evaluation of eutectic chloride molten salt for new generation of CSP plants. Part 1: thermal treatment assessment *J. Energy Storage* **27** 101125

[38] *PV FAQs* https://nrel.gov/docs/fy05osti/36542.pdf (accessed 30 March 2024)

[39] Multijunction III–V Photovoltaics Research, Solar Energy Technologies Office https://energy.gov/eere/solar/multijunction-iii-v-photovoltaics-research (accessed 30 March 2024)

[40] Wiesenfarth M *et al* Current Status of Concentrator Photovoltaic (CPV) Technology, Version 1.3, April (2017) file:///C:/Users/Appa/Downloads/cpv-report-ise-nrel.pdf (accessed 31 March 2024)

[41] micro-CPV—Development of a Highly Concentrating CPV Module Based on Modern Micro-Production Technology https://ise.fraunhofer.de/en/research-projects/micro-cpv.html (accessed 30 March 2024)

[42] Sato D *et al* 2021 Stretchable micro-scale concentrator photovoltaic module with 15.4% efficiency for three-dimensional curved surfaces *Commun. Mater.* **2** 7

[43] Gray J 2003 The physics of the solar cell ed A Luque and S Hegedus *Handbook of Photovoltaic Science and Engineering* (London: Wiley) pp 61–112

[44] PV Education—Average Solar Radiationaccessed 31 March 2024

[45] PV Education—Solar Cell Efficiency (accessed 31 March 2024). *PV Education—Fill Factor. Archived* from the original on 8 May 2019 (accessed 3 March 2019).

[46] Pulfrey D L 1978 On the fill factor of solar cells *Solid-State Electron* **21** 519–20

[47] Emery K and Osterwald C 1987 Measurement of photovoltaic device current as a function of voltage, temperature, intensity and spectrum *Sol. Cells* **21** 313–27

[48] https://en.wikipedia.org/wiki/Concentrator_photovoltaics (accessed 1 April 2024)

[49] Baig H, Heasman K C and Mallick T K 2012 Non-uniform illumination in concentrating solar cells *Renew. Sustain. Energy Rev.* **16** 5890–909

[50] Elqady H I *et al* 2021 Concentrator photovoltaic thermal management using a new design of double-layer microchannel heat sink *Sol. Energy* **220** 552–70

[51] Zaghloul H *et al* 2021 Optimization and parametric analysis of a multi-junction high-concentrator PV cell combined with a straight fins heat sink *Energy Concers. Manag.* **243** 114382

[52] Gorjian S, Minaei S, MalehMirchegini L, Trommsdorff M and Shamshiri R R 2020 *Photovoltaic Solar Energy Conversion* ed S Gorjian and A Shukla (Academic) pp 191–235

[53] Gorjian S, Ebadi H, Najafi G, Singh Chandel S and Yildizhan H 2021 Recent advances in net-zero energy greenhouses and adapted thermal energy storage systems *Sustain. Energy Technol. Assess.* **43** 100940

[54] Gorjian S *et al* 2022 Progress and challenges of crop production and electricity generation in agrivoltaic systems using semi-transparent photovoltaic technology *Renew. Sustain. Energy Rev.* **158** 112126

[55] Amaducci S, Yin X and Colauzzi M 2018 Agrivoltaic systems to optimise land use for electric energy production *Appl. Energy* **220** 545–61

[56] Marrou H, Guilioni L, Dufour L, Dupraz C and Wery J 2013 Microclimate under agrivoltaic systems: is crop growth rate affected in the partial shade of solar panels? *Agric. For. Meteorol.* **177** 117–32

[57] Marrou H, Dufour L and Wery J 2013 How does a shelter of solar panels influence water flows in a soil–crop system? *Eur. J. Agron.* **50** 38–51

[58] Goetzberger A and Zastrow A 1982 On the coexistence of solar-energy conversion and plant cultivation *Int. J. Sol. Energy* **1** 55–69

[59] Neupane Bhandari S *et al* 2021 Economic feasibility of agrivoltaic systems in food-energy nexus context: modelling and a case study in Niger *Agronomy* **11** 1906

[60] Trommsdorff M *et al* 2021 Combining food and energy production: design of an agrivoltaic system applied in arable and vegetable farming in Germany *Renew. Sustain. Energy Rev.* **140** 110694

[61] https://nrel.gov/news/program/2022/growing-plants-power-and-partnerships.html 18 August 2022 (accessed 17 December 2024)

[62] Pimentel Da Silva G D and Branco D A C 2018 Is floating photovoltaic better than conventional photovoltaic? Assessing environmental impacts *Impact Assess. Proj. Appraisal* **36** 390–400

[63] Ravi S *et al* 2016 Colocation opportunities for large solar infrastructures and agriculture in drylands *Appl. Energy* **165** 383–92

[64] Niblick B and Landis A E 2016 Assessing renewable energy potential on United States marginal and contaminated sites *Renew. Sustain. Energy Rev.* **60** 489–97

[65] McKuin B *et al* 2021 Energy and water co-benefits from covering canals with solar panels *Nat. Sustain.* **4** 609–17

[66] Ravi S, Lobell D B and Field C B 2014 Tradeoffs and synergies between biofuel production and large solar infrastructure in deserts *Environ. Sci. Technol.* **48** 3021–30

[67] Hernandez R R, Hoffacker M K and Field C B 2014 Land-use efficiency of big solar *Environ. Sci. Technol.* **48** 1315–23

[68] Burkhardt J J, Heath G A and Turchi C S 2011 Life cycle assessment of a parabolic trough concentrating solar power plant and the impacts of key design alternatives *Environ. Sci. Technol.* **45** 2457–64

[69] Lovich J E and Ennen J R 2011 Wildlife conservation and solar energy development in the Desert Southwest, United States *BioScience* **61** 982–92

[70] Wilshire H G, Nielson J N and Hazlett R W 2008 *The American West at Risk: Science, Myths, and Politics of Land Abuse and Recovery* (Oxford University Press)

[71] William P K *et al* 2007 Wind energy development and wildlife conservation: challenges and opportunities *J. Wildl. Manage.* **71** 2487–98

[72] Bennun L *et al* 2021 *Mitigating Biodiversity Impacts Associated with Solar and Wind Energy Development: Guidelines for Project Developers* (IUCN, Global Business and Biodiversity Programme, The Biodiversity Consultancy)

[73] Barron-Gafford G A *et al* 2016 The photovoltaic heat island effect: larger solar power plants increase local temperatures *Sci. Rep.* **6** 35070

[74] FAA Issues Policy on Solar Projects on Airports https://faa.gov/newsroom/faa-issues-policy-solar-projects-airports (accessed 1 April 2024).

[75] Lin L *et al* 2023 Influence of outdoor conditions on PV module performance – an overview *Mater. Sci. Eng. Int. J.* **7** 88–101

IOP Publishing

Recent Advances in Solar Cells

N M Ravindra, Leqi Lin and Priyanka Singh

Chapter 7

Instantaneous performance measurements of PV modules

The identification and tracking of the key performance indicators of photovoltaic (PV) modules, in real-time, allow for a quick assessment and evaluation of their properties, performance, reliability, temperature and health. The procedure and the associated equipment for the instantaneous performance measurements of solar modules are critical aspects of the solar PV infrastructure. The equipment that is required to perform the instantaneous monitoring of PV modules includes a variety of sensors and diagnostic tools for the evaluation of their basic electrical, electronic, optical and structural characteristics. The implementation of these procedures will ensure that: (a) the PV modules are working efficiently, and (b) corrections are applied in real-time in the event that the diagnostic procedures reveal potential issues.

7.1 Equipment required for instantaneous performance testing

Performance metrics of PV modules can be classified into the following categories: (a) instantaneous, (b) short-term, and (c) long-term assessment periods. Instantaneous performance measurement of PV modules refers to their real-time assessment in outdoor conditions. This measurement technique is utilized to assign ratings to the modules (STC—Standard Test Conditions, NOCT—Nominal Operating Cell Temperature, Temperature Coefficient) and to estimate degradation rate after specific term/duration of operation in the field. Besides, this technique provides immediate feedback that a user could utilize to optimize the PV module condition without delay. A list of equipment that is required to test the instantaneous performance of PV modules in outdoor conditions include I–V tracer, temperature sensor, solar radiation measurement instruments such as solarimeters and pyranometers, which are very important components to obtain basic performance-related conditions. Measurements of the angle of incidence, implementation of

doi:10.1088/978-0-7503-5994-8ch7

mesh to control the transmissivity, cooling and heating design and spectroradiometer are critical to maintain the long-term working operation of the PV modules.

7.1.1 The *I–V* tracer

The *I–V* tracer is a device with a variable electronic load to trace the current–voltage characteristics of a PV module within a fraction of a second. In order to utilize it, the *I–V* tracer should have a good number of load points to measure the current as a function of the voltage. For measurements and subsequent drawing of *I–V* curves of an individual module, the current and voltage measurement range should not be very high compared to module's actual current and voltage values. *I–V* tracer with voltage range from 10 to 1000 *V* can be used to measure the *I–V* curve of an individual module and the current range should be within 20–30 *A*. The uncertainty in the measurement of power, current and voltage should be within 1%. It is preferred to use a Kelvin or four wire method to measure the *I–V* curve to obtain more accurate results.

7.1.2 Temperature sensor

The temperature sensor is used to measure the module temperature during testing. Usually Pt-100 or Pt-1000 temperature sensors are used because of their accuracy and temperature measurement range. Three sensors are attached to the back side of the module (using conductive aluminum adhesive tape) to obtain the spatial distribution of temperature, in real time, in order to measure the temperature coefficient of the PV module.

7.1.3 Radiation measuring sensor

There are two types of irradiance sensors that are available; they can measure the radiation in the plane of the module. These are thermopile-based and solar cell-based methods. Thermopile sensors can measure irradiance in the wavelength range of 280–3000 nm. Silicon solar cell-based sensors can measure only in the wavelength range of 300–1100 nm. For performance measurement of PV module, usually a reference silicon solar cell is used. However, for spectral correction of effective irradiance, mismatch factor (MMF) can be used for different PV technologies.

7.1.4 Measurement setup of angle of incidence

The performance of PV module also depends on the angle of incidence (AOI) of the incident light. With the increase in AOI, the effective irradiance decreases and hence the power output. The AOI needs to be measured in the plane of the module, as shown in figure 7.1(a). AOI can be measured by utilizing a camera or by performing the measurements of the heights of shadows of a known object in the same plane. AOI can also be determined by using the methodology, as illustrated in section 7.1.5.

Figure 7.1. (a) Shadow length measurement setup (b) mesh with 60% light transmissivity (c) spectroradiometers installed at a fixed tilt equal to latitude. Reproduced with permssion from [1], copyright (2023) Lin *et al.*

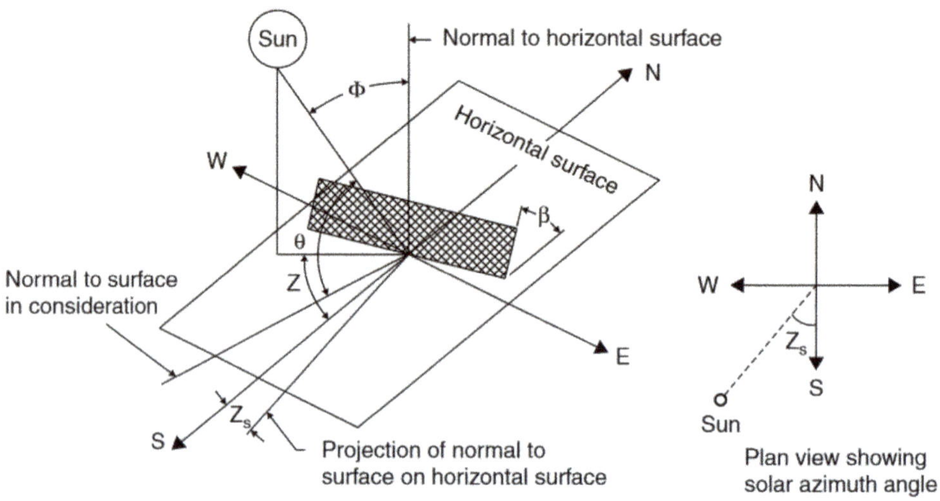

Figure 7.2. Solar angles diagram. Reprinted from [4], copyright (2009), with permission from Elsevier.

7.1.5 Determination of angle of incidence

The solar angle of incidence, θ, is the angle between the rays from the sun and the normal on a surface. The procedure for the determination of the solar angle of incidence for horizontal, vertical and tilted surfaces has been explained in detail by Keith and Kreider [2], Duffie and Beckman [3] and Kalogirou [4]. An illustration of the solar angle diagram is presented in figure 7.2 [4].

The incidence angle, θ, and the zenith angle, ϕ, are the same for a horizontal plane [4].

The final set of equations for various surfaces, based on the analysis in the literature [2–4] are as follows:

For horizontal surfaces, surface tilt angle from the horizontal $\beta = 0$, $\theta = \phi$;

$$\sin(\alpha) = \cos(\phi) = [\sin(\varphi)\sin(\delta)] + [\cos(L)\cos(\delta)\cos(h)] \tag{7.1}$$

where,

α is the solar altitude angle,
β is the surface tilt angle from the horizontal,
ϕ is the solar zenith angle,
L is the latitude,
δ is the declination,
h is the hour angle,
Z_s is the surface azimuth angle, the angle between the normal to the surface from true south, westward is designated as positive.

Illustrations of the definitions of various angles of the sun with respect to the earth and the apparent daily path are presented in figures 7.3 and 7.4, respectively.

$$\phi + \alpha = \Pi/2 = 90° \tag{7.2}$$

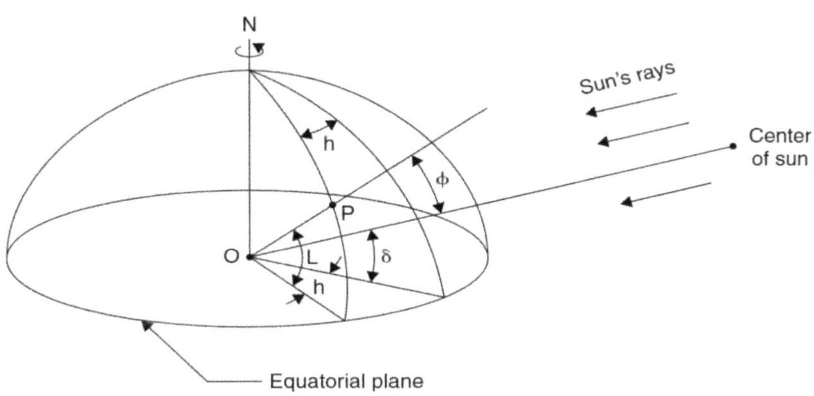

Figure 7.3. Definition of latitude, hour angle, and solar declination. Reprinted from [4], copyright (2009), with permission from Elsevier.

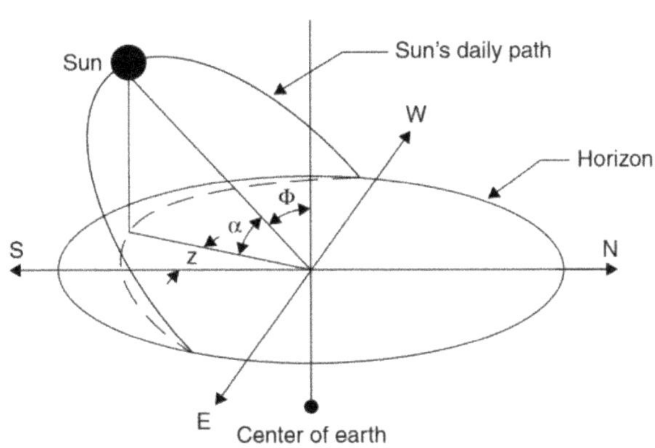

Figure 7.4. Apparent daily path of the sun across the sky from sunrise to sunset. Reprinted from [4], copyright (2009), with permission from Elsevier.

For vertical surfaces, $\beta = 90°$,

$$(\theta) = \cos^{-1}\left\{[-\cos(L)\sin(\delta)\cos(Z_s)] + [\sin(L)\cos(\delta)\cos(h)\cos(Z_s)] + [\cos(\delta)\sin(h)\sin(Z_s)]\right\} \tag{7.3}$$

For a south-facing, tilted surface in the Northern Hemisphere, $Z_s = 0°$;

$$(\theta) = \cos^{-1}\{[\sin(L - \beta)\sin(\delta)] + [\cos(L - \beta)\cos(\delta)\cos(h)]\} \tag{7.4}$$

For a north-facing, tilted surface in the Southern Hemisphere, $Z_s = 180°$,

$$(\theta) = \cos^{-1}\{[\sin(L + \beta)\sin(\delta)] + [\cos(L + \beta)\cos(\delta)\cos(h)]\} \tag{7.5}$$

A detailed discussion of the various solar angles along with the general equations and case studies for various locations has been presented in chapter 2—Environmental characteristics *Solar Energy Engineering: Processes and Systems* by Soteris Kalogirou—1st edn [4].

7.1.6 Mesh with different transmissivity

For the measurement of the performance of a PV module, as a function of irradiance at a fixed temperature, a mesh with different transmissivity is required. These data can be used to estimate the series resistance of the module. A mesh greater than the size of the module with transmissivity of 80% and 60% is required. There should not be any change in the spectrum of the transmitted light. A proper gap between the module and the mesh is required during the testing at different irradiance level. Figure 7.1(b) shows the mesh with 60% light transmissivity.

7.1.7 Arrangement for cooling or heating up the PV module

For the measurement of the temperature coefficient of the PV module, its performance needs to be measured at a fixed intensity and at different temperatures. The experimental setup should have the necessary facility to cool down or heat up the PV module beyond the ambient temperature. The module can be placed in an air-conditioned room and the module temperature is controlled to be at 15 °C. The back side of the module should be insulated; it is easy to utilize polymer materials in the back side to increase the thermal stability of the module.

7.1.8 Spectroradiometer

A spectroradiometer is used to measure the spectrum of the incident light. This data can be used to estimate the mismatch factor and can calculate the effective irradiance for the module technology being tested. The spectroradiometer should be capable of measuring the spectrum of the incident light in the wavelength range from 280 to 1700 nm. Figure 7.1(c) shows the spectroradiometers installed at fixed tilt equal to latitude.

7.2 Procedure for instant performance measurements

Performing instantaneous performance measurements of solar PV modules involves a systematic procedure to ensure accuracy and reliability.

7.2.1 Weather check

The irradiance should be more than 800 $W \cdot m^{-2}$ without any rapid cloud movement. A suitable time to perform the measurement is from 11:00 AM to 2:00 PM. This can be verified by the irradiance level before and after the *I–V* curve tracing. The irradiance checks help to optimize the perform-efficiency of the PV module and adjust the working conditions simultaneously. *I–V* curve tracing is done by cleaning the module first and connecting the *I–V* tracer before placing the irradiance sensor in the same plane of the module.

7.2.2 AOI correction

For the estimation of the AOI, the real-time measured data is used. The angle of incidence, θ, can be calculated in the plane-of-module using equation (7.6) [2–4].

$$(\theta) = \cos^{-1}\{[\sin(\delta)\sin(L - \beta) + [\cos(\delta)\cos(h)\cos(L - \beta)]\} \tag{7.6}$$

The effect of AOI on the performance of the PV module needs to be estimated for the distribution of AOI for the entire year. In this section, the methodology reported by King *et al* is used [5].

I_{scr} is the PV module short-circuit current at STC (A) and can be estimated by using equations (7.7) and (7.8):

$$I_{sc} = \frac{(I_{sc} \times E_o)}{[E_{poa} \times (1 + \alpha_{sc}(Tc - 25))]} \tag{7.7}$$

The relative optical response is given as in equation (7.8):

$$f_2(AOI) = \frac{\left[\dfrac{(I_{sc} \times E_o)}{[E_{poa} \times (1 + \alpha_{sc}(Tc - 25))]}\right] - ((E_{poa} - E_{dni} \times \cos(AOI)))}{[E_{dni} \times \cos(AOI)]} \tag{7.8}$$

where, E_{dni} = Direct normal solar irradiance ($W \cdot m^{-2}$); E_{poa} = Global solar irradiance on the plane-of-array (module) ($W \cdot m^{-2}$); E_o = Reference global solar irradiance, typically 1000 $W \cdot m^{-2}$; AOI = Angle between solar beam and module normal vector (degrees); T_c = Measured module temperature (°C); α_{sc} = Short-circuit current temperature coefficient (1/°C); I_{sc} = Measured short-circuit current (A).

Solar irradiance captured and used by the module is known as the effective irradiance. The effective irradiance is technology specific. The effective irradiance due to the AOI on the PV module can be calculated using equation (7.9):

$$E_e = \frac{[E_{dni} \times \cos(\text{AOI}) \times f_2(\text{AOI}) + f_d \times (E_{poa} - E_{dni} \times \cos(AOI))]}{E_o} \tag{7.9}$$

where, f_d = Fraction of diffuse irradiance used by module, typically assumed to be = 1.

7.2.3 Spectral correction

In the event that the radiation measuring device requires spectral correction, then the incident spectrum of light needs to be measured. By calculating the spectral MMF, the effective irradiance in terms of the technology specific spectral content of the incident irradiation can be estimated, as shown in figure 7.5. The MMF of a PV technology can be calculated by using the equation (7.10) [6].

$$\mathrm{MMF} = \frac{\int SR(\lambda)E_{AM1.5G}(\lambda)d\lambda}{\int E_{AM1.5G}(\lambda)d\lambda} \frac{\int E(\lambda)d\lambda}{\int SR(\lambda)E(\lambda)d\lambda} \tag{7.10}$$

where, $SR(\lambda)$ is the relative spectral response of the PV module and $E_{AM1.5G}$ is the standard AM1.5G spectrum. The effective solar irradiance (G_{eff}), corrected by the effect of the solar spectrum can be calculated by using equation (7.11):

$$G_2 = \frac{G}{MMF} \tag{7.11}$$

The installation of the module needs to be facing towards the sun with the angle of incidence as low as possible before placing the irradiance sensor in the plane of the module.

7.2.4 Temperature coefficient estimation

In order to mitigate the interruption and the associated errors in the generated potential, the module needs to be cooled before measurement. As described earlier, the temperature sensors are installed in three different positions on the back sheet (top, mid and bottom corner). The data of the I–V curve is traced at 15–30 s intervals at a fixed irradiance and at stabilized module temperature. Normally, the temperature coefficients

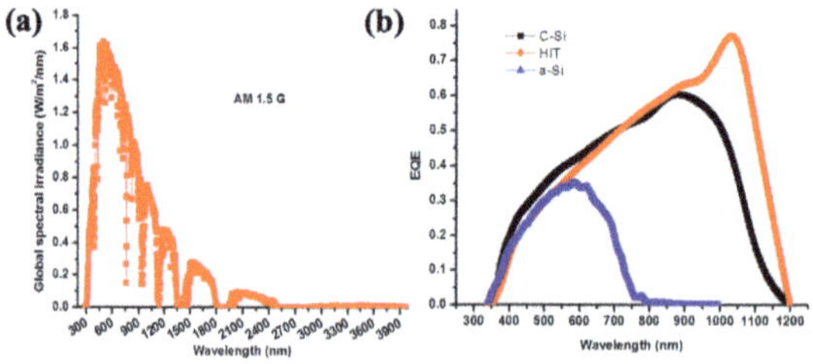

Figure 7.5. (a) AM 1.5 global spectrum as per IEC 60904-3. Reproduced with permission from [7]. The author thanks the International Electrotechnical Commission (IEC) for permission to reproduce Information from its International Standards. All such extracts are copyright of IEC, Geneva, Switzerland. All rights reserved. Further information on the IEC is available from http://www.iec.ch. IEC has no responsibility for the placement and context in which the extracts and contents are reproduced by the author, nor is IEC in any way responsible for the other content or accuracy therein. IEC 60904-3 ed. 4.0. Copyright (2019) IEC Geneva, Switzerland. (b) Spectral responses of PV modules. Reproduced with permission from [1], copyright (2023) Lin *et al.*

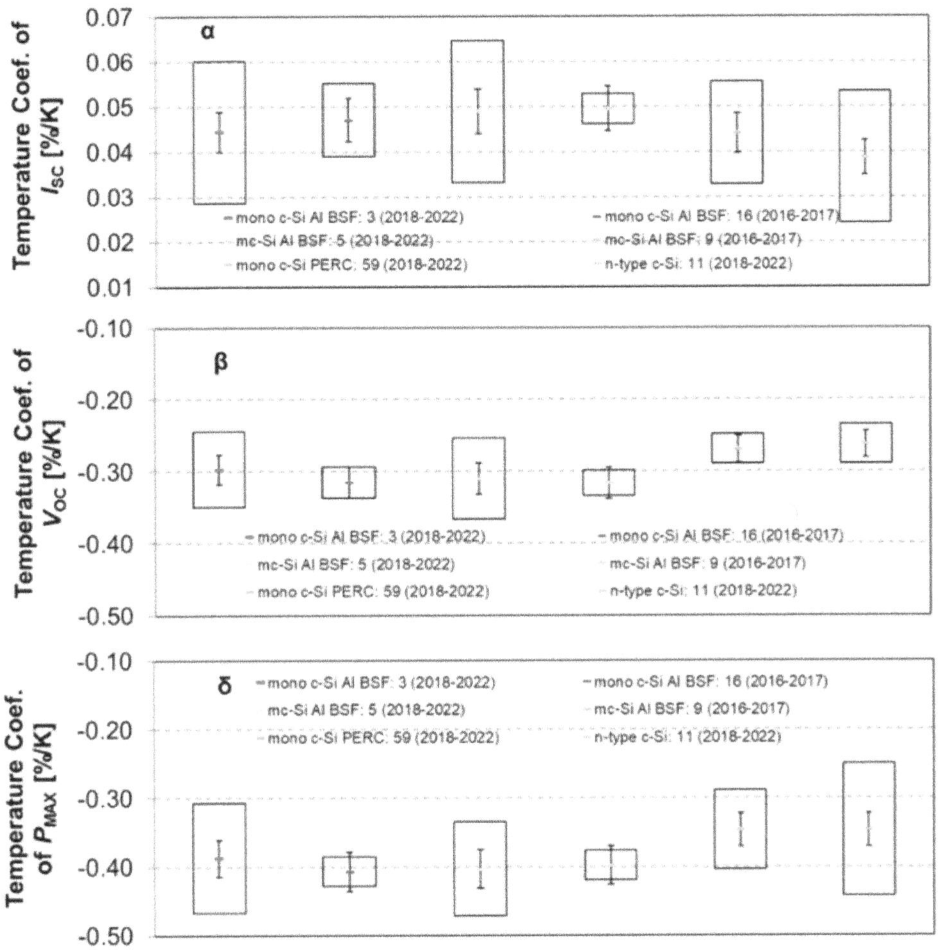

Figure 7.6. I_{sc}, V_{oc}, $P_{\max}$ as the function of module temperature. Reproduced from [8] Cc BY 4.0.

are measured in terms of the fundamental solar module performance parameters, I_{sc}, V_{oc}, $P_{\max}$. Several precautions need to be taken while performing these measurements. These precautions include the following: the total irradiance is at least as high as the upper limit of the range of interest; the irradiance variation caused by short-term oscillations should be less than 2% of the total irradiance as measured by the reference device and the wind speed should be less than 2 $\text{m}\cdot\text{s}^{-1}$. By performing these measurements, the variation in I_{sc}, V_{oc}, $P_{\max}$ as a function of the temperature are obtained, as shown in figure 7.6 [8]. Subsequently, percentage temperature coefficients are estimated by comparing them with the corresponding values of I_{sc}, V_{oc}, $P_{\max}$ at 25 °C [9].

7.2.5 Series resistance estimation

In accordance with IEC 60891 [9], for the estimation of series resistance, three I–V curves at different irradiance levels with constant temperature and spectrum are required. A mesh is used to obtain different irradiance at the same instance without any change in spectrum and module temperature. In procedure 1 of the IEC 60891 method, the current and voltage are obtained from the I–V characteristics by utilizing equations (7.12) and (7.13) [9].

$$I_r = I_m + I_{sc}\left(\frac{G_r}{G_m} - 1\right) + \alpha(T_r - T_m) \tag{7.12}$$

$$V_r = V_m - R_s(I_r - I_m) - \kappa I_r(T_r - T_m) + \beta(T_r - T_m) \tag{7.13}$$

where, I_r is the rated current, I_m is the measured current, I_{sc} is the short-circuit current, R_s is the series resistance, V_r is the rated voltage, V_m is the measured voltage, k is the curve correction factor and α is the temperature coefficient of the module.

The translational procedure 1 of IEC 60891 can be mathematically formulated as in equations (7.14)–(7.16).

$$I_2 = I_1 + I_{sc} \times \left(\frac{G_2}{G_1} - 1\right) \tag{7.14}$$

$$V_2 = V_1 - R_s \times (I_2 - I_1) \tag{7.15}$$

$$P = V_2 \times I_2 \tag{7.16}$$

where, I_1 and V_1 are the pair of measured points on the I–V characteristics, I_2 and V_2 are the pair of points of the resulting corrected characteristics, G_1 is the irradiance measured with the reference; G_2 is the irradiance at the reference or other desired conditions.

For the calculation of series resistance, the I–V curve corresponding to lower radiance with constant module temperature is translated into higher radiance data with the same module temperature. Since I–V curves are at the same temperature, the temperature coefficient will not play any role. Changes in R_s in steps of 10 mΩ in the positive or negative direction are noted. The deviations in $P_{\max}$ values of the transposed I–V characteristics are determined and the appropriate R_s is determined when the $P_{\max}$ deviation is within ± 0.5 % or better [9], as shown in figure 7.7(a).

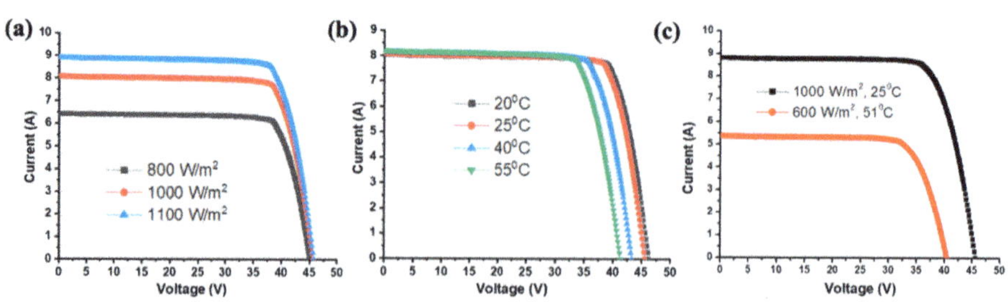

Figure 7.7. I–V curve of PV module (a) at different irradiance and at the same temperature (b) at constant irradiance but at different temperatures for estimation of curve correction factor (c) translated I–V curve. Reproduced from [1], copyright (2023) Lin *et al.*

7.2.6 Curve correction factor (κ) estimation

The current–voltage characteristics of the PV module at a constant irradiance and at different temperatures covering the range of interest of irradiance conditions are traced, which shall not differ by more than ± 1 % for the irradiance conditions at which the *I–V* measurements are performed. The *I–V* curves corresponding to all the temperatures are considered with the lowest module temperature being $\kappa = 0\ \Omega\ \mathrm{K}^{-1}$ in the translational equation. Starting from 0 mΩ K^{-1}, κ needs to be changed in steps of 1 mΩ K^{-1} in the positive or negative direction. The proper value of κ is determined when the deviation in the maximum output power values of the transposed *I–V* characteristics overlap within 0.5 % or better [9], as shown in figure 7.7(b).

7.2.7 Translation of *I–V* data to the desired irradiance and temperature condition

Besides the procedure 1 of IEC 60891 [9], there are two more procedures in the same IEC standard, which can also be used to translate *I–V* curves, as shown in figure 7.7(c). By using equations (7.17) and (7.18), the performance characteristics of PV modules can be estimated at the desired irradiance and temperature level. In tables 7.1 and 7.2, the parameters of measured and translated *I–V* curves are summarized. These parameters are based on the following equations:

Table 7.1. Electrical parameters of measured and translated *I–V* curves.

Conditions	I_{sc}	I_{mmp}	V_{oc}	V_{mmp}	$P_{\max}$	FF	R_s
Unit	A	A	V	V	W	%	Ohm
Measured at 600 W·m^{-2}, 51 °C	5.36	5.09	40.46	32.22	164.07	75.64	0.841
Translated to 1000 W·m^{-2}, 25 °C	8.81	8.43	45.63	36.30	306.14	76.14	0.583

Table 7.2. Maximum power (W) of a m-Si module as a function of irradiance and temperature as per IEC 61853-1.

		Parameter: maximum power ($P_{\max}$)			
Irradiance (W·m^{-2})	Spectrum	15 °C	25 °C	50 °C	75 °C
1100	AM1.5	NA	323.76	297.01	265.33
1000	AM1.5	303.74	295.06	270.77	243.03
800	AM1.5	240.66	233.35	212.56	194.42
600	AM1.5	178.76	173.29	158.76	146.12
400	AM1.5	115.99	113.20	103.48	NA
200	AM1.5	54.89	54.01	NA	NA
100*	AM1.5	25.84	25.59	NA	NA

$$I_2 = I_1 + I_{sc}\left(\frac{G_1}{G_2} - 1\right) + \alpha \times (T_2 - T_1) \tag{7.17}$$

$$V_2 = V_1 - R_s(I_2 - I_1) - k \times I_2 \times (T_2 - T_1) + \beta \times (T_2 - T_1) \tag{7.18}$$

where, I_2, V_2 are target current and voltage values, respectively; I_1, V_1 are measured current and voltage values, respectively; G_2, T_2 are target irradiance and temperature values, respectively; G_1, T_1 are measured irradiance and temperature values, respectively; α is the temperature coefficient of I_{sc} (A °C^{-1}); β is the temperature coefficient of V_{oc} (V °C^{-1}); R_s represents series resistance (Ω) values and κ is Curve correction factor (Ω °C^{-1}).

7.3 Conclusion

Instantaneous performance measurements of solar PV modules play a critical role in maintaining and enhancing the efficiency, reliability, stability and life expectancy of various components, systems and processes in photovoltaics from cell to module to farms. Advanced analytics and data visualization tools are often employed to make sense of the real-time data and facilitate quick decision-making in managing solar PV power plants. By following a structured procedure for instantaneous performance measurements, the industry practitioners can obtain the most precise data and use it to correct the entire system when required. This ensures that the PV systems are monitored effectively in real time.

This chapter was reproduced with permission from [1].

References

[1] Lin L *et al* 2023 Influence of outdoor conditions on PV module performance—an overview *Mater. Sci. Eng. Int. J.* **7** 88–101

[2] Keith F and Kreider J F 1978 *Hemisphere* (Washington, DC: Principles of Solar Engineering)

[3] Duffie J and Beckman W 1991 *Solar Engineering of Thermal Processes* (Wiley)

[4] Kalogirou S 2009 *Solar Energy Engineering: Processes and Systems* 1st edn (Academic)

[5] King D L, Kratochvil J A and Boyson W E 1997 *Conf. Record of the 26th IEEE Photovoltaic Specialists Conf.* pp 1113–6

[6] Magare D B *et al* 2016 Effect of seasonal spectral variations on performance of three different photovoltaic technologies in India *Int. J. Energy Environ. Eng.* **7** 93–103

[7] IEC 60904-3 2019 Photovoltaic devices—part 3: measurement principles for terrestrial photovoltaic (PV) solar devices with reference spectral irradiance data https://webstore.iec.ch/publication/61084

[8] Monokroussos C *et al* 2023 Energy performance of commercial c-Si PV modules in accordance with IEC 61853-1, -2 and impact on the annual specific yield *EPJ Photovolt.* **14** 6

[9] IEC 60891:2021 photovoltaic devices—procedures for temperature and irradiance corrections to measured *I–V* characteristics https://webstore.iec.ch/publication/61766

IOP Publishing

Recent Advances in Solar Cells

N M Ravindra, Leqi Lin and Priyanka Singh

Chapter 8

Long term performance measurements

The relentless pursuit of sustainable energy sources has propelled the evolution of solar cell technologies, with a growing emphasis and demands on their long-term performance, reliability and life expectancy. In this chapter, a brief introduction to the components and instruments that monitor the durability, stability, and advancements in materials and technologies, which are contributing to the extended operational lifespan of solar cells and solar panels, is presented. Through the analysis of environmental resilience, material degradation, and reliability under operational stress, the study unveils critical insights into the trends in longevity and life expectancy of solar cells/solar panels. The data analysis tools and comparisons are provided for a better evaluation of their long-term performance. The temperature dependence of solar cell/solar panel performance is briefly discussed. An introduction to the various solar panel components is presented. The various industry standards and the required warranties at the component and system level are briefly described.

8.1 Weather station system

8.1.1 Pyranometer

For long-term stability evaluation of PV modules under different outdoor conditions, continuous measurement and monitoring of weather-related parameters and performance of PV modules is required. In order to maintain uniformity in the intensity as well as the surroundings, the height of the weather station is the same as the height of the equipped PV test bed to collect the data. The interval of weather data recording should be less than 5 min to capture the variations properly. Weather station systems play a crucial role in measuring the environmental conditions that impact the performance of solar cells and solar panels. In table 8.1, a summary of the weather-related parameters and instruments that are required for the measurements is presented.

doi:10.1088/978-0-7503-5994-8ch8

Table 8.1. List of weather-related parameters to be measured and instrument required.

Parameter	Instrument to be used
Global irradiance in-plane to the module surface and at horizontal surface (0–1400 $W \cdot m^{-2}$)	Pyranometer (Thermopile or silicon solar cell-based)
Direct normal Irradiance (0–1400 $W \cdot m^{-2}$)	Pyrheliometer
Spectrum (Wavelength: 0.28 to 1.7 μm)	Spectroradiometer
Ambient temperature (−40 to +60 °C)	Digital thermometer
Wind speed and direction (For wind speed 0–99 mph, For wind direction 0–359°)	Anemometer/Wind monitor
Atmospheric pressure (500–1100 hPa)	Barometer
Rain (0–9999 mm)	Rain gauge
Humidity (10–100%)	Humidity probe
Angle of incidence of sunlight (−90 to +90°)	Sensor of angle of incidence

Pyranometer measures the solar irradiance which is critical in the assessment of the available solar energy for conversion to useful electricity in real-time conditions. Thermopile or silicon solar cell-based devices are the two of the most widely used sensors with different operating conditions. Thermopile pyranometers measure the total irradiance incident on a flat surface, and can quantify the irradiance from all the directions [1]. The gradient of the converted incident solar radiation to heat is measured across an area between a hot and cold point. This temperature difference is proportional to the incident irradiance, as illustrated in equation (8.1). Such devices measure irradiance with a spectral response from ultraviolet (UV) to the mid infrared (MIR), i.e., wavelengths ranging from 280 to 2800 nm [2].

$$V_{\text{out}} = (N)(S)(T_x - T_{\text{Ref}}) \tag{8.1}$$

where, V_{out} is the output voltage of the thermopile, S is the Seebeck coefficient, N is the number of thermocouples, T_x and T_{Ref} are the temperatures on positions x and Ref on the panel, respectively.

Silicon based pyranometers consist of a silicon photodiode embedded behind a diffuser, which can only measure the irradiance induced response in a narrow wavelength range of 300–1100 nm. The photocurrent produced by the semiconductor diode is proportional to the amount of received irradiance [2]. In general, the calibration of pyranometers requires simultaneous measurements of solar irradiance measured by a broadband thermopile reference (Ref) and the unit under test (*UUT*) [3, 4]. The responsivity, *R*, of the pyranometer is determined by:

$$R = UUT\ (\mu\text{V})/\text{Ref}\ (\text{W. m}^{-2}) \tag{8.2}$$

For a silicon photodiode-based pyranometer, this calculation represents the energy collected by the photodiode in response to the total energy in the solar spectrum [3, 4].

Figure 8.1 shows the different types of sensors for irradiance measurements.

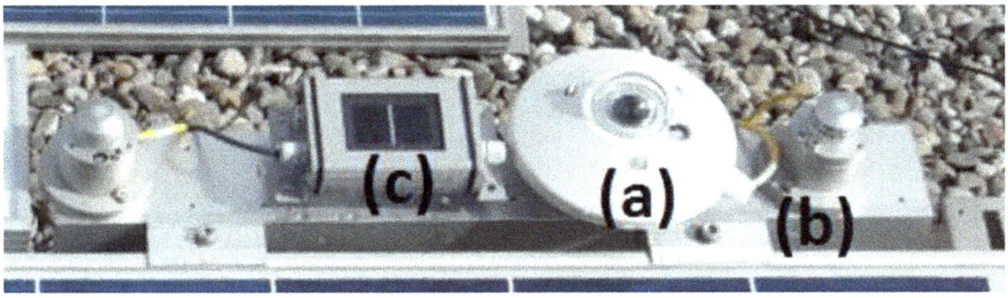

Figure 8.1. (a) Thermopile pyranometer, (b) silicon diode pyranometer, and (c) working-class PV reference cell, which were all tested, in this case, at the measurement location (Delft, The Netherlands). All the instruments were installed at one mounting structure with a 30° tilt angle facing south (182°). [1] John Wiley & Sons. Copyright 2023 The Authors. Progress in Photovoltaics: Research and Applications published by John Wiley & Sons Ltd.

8.1.2 Environmental resilience

Environmental resilience refers to the ability of a system or infrastructure to withstand and bounce back from the impacts of climate change. The utilization and incorporation of sensors in the solar panel boosts the robustness and energy reliability of the modules. By alleviating strain on the grid, solar panels can accumulate excess energy produced during peak sunlight conditions and stored in the battery/supercapacitor storage systems. This stored energy becomes valuable during low sunlight periods or power outages. Consequently, this setup establishes a dependable backup power solution, diminishing reliance on the conventional power grid and enhancing energy security in times of emergencies or natural disasters.

8.1.3 Reliability under thermal stress

The operation of solar cell panels, under high temperatures, is not advisable due to the decreasing efficiency and performance with increasing temperature [5]. Examples of solar cells operating under thermal stress are illustrated in figure 8.2, with considerations of space and terrestrial operational conditions. In space, solar cells are usually operating under high illumination and high temperatures since they are outside the earth's atmosphere. Terrestrial solar cells generally switch between grid-connected and off-grid conditions and are under thermal stress due to the constant exposure to the sun. Solar cells operating in space missions, close to the sun, may be considered to be under similar stress conditions as concentrating photovoltaics (CPVs) on Earth. However, in space, the thermal energy can only be dissipated by radiation, and space crafts may even be subjected to additional planetary infrared radiation [6]. On earth, hybrid solar cell systems are able to achieve better solar-to-thermal energy conversion. Hybrid photovoltaic-thermal (PV-T) collectors combine PV cells with a solar-thermal converter in which thermal energy is harvested by a solar absorber exchanging heat with a moving fluid [9]. Hybrid photovoltaic-thermoelectric (PV-TE) devices complement solar-to-thermal energy systems with solid-state thermal-to-electrical energy conversion, while, hybrid photovoltaic-

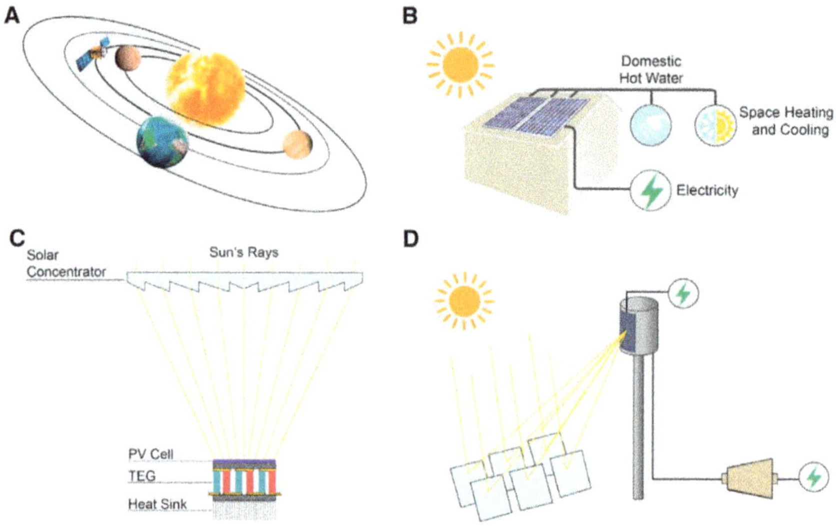

Figure 8.2. (A) Space-based solar cells on near-solar missions, (B) photovoltaic-thermal (PV-T), (C) photovoltaic- thermoelectric (PV-TE), and (D) photovoltaic-thermal concentrated solar power (PV-CSP) systems. Reprinted from [8], copyright (2020), with permission from Elsevier.

thermal concentrated solar power (PV-CSP) systems generate electricity with solar cells and a solar-to-thermal energy converter combined with a heat engine [10].

8.2 Performance metrics

8.2.1 Test bed with *I–V* tracer and inverter

For the purpose of continuous performance measurements, different types of test beds are designed. In this section, two different types of measurement setup are explained, which are test bed with *I–V* tracer and test bed with inverter. The monitoring of PV module is critical for better detection and recording of their abnormality. Two types of exposure procedures can be used in the reliability testing of module using *I–V* tracer, as shown in figure 8.3(a–b) [9]. The first, module exposure test, is a procedure in which each module is exposed without loads, and the other, system exposure test, requires that the array consisting of several modules is exposed under maximum power point tracking (MPPT) control with loads [10]. In each procedure, the *I–V* characteristics of the modules and that of the array are measured in a certain interval (5–10 min) [11]. In the module exposure test, modules are in loaded condition during stand-by condition. The modules are connected to the load in parallel. In the system exposure test, the operation current, under MPPT control, flows in the modules during the stand-by condition, and it is a similar condition when the modules are used. Moreover, in the system exposure test, the string voltage potential should be similar as they are used in the MWp scale power plants. It may cause a potential-induced degradation in the system exposure test. The procedure for the system exposure test has the possibility to cause failures, such

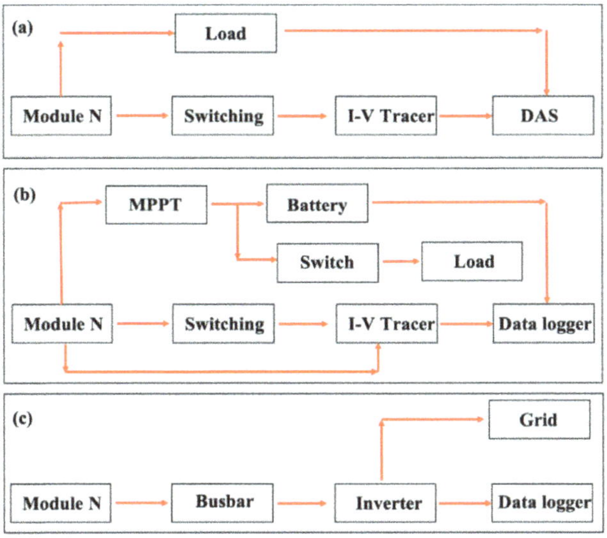

Figure 8.3. (a) Grid-connected power plant with *I–V* tracer, (b) grid-connected power plant with *I–V* tracer and battery connected power plant with *I–V* tracer, (c) grid-connected PV power plant with inverter.

as hot-spots, insulation failure and delamination rather than that in the module exposure test.

In grid-connected systems, the power conditioning system is connected which can control the MPPT and grid connection. In the case of a stand-alone measurement system, the PV array is MPPT controlled by MPPT unit, and the power is stored in battery during stand-by condition by the charge controller. The power stored in the battery is used for the measurements. If the power exceeds the battery capacity, the excess power will be consumed by the electrical load.

In the case of a test bed with inverter module, it can be exposed in both ways. In the case of system exposure, the array of several modules is connected to the inverter similar to the connections in a power plant, as shown in figure 8.3(c). The inverter should be MPPT operated. The string voltage of the module should be similar to the case when they are used. Megawatt power plants can also be used as test beds if the system voltage needs to be maintained. The measurement of maximum current and voltage can be performed with the help of MPPT controller within certain interval of time (six times a second). For the measurement of short-circuit current and open-circuit voltage, usually two reference modules are used with relay. This can be referred to as module exposure test. The maximum current and voltage measurement can also be performed with the help of MPPT controller. The MPPT samples the output of the PV modules and selects the most appropriate operating point to obtain maximum power under the prevailing environmental conditions. The PV performance data thus collected should include the MPPT efficiency. Operational data needs to be recorded for a sufficient period of time to allow the PV modules to complete any initial light induced degradation, and to collect a minimum of one year of operation data in order to visualize the impact of seasonal weather variation in each PV system.

8.2.2 Performance ratio (*PR*)

For a long-term performance evaluation and degradation rate of PV power plants, parameters such as the performance ratio and capacity utilization factor are used. The performance ratio (*PR*) represents the performance of the PV modules for outdoor conditions compared to their performance under standard test conditions (STCs) [12]. These are illustrated in equations (8.3)–(8.5), which are calculated as the final PV system yield, Y_f, divided by the reference yield (Y_r). *PR* depends mainly on the module temperature. For reducing the seasonal effect and the influence of temperature, *PR* should be normalized with respect to temperature. This can be estimated by normalizing the instantaneous output to 25 °C from the measured module temperature using the power temperature coefficient. Temperature normalized power can be estimated by using equations (8.4) and (8.5).

$$PR = \frac{Y_f}{Y_r} \tag{8.3}$$

where Y_f (dimensionless quantity) is the final PV system yield, Y_r is the reference yield.

$$Y_f = \frac{E}{P_o} \tag{8.4}$$

where, the PV system yield, Y_f (kWh kW^{-1}) or (hours) is the net energy output E divided by the DC power P_o of the installed PV array.

The reference yield, Y_f (hr) represents an equivalent number of hours at the reference irradiance. It is a function of the location, orientation, and tilt angle of the PV array, and changes from day-to-day, month-to-month and year-to-year as well as the weather variability.

$$Y_r = \frac{H_t}{G} \tag{8.5}$$

where the reference yield, Y_r (hr) is the total in-plane solar insolation H_t (kWh) divided by the PV's reference G (STC) irradiation (kW). *PR* depends mainly on the module temperature [13]. For reducing the seasonal effect and temperature effect, *PR* should be normalized with respect to temperature. This can be estimated by normalizing the instantaneous output to 25 °C from the measured module temperature using the power temperature coefficient.

$$P_{\text{norm}} = P + (T_m - 25) \times \gamma \tag{8.6}$$

where, P_{norm} is the normalized power, T_m is the module temperature and γ is the power temperature coefficient. The total energy is estimated by the sum of the normalized power for the desired interval as in equation (8.7).

$$E = \sum P_{\text{norm}} \tag{8.7}$$

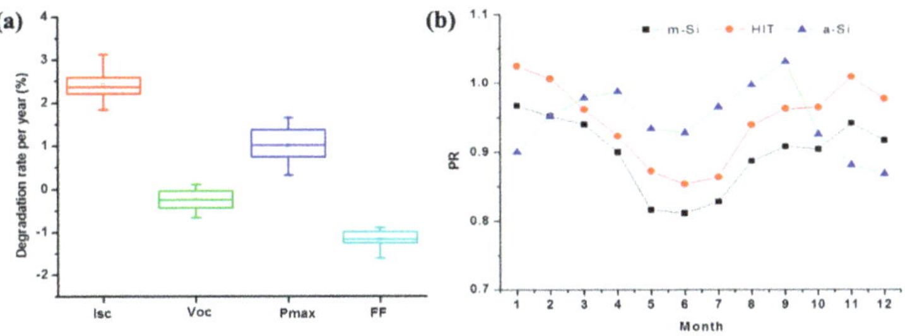

Figure 8.4. (a) Distribution of instantaneous degradation rate of PV modules in a power plant in India after three years of operation and (b) Distribution of *PR* in a year of three different technologies: m-Si (multicrystalline silicon), HIT (heterojunction with intrinsic thin layer and a-Si (amorphous silicon) solar cells.

8.2.3 Capacity utilization factor (*CUF*)

CUF is the ratio of the energy output from a solar PV plant over the time of interval to the maximum possible output from it for the same interval of time under ideal conditions [14, 15].

$$CUF = \frac{\text{Total energy output for the desired time period (kWh)}}{\text{Installed plant capacity (kW)} \times 24\text{h} \times \text{Time period (days)}} \tag{8.8}$$

In general, *CUF* of a PV power plant varies from 15% to 20% and it depends on the climatic conditions at the site [16]. In the case of concentrated PV systems, a maximum of around 35% *CUF* is observed in the field.

8.2.4 Degradation rate estimation of PV module

Degradation rate can be estimated by using equation (8.9) [17].

$$\text{Percentage of degradation} = \frac{(\text{initial } P_{\max} - \text{final } P_{\max}) \times 100\%}{\text{initial } P_{\max}} \tag{8.9}$$

There are various methodologies for estimating the degradation rate. It can be determined by both continuous and discrete data sets. In case of continuous data category, the degradation rate can be calculated by using performance ratio or by using the energy estimation method such as the PVUSA [18]. As an illustration, figure 8.4 shows the distribution of instantaneous degradation rate of PV modules in a power plant in India after three years of operation and distribution of *PR* in a year of three different technologies.

8.3 Data analysis tools and comparative analysis

8.3.1 Data analysis tools

The data analysis tools consist of hardware and software for handling weather sensors, estimating, averaging, integrating and tabulating weather-related parameters, as shown in figure 8.5. The embedded device collects all the data from weather sensors, forms tables in a database using a desired database package. It is possible to

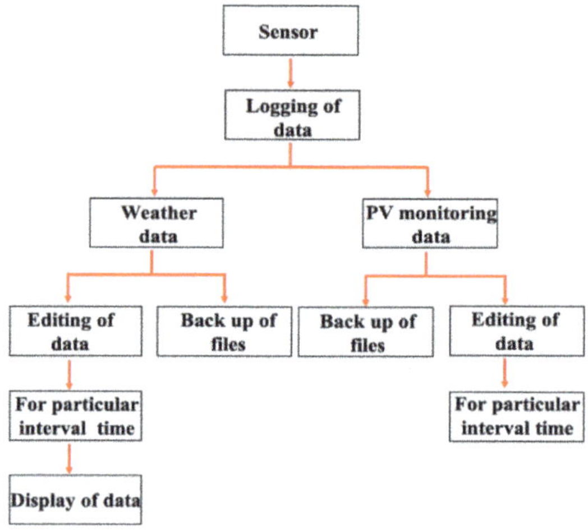

Figure 8.5. Flow chart of data collection.

create CSV files containing individual tables, as well as the entire database. The data can be collected over a wide duration and data for a specific period can be easily selectable for a specific period. The software generally includes programs for acquiring *I–V* measurements of PV arrays and modules as a function of time. The data is collected for the entire 24 h. The heart of the data logging facility is the timestamp facility, to help the user to view and compare *I–V* curves over selected time period for different technology modules. This is achieved by using indexing feature of database tables [19], so that the insertion and retrieval of data can be fast, even if the span of data is between years. It is estimated that for a day, about 230 KB are required for weather data and about 3 MB is required for storing PV monitoring data for one technology per day.

Besides, the reliability of the system should be high enough to operate in harsh environments. The design should support high channel-to-channel isolation, noise rejection, low-cost, high-density I/O, analog, digital modules, surge suppression and signal conditioning. Moreover, the installation and expansion should be done easily. For the analysis of data of PV module, the software should be able to estimate series resistance, curve correction factor and translation of *I–V* data to desired irradiance and temperature conditions.

8.4 Temperature dependence of solar cell/solar panel performance

The fundamentals of the physics of the temperature dependence of solar cells have been described in earlier chapters. The current–voltage characteristics of a (a) typical silicon solar cell; (b) solar panel are illustrated in figure 8.6 [20] and figure 8.7 [20], respectively.

As an example, the temperature dependent characteristics of a *Canadian Solar* BiHiKu6 530W-550W Bifacial MONO PERC (Passivated Emitter and Rear Contact) panel [21] are as follows:

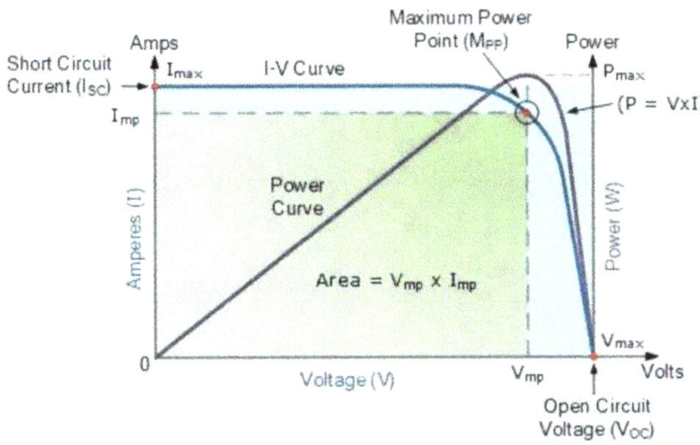

Figure 8.6. Current–voltage characteristics of a typical silicon solar cell. Reproduced with permission from [20]. Alternative Energy Tutorials, copyright (2010–2024) all rights reserved.

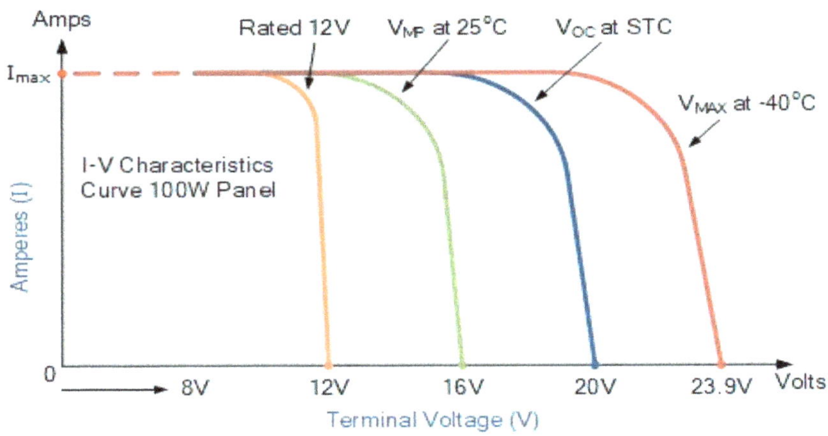

Figure 8.7. Current–voltage characteristics of PV panels illustrating the temperature dependence of the open-circuit voltage, V_{oc}. Reproduced with permission from [20]. Alternative Energy Tutorials, copyright (2010–2024) all rights reserved.

Specification data
Temperature coefficient (P_{max}) −0.34%/°C
Temperature coefficient (V_{oc}) −0.26%/°C
Temperature coefficient (I_{sc}) 0.05 %/°C
Nominal module operating temperature 41 °C ± 3 °C

In a solar cell, V_{oc} is more sensitive to temperature while the change in I_{sc} is small.

The most efficient solar panels are summarized in figure 8.8 (MBB = Multi-busbar; Topcon = Tunnel oxide passivated contact; HJT = Heterojunction technology).

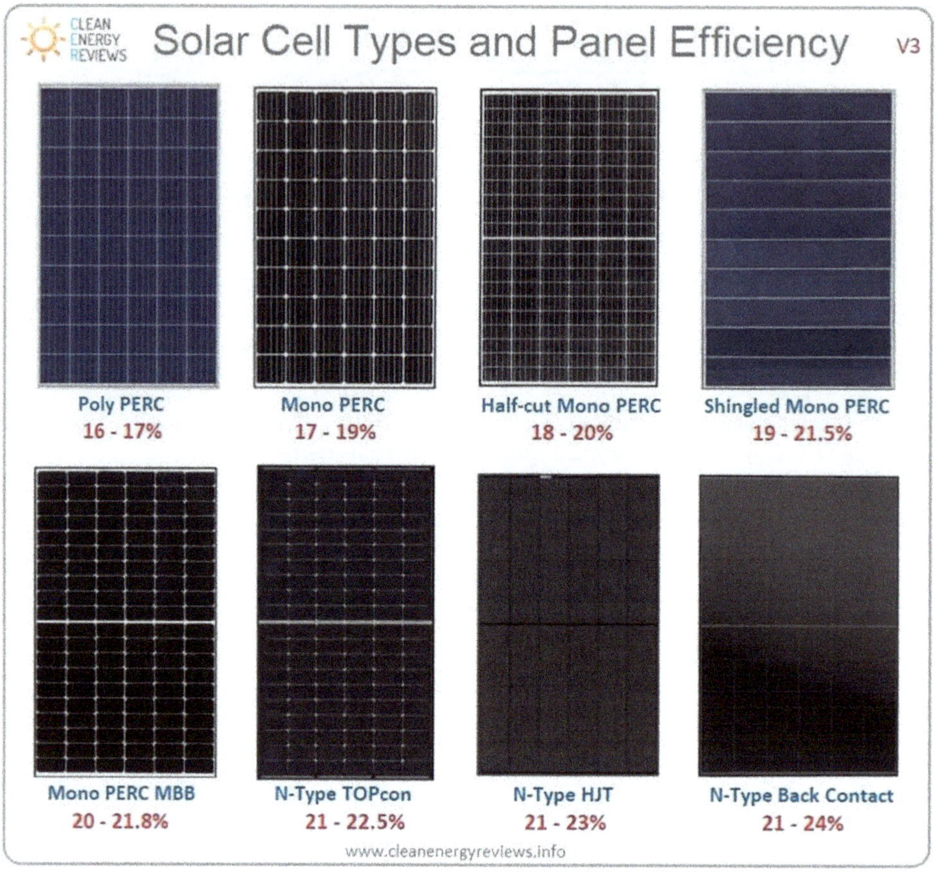

Figure 8.8. Most efficient solar panels in 2024. Reproduced with permission from [22].

As can be seen in figure 8.8, the solar panel efficiency is influenced by factors such as the cell type, design and configuration. As of today, the n-type interdigitated back contact (IBC) cell is the most efficient with 24% efficiency [22].

The most efficient residential solar panels, along with their performance specifications, are summarized in table 8.2. As can be seen in the table, the temperature coefficients are generally in the range of −0.24 to −0.35%/°C [23].

8.5 Solar panel components and cost considerations

An illustration of cells, modules, panels and arrays is shown in figure 8.9 [24]. Cells, connected in series yield higher voltages; connecting them in parallel will produce higher currents. Cells are connected and sealed in an environmentally protective laminate to form modules. Modules are assembled and pre-wired to form panels, which are subsequently integrated to produce arrays.

A schematic of the process steps involved in the fabrication of PV modules is presented in figure 8.10 [25]. In the first step, i.e., cell interconnection, solar cells in one

Table 8.2. A summary of most efficient residential panels [23].

Company	Panel and capacity	Max. efficiency rating	Wattage	Temperature coefficient
Maxeon	Maxeon 7	23.5%–24.1%	435–445	−0.27
Canadian solor	TopHiKu6 CS6.1–54TD	21.8%–23%	445–470	−0.29
REC	REC Alpha Pure RX	21.6%–22.6%	450–470	−0.24
Jinko solor	Tiger Neo N-type 54HL4R-(V)	21.52%–22.52%	430–450	−0.29
Qcells	Q.Tron BKL M-G2+	21.3–22.5%	415–440	−0.3
Panasonic	Evervolt HK2 Black	21.7%–22.2%	420–430	−0.24
Silfab	SIL-420/430 QD	21.5%–22.1%	420–430	−0.29
JA solor	JAM54S30/GR	20.5%–21.8%	400–425	−0.35
ZNshine solor	ZXM7-SH120 Series	20.35%–21.55%	440–465	−0.35
Longi	Hi-MO 5 m	21.30%	410–420	−0.34
Trina Solor	Vertax S TSM-DE09C.07	19.8%–21.1%	380–405	−0.34

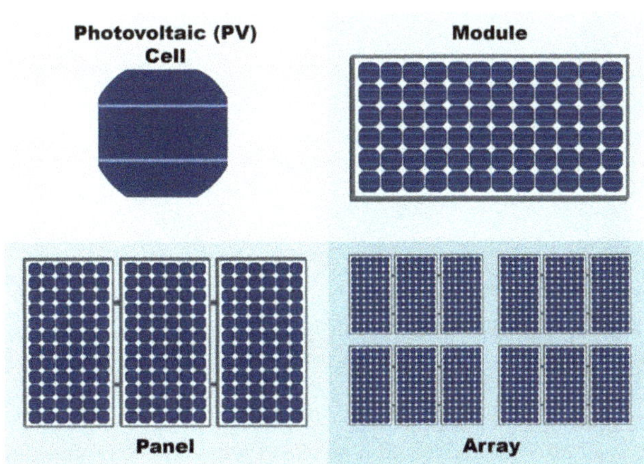

Figure 8.9. Illustration of PV cell, module, panel and array. Reproduced from [34] CC BY 4.0.

column of the PV module are soldered either manually or by a tabber and stringer machine. Electroluminescence imaging technique is utilized to inspect these strings to identify defects in the production process. In the next step, these strings are placed on top of a glass sheet with a layer of EVA–ethylene vinyl acetate and the strings are interconnected. Another layer of EVA is put on top of the interconnected strings and a backsheet, typically tedlar, is placed on top of the EVA and the entire stack is placed in a laminator; it is heated for 10–15 min at ∼200 °C. Subsequently, the junction box and the frame are added to the stack to complete the module. Quality checks are performed, after each process step, to ascertain a high-quality product.

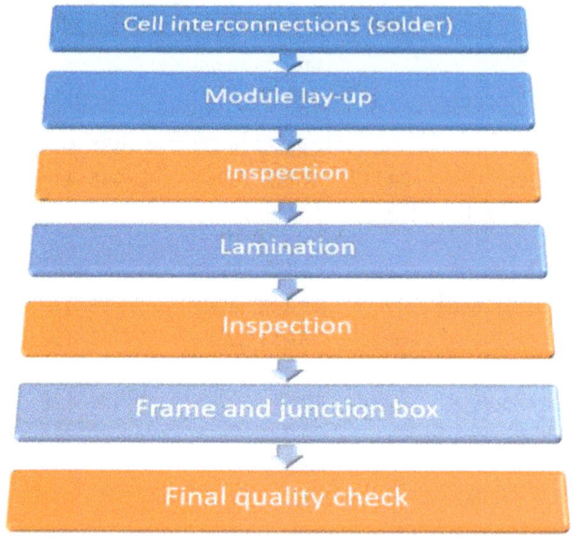

Figure 8.10. Schematic process flow of the fabrication of PV modules. Reproduced with permission from [25].

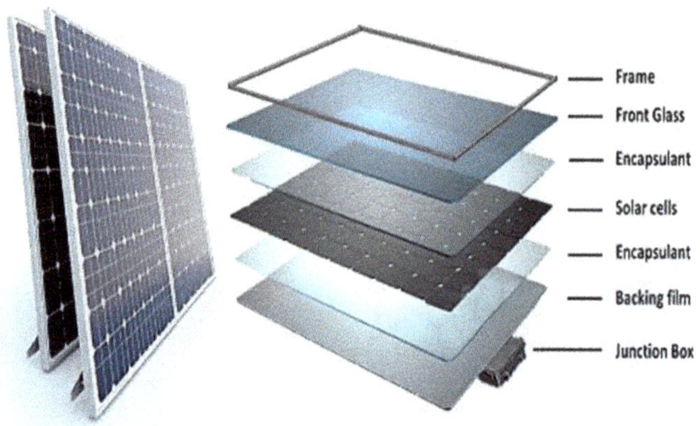

Figure 8.11. An example of a fully assembled glass on glass PV module. Reproduced from [26] CC BY 4.0.

The fully assembled solar module (dual glass—glass on glass) and the associated parts in the module are illustrated in figure 8.11 [26]. In order to reduce reflections, the solar glass in the PV module is textured. The use of this dual glass approach will strengthen the solar panel by contributing to the following factors: (a) reduce the potential for the panel to become deformed, (b) reduce the possibility for formation of microcracks in the cells, (c) enhance thermal stability and (d) increase yields from PV modules.

The National Renewable Energy Laboratory (NREL) has been performing detailed cost analysis of the manufacture of c-Si solar cells and solar panels since 2010. The results of the NREL model is presented in figure 8.12 [27]. In accordance with this model, the anticipated cost of c-Si modules will be $0.18/W in 2025.

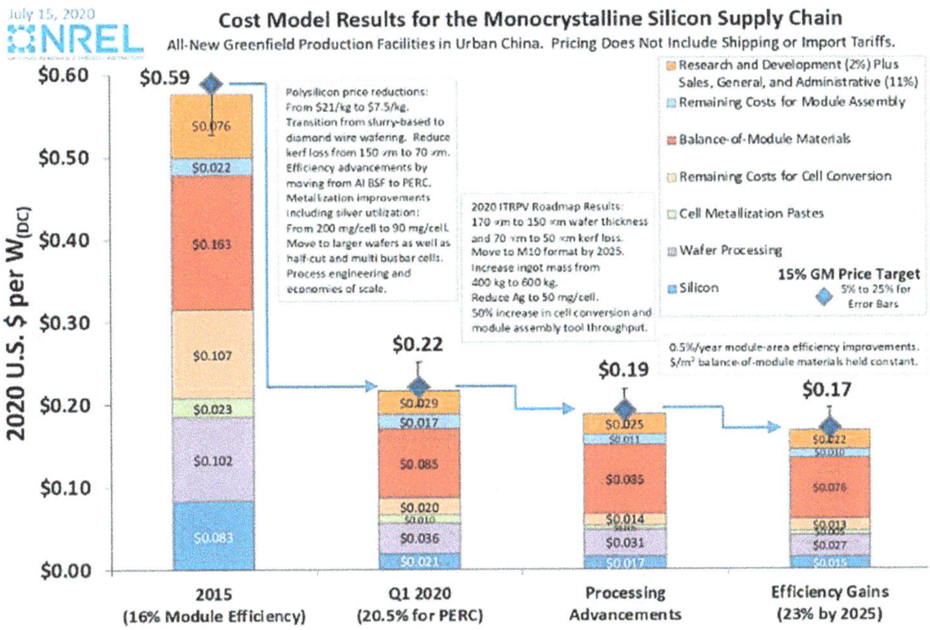

Figure 8.12. Summary of results for 2020 benchmark and future module costs analysis of monocrystalline silicon PV. Reproduced with permission from [27], credit: NREL.

8.6 Standards

The solar industry as well as several international and national organizations have established standards that address PV system component safety, design, installation and monitoring. The standards for PV include almost every stage of the PV industry such as: (a) materials and processes used in the production of PV panels, (b) testing methodologies, (c) performance standards, (d) design and installation guidelines.

Table 8.3 presents some of the major standards from various organizations. These standards are based on the summary as presented by USAID, International Electrotechnical Commission (IEC), Institute of Electrical and Electronics Engineers (IEEE), Underwriters Laboratories (UL), American Society for Testing and Materials (ASTM), and the International Organization for Standardization (ISO) [28–32]. ISO has several certifications for the solar PV industry. The ISO standards are internationally agreed-upon guidelines and requirements for energy efficiency and renewable energy solutions.

In general, in practical use and implementation, solar panel efficiency can be enhanced by the following approaches: (a) proper orientation and tilt—maximize interaction with the incident sunlight; (b) minimize shading—this will require avoiding physical obstructions such as buildings, trees, structures; (c) maximize the amount of sunlight captured by the panel—this may require tracking systems especially in solar farms; (d) maintain the panels clean. While north-facing roofs are the most unfavorable option for the installation of solar panels, panels that are

Table 8.3. Summary of organizations and standards for the PV industry.

Agency/standard	Subject matter/certification
International Electrotechnical Commission (IEC)	
Balance of system	
IEC 62093:2005	Balance-of-system components for photovoltaic systems–design qualification natural environments
IEC 62109–1:2010	Safety of power converters for use in photovoltaic power systems–Part 1: general requirements
IEC 62109–2:2011	Safety of power converters for use in photovoltaic power systems–Part 2: particular requirements for inverters
IEC 60269-6 ed1.0	Low-voltage fuses–part 6: Supplementary requirements for fuse-links for the protection of solar photovoltaic energy systems
Characteristics	
IEC 61727 ed2.0	Photovoltaic (PV) systems–characteristics of the utility interface
Commissioning	
IEC 62446–1:2016	Photovoltaic (PV) systems–requirements for testing, documentation, and maintenance–Part 1: grid-connected systems–documentation, commissioning tests and inspection.
Design	
IEC 62124 ed1.0	Photovoltaic (PV) stand-alone systems–design verification
IEC 62253 ed1.0	Photovoltaic pumping systems–design qualification and performance measurements
Installation	
IEC 60364-1 ed5.0	Low-voltage electrical installations–Part 1: fundamental principles, assessment of general characteristics, definitions
IEC 60364-7-712:2017	Low-voltage electrical installations–Part 7–712: Requirements for special installations or locations–Solar photovoltaic (PV) power supply systems
Monitoring	
IEC 61724–1:2017	Photovoltaic system performance–Part 1: monitoring
IEC TS 61724–2:2016	Photovoltaic system performance–Part 2: capacity evaluation method
IEC TS 61724–3:2016	Photovoltaic system performance–Part 3: energy evaluation method
Performance	
IEC 62509 ed1.0	Battery charge controllers for photovoltaic systems–performance and functioning
Rural Electrification	
IEC TS 62257–1:2015	Recommendations for renewable energy and hybrid systems for rural electrification–Part 1: general introduction to IEC 62257 series and rural electrification

Safety	
IEC 61730–1:2016	Photovoltaic (PV) module safety qualification–Part 1: requirements for construction
IEC 61730–2:2016	Photovoltaic (PV) module safety qualification–Part 2: requirements for testing
Terms	
IEC TS 61836:2016	Solar photovoltaic energy systems–terms, definitions and symbols
Testing	
IEC 61215–1:2016	Design qualifications and type approval Part 1: testing requirements (all chemistries)
IEC 61215–2:2016	Design qualifications and type approval Part 2: testing procedures (all chemistries)
IEC 62116:2014	Utility-interconnected photovoltaic inverters–test procedure of islanding prevention measures
Standard for PV module safety	
IEC 61730	Determination of safety of solar module for installation. For instance, it should not show any electrical, mechanical, or thermal hazards
Ammonia corrosion testing of PV modules	
IEC 62716	Tests for a solar module's resilience against ammonia damage
Institute of Electrical and Electronics Engineers (IEEE)	
Interconnection	
IEEE 1547.2–2008	IEEE application guide for IEEE Std 1547(TM), IEEE Standard for interconnecting distributed resources with electric power systems
IEEE 1547.3–2007	IEEE guide for monitoring, information exchange, and control of distributed resources interconnected with electric power systems
IEEE 1547–2018	IEEE standard for interconnection and interoperability of distributed energy resources with associated electric power systems interfaces
Performance	
IEEE 1526–2003	IEEE recommended practice for testing the performance of stand-alone photovoltaic systems
Sizing	
IEEE 1562–2007	IEEE guide for array and battery sizing in stand-alone photovoltaic (PV) systems
Underwriters Laboratories (UL)	
Balance of System	
UL-2703, 1st Edition	Standard for mounting systems, mounting devices, clamping/retention devices, and ground lugs for use with flat-plate photovoltaic modules and panels

(*Continued*)

Table 8.3. (*Continued*)

Agency/standard	Subject matter/certification
UL 1741	Standard for inverters, converters, controllers and interconnection system equipment for use with distributed energy resources
UL-1699B	Standard for photovoltaic (PV) DC arc-fault circuit protection
UL-4703	Standard for photovoltaic wire
UL-854	Standard for service–entrance cables
UL-4248-19	Fuseholders–Part 19: PHOTOVOLTAIC
UL-6703	Standards for connectors for use in photovoltaic systems
UL-3730	Standard for photovoltaic junction boxes
UL-489B, 1st Edition	Molded-case circuit breakers, molded-case switches, and circuit-breaker enclosures for use with photovoltaic (PV) systems
Concentrated	
UL 8703, 3rd Edition	Concentrator photovoltaic modules and assemblies
Flat-Plate PV	
UL 1703, 3rd Edition	Standard for flat-plate photovoltaic modules and panels
Mounting	
UL 790, 8th Edition	Standard for standard test methods for fire tests of roof coverings
UL 1897, 7th Edition	Standard for uplift tests for roof covering systems
Testing	
UL-SU 5703	Determination of the maximum operating temperature rating of photovoltaic (PV) backsheet materials
American Society for Testing and Materials (ASTM)	
Terms	
ASTM E772-15	Standard terminology of solar energy conversion
Testing	
ASTM E2848-13(2018)	Standard test method for reporting photovoltaic non-concentrator system performance
ASTM E927-19	Standard classification for solar simulators for electrical performance testing of photovoltaic devices [29]
International Organization for Standardization (ISO)	
ISO 50001	Set of standards for implementing Energy management systems (EnMS) in an organization [30, 31]
ISO 9060:2018	Specification and classification of instruments for the measurement of hemispherical solar and direct solar radiation [32]

ISO 9001	Implementation of quality management system in an organization
ISO 14001	Establishment of an environmental management system in an organization
ISO 41001	Implementation of facility management system in an organization
Recommended Testing and Certifications for Solar Panels	
UL/IEC 61730	Evaluation of safety of PV modules and outlines the requirements of PV construction
UL/IEC 61730–1	Photovoltaic compliance testing
UL/IEC 61730–2:2023	Photovoltaic (PV) module safety qualification–Part 2: Requirements for testing
UL/IEC 61215–1–2	PV module standards
IEC 61215–2, 4.17	Hail Test, 2′ Dia., 27 m s^{-1}
ISO 62716	Ammonia Test
IEC 6170	Salt spray test (severity: level-3, level-6)
EN 60068–2–68	Sand/dust storm test
IEC 62782/IEC 6121	Dynamic mechanical load test
IEC 61853–1	Irradiance and temp performance and power rating
IEC TS 63342:2022	C-Si photovoltaic (PV) modules–light and elevated temperature induced degradation (LETID) test–Detection
IEC TS 62804–1–1:2020	Photovoltaic (PV) modules–test methods for the detection of potential-induced degradation–Part 1–1: crystalline silicon–delamination
CEC Listed	Clean energy council
FSEC Registered	Florida solar energy center

Table 8.4. Summary of solar panel warranties.[a]

Warranty	Coverage	Conditions
Product	Solar panel replacement–If the panel/s malfunction Generally ⩾ 10 Years; Best warranties—25 Years	Defects in material or workmanship
Power	Performance generally 25–30 years Best panels: 85% productivity after 25 Years	Maximum degradation rate specified by manufacturer 2–3% Degradation—Year 1 ⩽0.5% per year after Year 1

(*Continued*)

Table 8.4. (*Continued*)

Warranty	Coverage	Conditions
Installation	Manufacturer's warranty covers solar panels (material defects, workmanship defects, loss of power output) Installation warranty = 5–25 Years	Installation warranty = System warranty. racking, wiring etc
Integrated	Full: product, performance and installation	Manufacturer with its own installation services
Not covered	System components = batteries, inverters, racking Batteries, inverters = 10–12 years Microinverters = 25 years	System components have their own warranties—shorter period
Factors to consider	Solar panel warranty transferable Inclusion of labor in solar product warranty Warranty under extreme weather, fire	Panel/product replacement with product warranty; Labor costs may not be covered. Homeowner's insurance coverage?
Maintaining warranty	Self-installation, Any modification by self = > Cautious approaches to cleaning Reporting problems in a timely manner	Loss of warranty

[a] Based on [33].

south-facing perform with the greatest production of PV-enabled electrical power. Panels that are east- or west-facing also perform well.

8.7 Warranties

Solar panels represent multi-faceted functionalities. While they are primarily products, their function is to provide power. In order to maximize the power obtained from solar panels, they need to be installed properly. A summary of the various warranties associated with solar panels is presented in table 8.3.

In general, most homeowner's insurance policies cover solar energy systems since they are considered to be a permanent attachment to the property.

8.8 Conclusion

The durability and stability assessments provide detailed information on the resilience of solar cells and solar panels in various environmental conditions, showcasing their potential for long-term viability. The long-term performance of solar panels is not merely a measure of endurance but a testament to the dynamic evolution of materials and technologies. By understanding the complexities and embracing the innovations highlighted herein, the way for a brighter, more sustainable energy future can be thought through.

References

[1] Karki S *et al* 2021 Performance evaluation of silicon-based irradiance sensors versus thermopile pyranometer *IEEE J. Photovolt.* **11** 144–9

[2] Rösemann R, Lee C and Kipp Z 2011 *A Guide to Solar Radiation Measurement: From Sensor to Application: An Overview of the State of the Art: UV, Visible, Infrared* 2nd edn (Delft: Kipp & Zonen)

[3] Sengupta M, Gotseff P and Stoffel T 2012 Evaluation of photodiode and thermopile pyranometers for photovoltaic applications *27th European Photovoltaic Solar Energy Conf. and Exhibition (Frankfurt, Germany, September 24–28)*

[4] Sengupta M, Gotseff P, Myers D and Stoffel T 2012 Performance testing using silicon devices–analysis of accuracy *Presented at the 2012 IEEE Photovoltaic Specialists Conf. (Austin, Texas) (June 3–8)*

[5] Gautam N K and Kaushika N D 2002 Reliability evaluation of solar photovoltaic arrays *Sol. Energy* **72** 129–41

[6] Loehberg A *et al* 2017 The bepicolombo mercury planetary orbiter (MPO) solar array design, major developments and qualification *E3S Web Conf* **16** 04006

[7] Zondag H A 2008 Flat-plate PV-thermal collectors and systems: a review *Renew. Sustain. Energy Rev.* **12** 891–959

[8] Vaillon R, Parola S, Lamnatou C and Chemisana D 2020 Solar cells operating under thermal stress *Cell Rep. Phys. Sci.* **1** 100267

[9] Sayyad J and Nasikkar D 2021 Design and development of low cost, portable, on-field I–V curve tracer based on capacitor loading for high power rated solar photovoltaic modules *IEEE Access* **9** 70715–31

[10] Quiroz J, Stein J, Carmignani C and Gillispie K 2015 In-situ module-level I–V tracers for novel PV monitoring *2015 IEEE 42nd Photovoltaic Specialist Conf. (PVSC) (New Orleans)* pp 1–6

[11] Aboagye B, Gyamfi S, Ofosu E A and Djordjevic S 2021 Degradation analysis of installed solar photovoltaic (PV) modules under outdoor conditions in Ghana *Energy Rep.* **7** 6921–31

[12] Timothy Dierauf A G, Kurtz S, Becerra Cruz J L, Riley E and Hansen C 2013 *Weather-Corrected Performance Ratio* (National Laboratory of the U.S. Department of Energy, Office of Energy Efficiency and Renewable Energy).

[13] Chandra S, Agrawal S and Chauhan D 2018 Effect of ambient temperature and wind speed on performance ratio of polycrystalline solar photovoltaic module: an experimental analysis *Int. Energy J.* **18** 171–80

[14] Tripathi A K 2020 Factors affecting capacity utilization of thermal power (coal) plants in India https://www.preprints.org/manuscript/202008.0414/v1

[15] Zhang J *et al* 2020 Measuring the capacity utilization of China's transportation industry under environmental constraints *Transp. Res. Part D: Transp. Environ.* **85** 102450

[16] Manoj Kumar N *et al* 2020 Operational performance of on-grid solar photovoltaic system integrated into pre-fabricated portable cabin buildings in warm and temperate climates *Energy Sustain. Develop.* **57** 109–18

[17] Phinikarides A, Kindyni N, Makrides G and Georghiou G E 2014 Review of photovoltaic degradation rate methodologies *Renew. Sustain. Energy Rev.* **40** 143–52

[18] Daher D H, Gaillard L and Ménézo C 2022 Experimental assessment of long-term performance degradation for a PV power plant operating in a desert maritime climate *Renew. Energy* **187** 44–55

[19] Magare D, Sastry O, Gupta R, Kumar A and Sinha A, *Proc. of 27th European Photovoltaic Solar Energy Conf.* 3259–62

[20] https://alternative-energy-tutorials.com/photovoltaics/temperature-coefficient.html (accessed 1 July 2024)

[21] Canadian Solar, BiHiKu6 530W-555 W Bifacial MONO PERC Panel Specifications (accessed 1 July 2024)

[22] https://cleanenergyreviews.info/blog/most-efficient-solar-panels (accessed 1 July 2024)

[23] https://cnet.com/home/energy-and-utilities/most-efficient-solar-panels/(accessed 1 July 2024)

[24] https://energyresearch.ucf.edu/consumer/solar-technologies/solar-electricity-basics/cells-modules-panels-and-arrays/ (accessed 2 July 2024)

[25] https://pv-manufacturing.org/solar-cell-manufacturing/pv-module-manufacturing/ (accessed 2 July 2024)

[26] Chiang P F, Han S, Claire M J, Maurice N J, Vakili M and Giwa A S 2024 Sustainable treatment of spent photovoltaic solar panels using plasma pyrolysis technology and its economic significance *Clean Technol.* **6** 432–52

[27] Woodhouse M *et al* 2021 Research and development priorities to advance solar photovoltaic lifecycle costs and performance *Technical Report NREL/TP-7A40–80505, National Renewable Energy Laboratory*

[28] https://usaid.gov/energy/powering-health/technical-standards/photovoltaic-systems (accessed 2 July 2024)

[29] https://webstore.ansi.org/industry/solar-energy (accessed 2 July 2024)

[30] https://siscertifications.com/iso-certification-for-solar-industry/ (accessed 2 July 2024)

[31] https://siscertifications.com/iso-50001-certification/ (accessed 2 July 2024)

[32] https://iso.org/standard/67464.html (accessed 2 July 2024)

[33] Leonardo D 2024 A guide to understanding solar panel warranties ed S Lopez https://marketwatch.com/guides/solar/solar-panel-warranty/ (accessed 2 July 2024)

[34] Olayiwola T N, Hyun S-H and Choi S-J 2024 Photovoltaic modeling: A comprehensive analysis of the I–V characteristic curve *Sustainability* **16** 432

IOP Publishing

Recent Advances in Solar Cells

N M Ravindra, Leqi Lin and Priyanka Singh

Chapter 9

Influence of weather conditions

The performance of solar cells/panels is, in general, a function of the weather conditions. The weather conditions include the following: humidity, irradiance, mechanical stress and temperature. These parameters affect the durability, efficiency, performance, stability, and life expectancy of solar cells/panels. Variables such as high temperature, humidity and mechanical stress have a negative impact on the short-circuit current and open-circuit voltage, whereas, with increasing irradiance, a linear increasing trend in the power conversion efficiency is observed. Examples of performance of solar cells/panels with changes in weather-related conditions are briefly summarized in this chapter. Case studies of space solar cells and concentrator systems are presented.

9.1 Effect of irradiance

This section was reproduced with permission from [48].

Irradiance refers to the power of electromagnetic radiation per unit area incident on a surface. A higher irradiance usually leads to higher conversion efficiency of cells/panels from sunlight to electricity. The impact of irradiance on the solar cell performance is described with respect to mainly three factors: I–V curve, efficiency and energy conversion. Considerations are given to the actual irradiance flux as function of weather and related changes throughout the day. Many research groups have worked on the performance evaluation of solar cells/panels under varying weather conditions. Besides, the efficiency of solar cells, under outdoor conditions, is different from that at STC (Standard Test Conditions, as per IEC 61538, Cell temperature = 25 °C, Irradiance = 1000 W m^{-2}, Air Mass = 1.5). This is due to factors such as humidity, irradiance and particle radiation (in space applications). During tests, solar panels, under the influence of varying irradiance, are monitored by using a real-time irradiance monitor, and system designs are employed for the maximization of power output from sunlight.

doi:10.1088/978-0-7503-5994-8ch9

The influence of irradiance on the performance efficiency of solar modules has been broadly reported in the literature. Buday *et al* [1] found out that the CIGS modules demonstrated linear increase in $P_{\max}$ and J_{sc}; FF and V_{oc} exhibit a rapid increase from dark condition to light condition and undergo a steady but gradually decreasing trend after 600–800 $W{\cdot}m^{-2}$ over a full range of increasing irradiance, as shown in figure 9.1. In the new generation solar cell modules, such as CIGS, (dye-sensitized) DSC and (organic thin films–organic photovoltaics) OPV, the experimental results show that the efficiency η and FF decrease when the irradiance increases from 200 to 1000 $W{\cdot}m^{-2}$, which may be ascribed to the variation in series resistance in these thin film solar cells [2]. V_{oc} increases with increasing irradiance due to the higher photo-generated current density [2]. In figure 9.2 [3–7], in comparing the PSC-2019-Tress (perovskite solar cell) [6] and HIT-2019-Tress (heterojunction with intrinsic thin layer) [6] regarding the irradiance dependence of *PCE*, it is observed that PSC-2019-Tress [6] decreased less than the HIT-2019-Tress [6] from 10–1000 $W{\cdot}m^{-2}$. This is due to the higher V_{oc} than that of the HIT-2019-Tress [6], leading to smaller relative changes in *PCE* when V_{oc} decreased by a

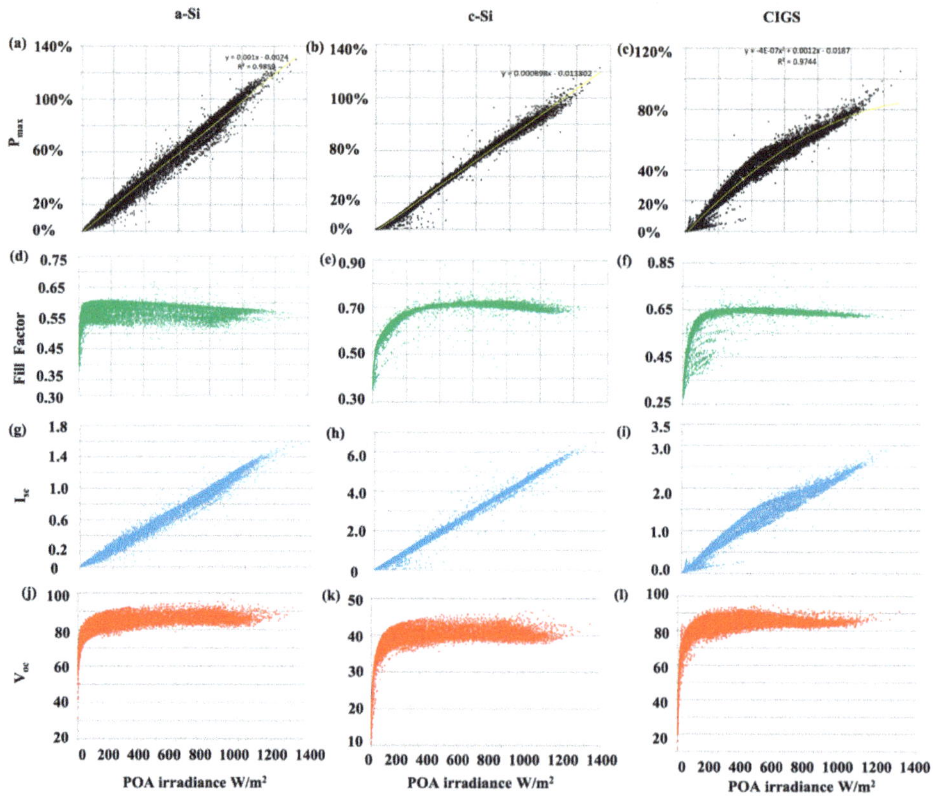

Figure 9.1. a-Si, c-Si and CIGS module (a)–(c) $P_{\max}$, (d)–(f) fill factor *FF*, (g)–(i) I_{sc} and (j)–(l) V_{oc} versus incident light intensity (POA irradiance) in the range of 0–1400 W m^{-2} in Taxila, Pakistan. The figure is regenerated from reference [1] with permission.

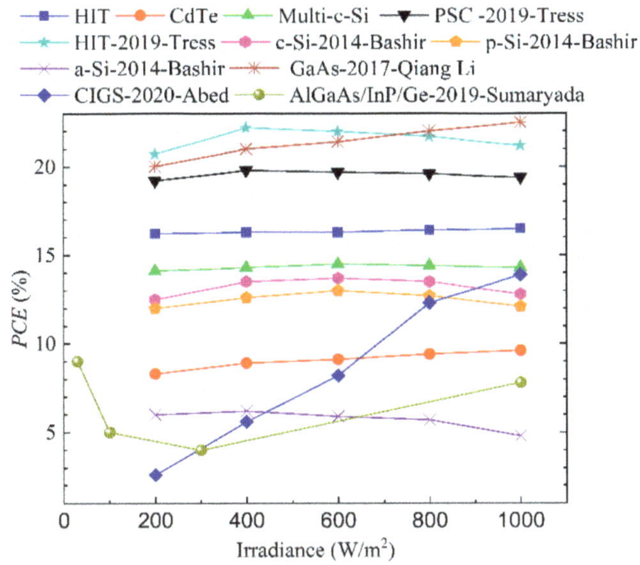

Figure 9.2. A comparison of *PCE* with irradiance for different solar cells; the data is extracted from reference [3–7].

constant value; the larger V_{oc} is ascribed to larger E_g of PSC. Besides the HIT and PSC devices, silicon-based devices (e.g., m-Si, p-Si and c-Si), especially in the case of a-Si solar cells, a decrease in *PCE* is observed in the irradiance range of 500–1000 $W \cdot m^{-2}$, in which most electricity is produced. However, for the thin film and III–V solar cells, such as GaAs, CIGS, CdTe and AlGaAs/InP/Ge, they have a higher efficiency at higher irradiance. It is also reported that tandem cells and multi-junction layer cells, such as perovskite/c-Si and AlGaAs/InP/Ge have higher *PCE* as the irradiance increases from 1 sun to ~70 suns, which is ascribed to their larger E_g and higher long-term thermal stability. Different light intensity not only changes the photo-generated current density, but also has a significant impact on the PV parameters, such as R_s, R_{sh}, *FF*, diode ideality factor *A* and reverse saturation current density I_0 [3].

9.2 Effect of humidity

This section was reproduced with permission from [48].

Besides the influence of irradiance, the exposure of PV modules to moisture will lead to oxidation and instability of materials and devices, leading to the degradation of the device eventually. Thus, the resistance test and influence of humidity are key factors for a successful performance of solar modules in outdoor conditions. The relative humidity of air (RH) which varies with respect to water vapor saturation pressure is mainly affected by the temperature. For PSCs, the degradation is ascribed to the hydrolysis reaction, in which the organic species in the perovskite decomposes and releases gas phase HI and CH_3NH_2 when reacting with water [8]. It has been observed that when the PV module is exposed to high RH conditions, the

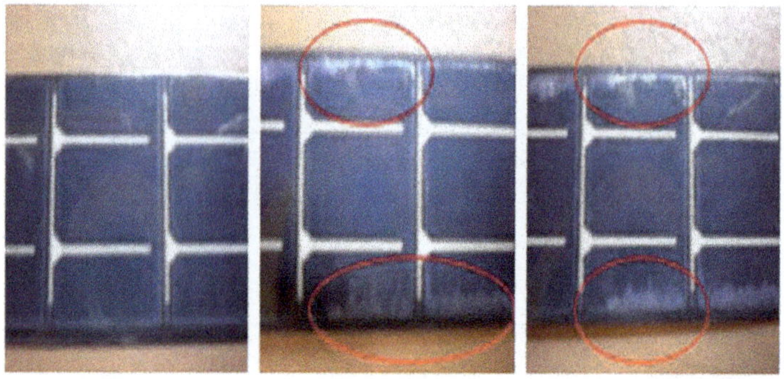

(a) Initial **(b) after 24 hr test** **(c) after 120 hr test**

Figure 9.3. The encapsulant delamination of solar cell device after 24, 120 h humidity test. The figure is regenerated from reference [9], copyright (2010) with permission from Elsevier.

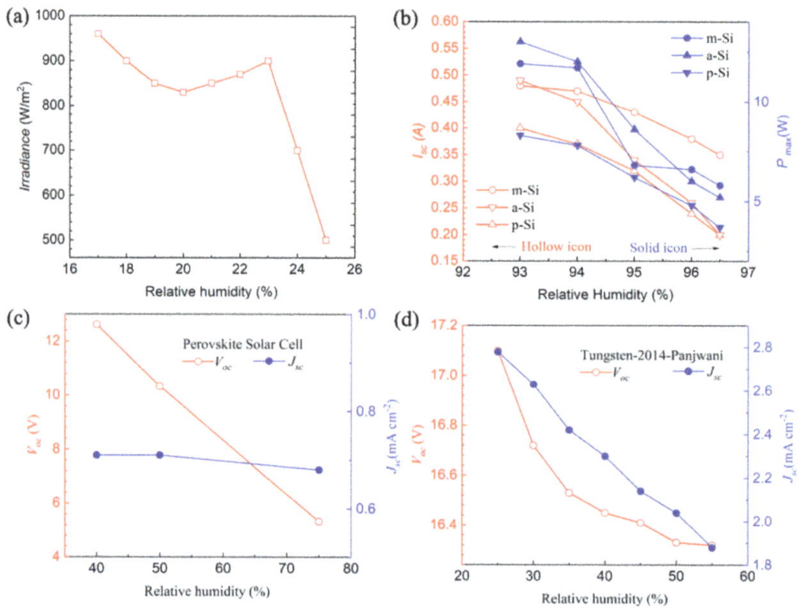

Figure 9.4. (a) Irradiance-RH relations at the low RH range for a-Si module, the figure is regenerated with permission from reference [9]. (b) I_{sc}-RH-$P_{\max}$ relation at high humidity for m-Si, a-Si and p-Si modules. The data is extracted from reference [14]. (c) V_{oc}-RH-J_{sc} for perovskite solar cells and (d) for tungsten [15]. Reproduced with permission from [48], copyright (2023) Lin *et al.*

water will penetrate into the cell which leads to delamination and corrosion of the encapsulant [9]; this is shown in figure 9.3. Panjwani *et al* [10] found that humidity reduces the utilization of solar energy from 70% to 55%–60% from the sun. Mekhilef *et al* [11] also reported a decrease in irradiance level with increasing relative humidity on a-Si module, as shown in figure 9.4(a). This may be due to the refraction and reflection of the incident light, and hence the change in irradiance from the water molecules. The PV modules exhibit high efficiency at low RH, which implies that high RH adversely affects the performance of PV parameters [12]. In figure 9.4(b), p-Si shows a sharper decrease than a-Si and m-Si in both I_{sc} and $P_{\max}$, which indicates that p-Si is much more sensitive to the variation in humidity. Besides the silicon-based PV modules, for the thin film solar cell and tungsten, PV parameters (e.g., $P_{\max}$, V_{oc} and J_{sc}) decrease with increasing RH, as shown in figures 9.4(c) and 9.4(d). The high RH also facilitates the growth of fungi and other algae, which will possibly accelerate the degradation of solar cell devices [13].

9.3 Influence of mechanical stress

The mechanical stability of PV modules is of great importance for installation and long-term utility in the field, especially in areas with heavy or continuous snow days. It is also inevitable for PV modules to be exposed to heavy loads and high pressure during transportation and strong wind through their lifetime. Moreover, during their assembly and production, the lamination and soldering of solar cells can cause cracks which will turn into security risks when PV modules experience large temperature cycling due to the differences in coefficient of thermal expansion of materials [16]. An example of the results of imaging of a stressed and degraded Si mini module is presented in figure 9.5 [17].

In order to address factors such as high load and weather conditions, several tests are designed to examine and certify the safety and quality of PV modules. These required certifications from various organizations have been described in detail in chapter 8. IEC 61215 includes many parameters that contribute to the aging of

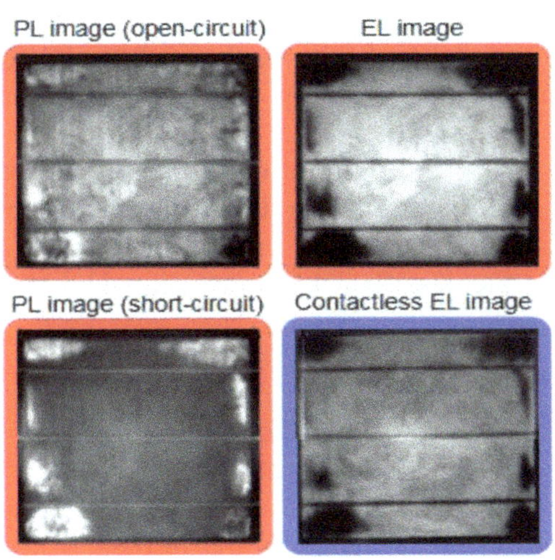

Figure 9.5. Imaging of a stressed and degraded Si mini module; PL images are collected in both open-circuit and short-circuit conditions. High resistivity areas (grid corrosion) are bright in PL and further highlighted bright in short-circuit. The EL image is collected by applying bias. This image was collected in a dark enclosure and is shown for comparison to the light- induced EL image below. Areas of high series resistance exhibit bright PL and dark EL. High carrier recombination regions are dark in both PL and EL. Reprocuced with permission from [17]. Courtesy: National Renewable Energy Laboratory.

terrestrial PV modules for long-term operation in open-air climates. It specifies the mechanical load tests to be performed on a module after damp heat exposure. Lee *et al* [16] reported the stress investigation method with uniform loading of 5400 Pa on silicon cells and non-tempered float glass by using finite element analysis in accordance with the regulation IEC 61215. The impact of snow and load during transportation and lamination can thus be thoroughly investigated, which provides insight into methods to improve the future designs to make the production of solar cells/panels and solar glass safer with high durability and enhanced life expectancy. Besides, the superposition of mechanical stress fields from the influence of production steps can also be studied by using finite element analysis [18]. Stress analysis of solar cells has become crucial nowadays with the advancements of manufacturing technologies and the increased role of IoT (Internet of Things) generation; researchers are able to make thinner wafers and thin films for different types of solar cells in different sizes [19].

The accurate understanding of wind pressure on PV modules is critical for the analysis of the performance of tilted solar panels [20]. The wind velocity topography changes with the altitude and angle; the wind load is very different for the panels at the ground level when compared to those that are installed on top of the roof [21]. Wind load includes the impact of vibration of the solar cells, in which, frequencies up to 14 Hz and an amplitude of 1.6 mm leads to a deflection ramp of $\sim$5200 mm min^{-1} [22]. The roughness factor and orography for different areas (urban, rural) changes the wind turbulence accordingly. Thus, it is necessary to have *in situ* measurement on the pressure loading for PV modules and the simulation models to help determine the proper position with ideal tilt angle for their installation. Besides, one can estimate and monitor the moment of structural damage and delamination inside the PV module. In order to determine the structural deformation in solar cells, Laha *et al* [23] proposed a methodology by using a shear stress transport model, based on a 3D Reynolds-averaged Navier–Stokes algorithm, with a wobbly solver. The results show that the structural deformation increases rapidly with the increase in wind pressure for silicon solar cells. The same methodology can be used for determining the structural deformation in third generation solar cells or tandem solar cells to be able to increase their life expectancy [23].

9.4 Influence of temperature

The influence of temperature on the performance efficiency of solar cells/modules has been reported extensively in the literature [24, 25]. The study of the impact of temperature on solar cell performance is important since, in terrestrial applications, solar cells are generally exposed to temperatures ranging from 15 °C to 50 °C, and to even higher (and lower) temperatures in space (+120 °C to −170 °C) and concentrator systems [26]. A comparison of real-word efficiency of commercial, space-qualified solar cells and research cells is presented in figure 9.6 [27]. As can be seen in the figure, the efficiency of commercial cells is $\sim$33% while that of research cells is $\sim$48%. Much work needs to be done in order to improve the efficiency of research cells to be compatible with the standards set forth by the space PV industry.

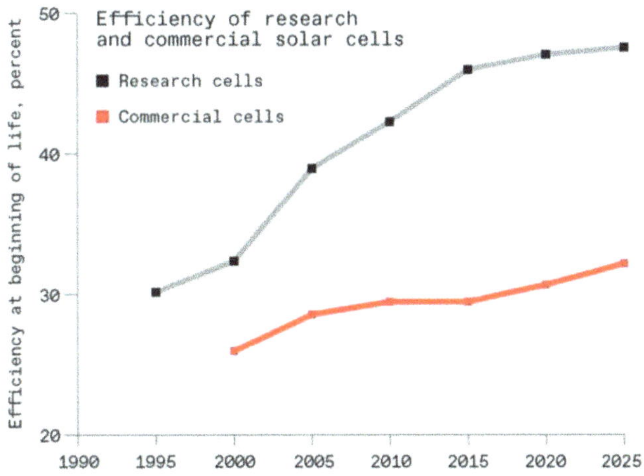

Figure 9.6. Real-world efficiency of commercial, space-qualified solar cells and research cells. (2024) IEEE, reprinted with permission from [27] .

For example, for high temperature near-sun operations, NASA requires [28, 29] that solar panels must have improved efficiency at high temperatures (for Mercury Orbiter, a sun-facing solar array with no reflectors, equilibrium temperature ≈450 °C) as well as improved lifetime at high temperatures. This is in addition to the preference for semiconductors with direct bandgap, improved bandgap tailoring to match the solar spectrum, superior tolerance to charged particle radiation and high temperature.

In the late 1950s, silicon solar cells were used to power satellites in space. Soviet space station, MIR launched in 1986, was equipped with 10 kW GaAs solar cells. Since then, multi-junction III–V compound semiconductor solar cells have replaced silicon p-n junctions for space solar cells. GaAs has several significant advantages over silicon for solar cell applications in space as well as in concentrator photovoltaic systems. Being a direct bandgap semiconductor, GaAs offers a high-efficiency, lightweight thin film PV option with a low temperature coefficient [30, 31] and excellent radiation hardness. An illustration of the comparison of the absorption coefficient of various semiconductors, as a function of wavelength, is presented in figure 9.7 [32].

Concentrated solar systems are particularly interesting from the perspective of the similarities in the influence of temperature on solar panel performance in space applications. An example of the physical schematic of a monolithic triple junction n-on-p solar cell deposited epitaxially on a substrate and (b) the electrical circuit equivalent diagram showing top, middle and bottom junction diodes and interconnecting upper and lower tunnel junctions, is shown in figure 9.8 [33].

Equation (9.1) illustrates the change in solar panel efficiency with temperature. The normalized temperature coefficient, $(1/\eta)(d\eta/dT)$, of solar panels is given by [27, 29]:

$$(1/\eta)(d\eta/dT) = (1/V_{oc})\,(dV_{oc}/dT) + (1/J_{sc})\,(dJ_{sc}/dT) + (1/FF)\,(dFF/dT) \qquad (9.1)$$

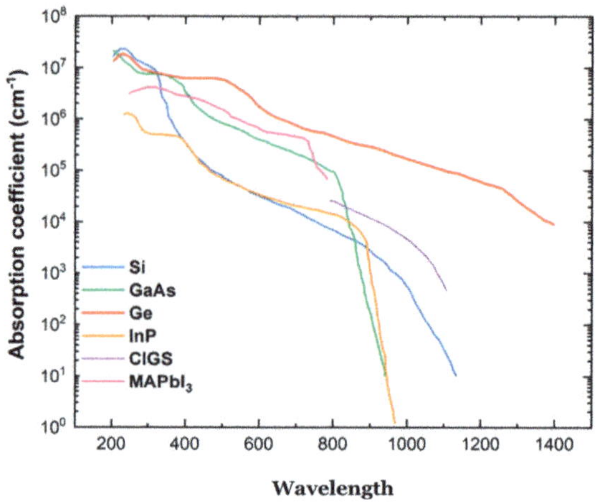

Figure 9.7. Comparison of the absorption coefficient versus wavelength for various semiconductors of interest to photovoltaics. Reproduced from [32] CC BY 4.0.

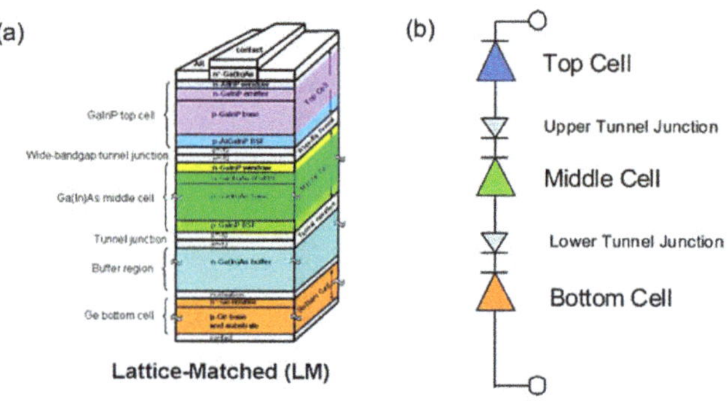

Figure. 9.8. Example of physical schematic of a monolithic triple junction n-on-p solar cell deposited epitaxially on a substrate. (b) The electrical circuit equivalent diagram showing top, middle and bottom junction diodes and interconnecting upper and lower tunnel junctions. Reproduced from [33] with permission from the Royal Society of Chemistry.

It has been shown earlier that V_{oc} decreases with increasing temperature *T*, whereas due to decreasing bandgap, J_{sc} increases slightly with increasing *T* [6, 24, 25] The effect of temperature on the solar cell performance has been investigated in detail from 273 K–523 K, in figure 9.9, with various material types of solar cells, such as Si, Ge, GaAs, CdTe, CdS and InP [24]. *FF* is calculated using the Green's formula with the values of measured $V_{oc.}$ The drastic impact of temperature on the *FF* can be observed in Ge cell, whereas the *FF* of CdS is the least sensitive to the change in temperature [24].

In the context of the development of high-temperature solar cells, the theoretical efficiency of a solar cell, as a function of bandgap, for temperatures in the range of 27 °C to 900 °C, is shown in figure 9.10 [29]. As can be seen in the figure, the optimal

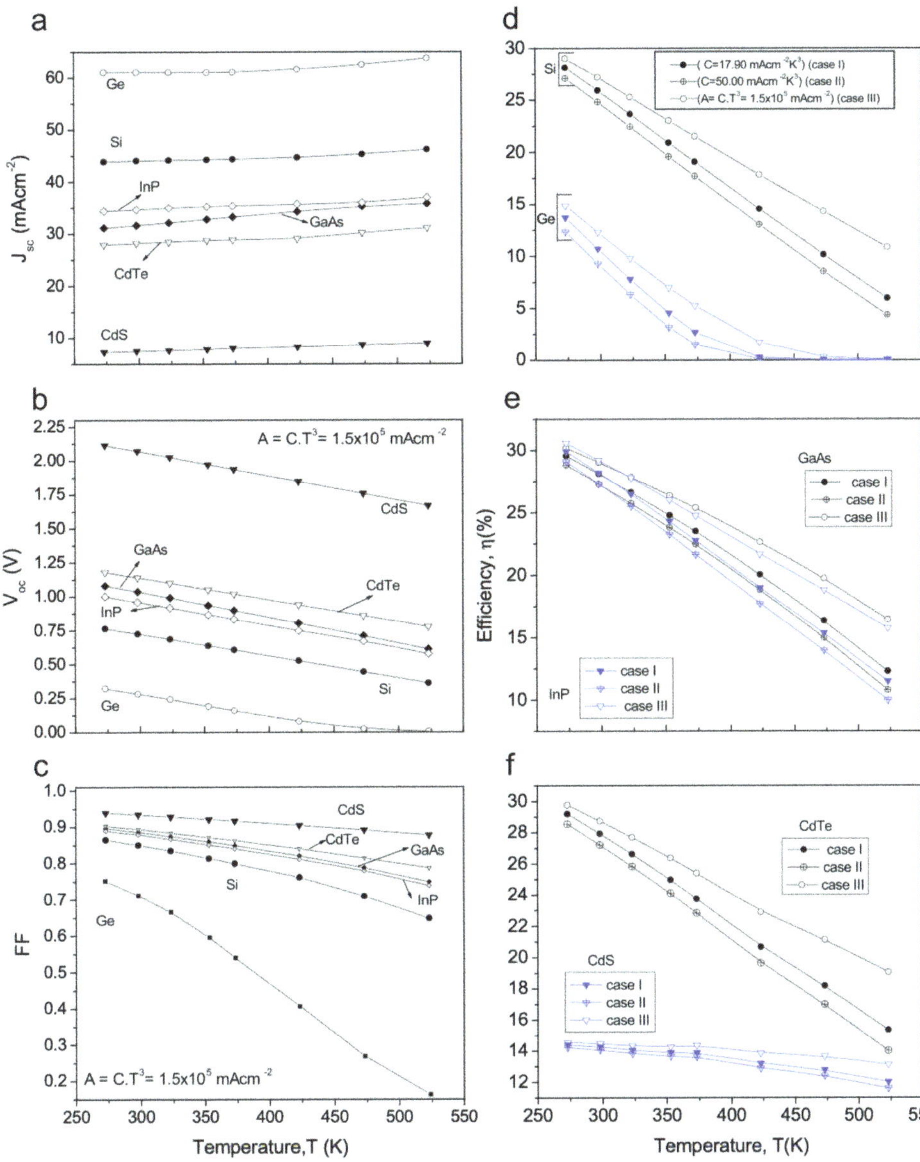

Figure 9.9. (a) J_{sc}, (b) V_{oc}, (c) FF and (d–f) efficiency for Ge, Si GaAs, CdTe, CdS and InP in the temperature range of 273 K–523 K. Reprinted from [24], copyright (2012) with permission from Elsevier.

bandgap for photovoltaic energy conversion shifts from 1.4 eV at room temperature (27 °C) to ~2.3 eV at 900 °C [29].

An example of the efficiency losses in solar panels, as a function of temperature, is presented in figure 9.11 [34]. As can be seen in the figure, the efficiency losses increase with increase in temperature.

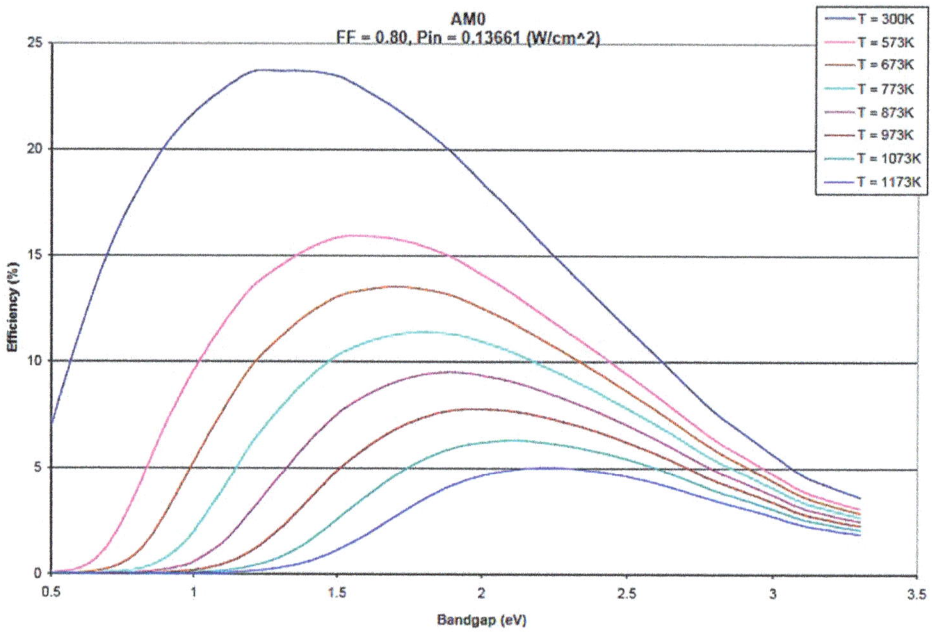

Figure 9.10. Theoretical efficiency of a solar cell as a function of bandgap, showing the shift in optimum bandgap for photovoltaic conversion from about 1.4 eV at room temperature (27 °C) to about 2.3 eV at 900 °C. Reproduced with permission from [29].

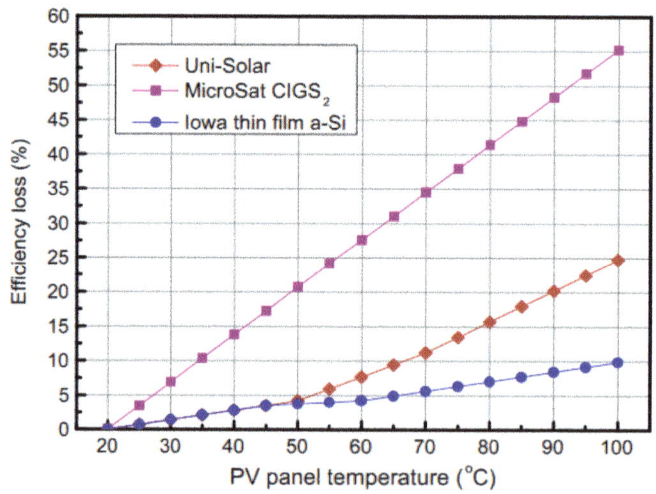

Figure 9.11. Efficiency losses as function of temperature in solar panels. Reprinted from [34], copyright (2016) with permission from Elsevier.

With a view to address the thermal management of concentrator photovoltaics as well as make solar PV technology affordable and cost-effective, a research team at IBM has proposed a high-concentration PV thermal system based on water-cooled solar panels [35]. Earlier, a team from IBM utilized a

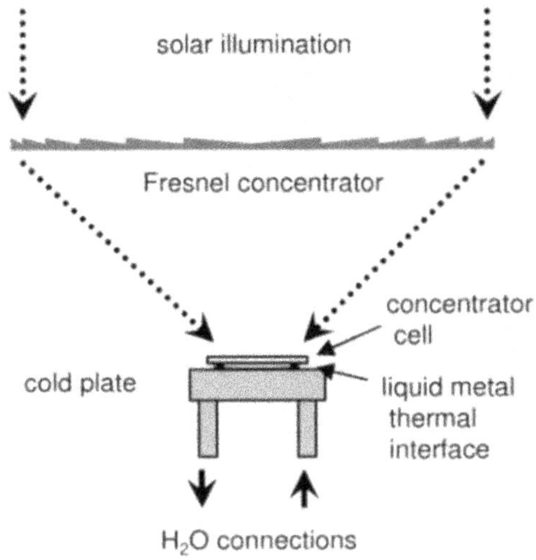

Figure 9.12. Schematic of test arrangement comprising a 2500 sun Fresnel concentrator, Spectrolab 3JT cell interfaced with liquid metal to a Mikros mini channel water-cooled plate. Copyright 2008, IEEE. Adapted with permission from [36].

liquid metal thermal interface with passive cooling for enhancing the PV operation of high-concentration systems beyond 2000 suns [36]. This is illustrated in figure 9.12 [36].

In recent years, PSCs have attracted significant interest among various solar cell types due to their unique optoelectronic properties. The effect of temperature on the performance of PSCs is of importance in order to obtain thermally stable and high efficiency PV systems for wide applicability from ocean to ground to even space. Meng *et al* [37] investigated the thermal stability of the mixed cation organic–inorganic lead halide perovskites $(FAPbI_3)_{1-x}MAPb(Br_{3-y}Cl_y)_x$ films from 298 K–523 K to reveal the performance parameters in such a wide temperature range [37]. Their results show that high temperatures can cause the decomposition of these perovskites into PbI_2 and introduce defects in the structures, especially when the temperature is higher than 150 °C. This formed PbI_2 as well as the defects hinder the transfer of carriers and bring aggravation in the form of electron–hole recombination that causes a sharp decay in device performance. However, the fabrication of PSCs almost invariably involves high-temperature annealing to eliminate the defects and cracks [38, 39]. Thus, it is also critical to understand the impact of anneal temperature on the performance of PSCs. As shown in figure 9.13(a), the perovskites gradually transform from the intermediate phase to the perovskite α phase during heat treatment. Figure 9.13(b) shows that the grain size is small at 100 °C, and increases with increasing annealing temperature and turns into sheet of grains when the temperature reaches 140 °C [40].

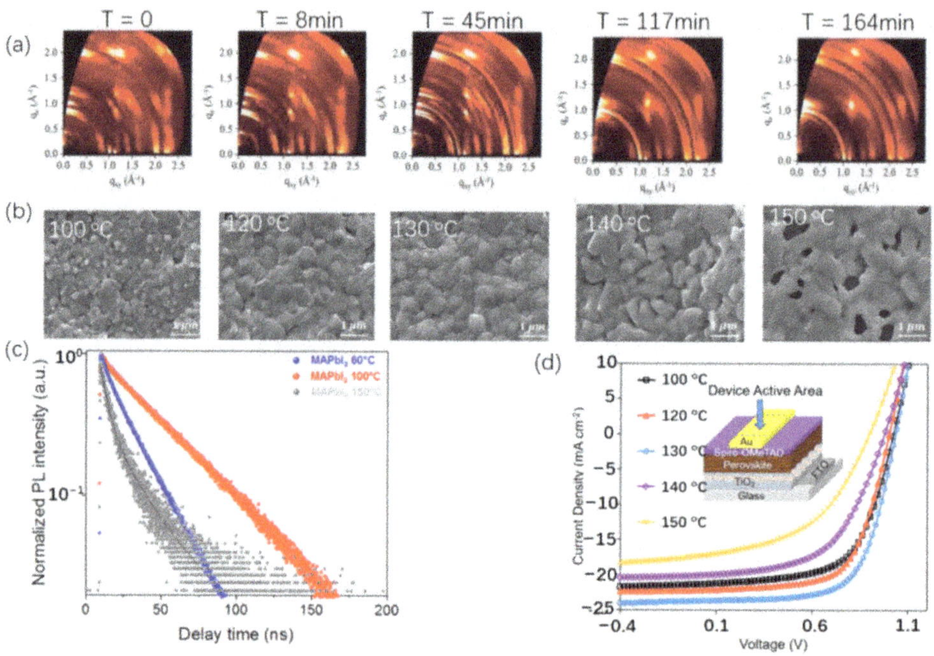

Figure 9.13. (a) The annealing process of PSCs at 80 °C with recording time from 8–164 min. (b) grain size evolution when the annealing temperature is changed from 100–150 °C. (c) and (d) the evolution of photoluminescence (PL) decay and *I–V* curves at different temperatures. Reprinted from [40] with the permission for AIP Publishing.

Moreover, with the advancement in technologies, flexible substrates with PSCs provide potential applications for wearable electronics such as cellphones, biomedical sensors and laptops. Flexible PSCs usually utilize polymers such as PEN (polyethylene naphthalate), PET (polyethylene terephthalate), PI (polyimide) and PDMS (polydimethylsiloxane) which have poor performance at high temperatures but exhibit good mechanical bending stability and good conformability for wearable electronics [41]. In figure 9.14, a comparison of several flexible substrates with traditional metal thin film electrodes is presented [42]. The first paper on PSCs with a structure of $CH_3NH_3PbI_3$/Spiro-OMeTAD/ MoOx/Au/MoOx reported a *PCE* of 2.7%. This led to significant interest and curiosity for continued development in this area of research [43]. The critical challenge for attaining high performance with flexible PSCs is limited to a tolerable temperature (restricting to 150 °C) [44, 45]. Thus, many approaches have been proposed to address these challenges in the literature without sacrificing too much solar cell performance. A *PCE* of 17.9% for flexible PSCs based on $FAPbI_{3-x}Br_x$ was obtained by employing toluene as the anti-solvent with fabrication temperature below 150 °C [46]. A new method proposes a strategy for replacing the usual anti-solvents with CH_3NH_3Br ($CH_3NH_3^+$, MA). The resulting perovskite precursor significantly reduces the formation energy of α-phase FA-perovskites to 40 °C [47].

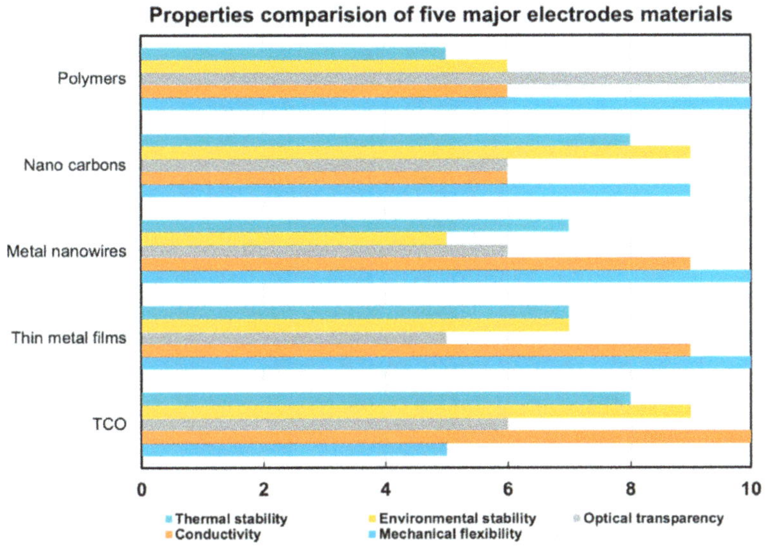

Figure 9.14. Comparison of relative physical properties of various electrode materials. The numerical number of the bar length indicates poor (5) to excellent (10) in meeting the requirements for a flexible transparent electrode. Reprinted from [42], copyright (2021) with permission from Elsevier.

9.5 Conclusion

A study of the effects of irradiance, temperature, humidity, and mechanical stress on solar cell/panel performance provides valuable insights for advancing the efficiency and durability of PV systems. The actual irradiance flux varies throughout the day. V_{oc} increases with the increasing irradiance due to high photo-generated current density. Besides the influence of irradiance, PV modules react with water which will lead to oxidation and instability in materials and cells, leading to the degradation of the device eventually. The relative humidity of air (RH) which varies with respect to water vapor saturation pressure is mainly affected by the temperature, where V_{oc} decreases with increasing T whereas J_{sc} increases slightly with increasing T. Additionally, the study emphasizes the significance of mechanical stress considerations in the design and deployment of solar PV systems. Structural integrity, vibration resistance, and mechanical robustness are paramount for ensuring the longevity and reliability of solar installations, particularly in harsh environmental conditions or during transportation and installation processes. A brief introduction to solar cells in space and concentrator systems is presented.

References

[1] Buday M S 2011 Measuring Irradiance, Temperature, and Angle Incidence Effects on Photovoltaic Modules in Auburn Hills, Michigan *Master's Practicum* University of Michigan

[2] Hishikawa Y *et al* 2013 *IEEE 39th Photovoltaic Specialists Conf. (PVSC)* pp 1417–22

[3] Li Q *et al* 2017 Comparative study of GaAs and CdTe solar cell performance under low-intensity light irradiance *Sol. Energy* **157** 216–26

[4] Bashir M A, Ali H M, Khalil S, Ali M and Siddiqui A M 2014 Comparison of performance measurements of photovoltaic modules during winter months in Taxila, Pakistan *Int. J. Photoenergy* **2014** 898414

[5] Sumaryada T, Rohaeni S, Damayanti N E, Syafutra H and Hardhienata H 2019 Simulating the performance of $Al_{0.3}Ga_{0.7}As$/InP/Ge multijunction solar cells under variation of spectral irradiance and temperature *Model. Simul. Eng.* **2019** 5090981

[6] Tress W *et al* 2019 Performance of perovskite solar cells under simulated temperature-illumination real-world operating conditions *Nat. Energy* **4** 568–74

[7] Abed A, Hussain H H and Kasim N 2020 Efficiency and performance improvement via using optical reflectors of on-grid CIGS PV solar system *Karbala Int. J. Modern Sci.* **6** 5

[8] Song Z *et al* 2016 Perovskite solar cell stability in humid air: partially reversible phase transitions in the PbI_2-CH_3NH_3I-H_2O system *Adv. Energy Mater.* **6** 1600846

[9] Tan C M, Chen B K E and Toh K P 2010 Humidity study of a-Si PV cell *Microelectron. Reliab.* **50** 1871–4

[10] Manoj Kumar Panjwani G B N 2014 Effect of humidity on the efficiency of solar cell (photovoltaic) *Int. J. Eng. Res. Gen. Sci.* **2** 499–503

[11] Mekhilef S, Saidur R and Kamalisarvestani M 2012 Effect of dust, humidity and air velocity on efficiency of photovoltaic cells *Renew. Sustain. Energy Rev.* **16** 2920–5

[12] Rahman M M, Hasanuzzaman M and Rahim N A 2015 Effects of various parameters on PV-module power and efficiency *Energy Convers. Manage.* **103** 348–58

[13] Tariq Ahmed Hamdi R, Abdulhadi S, Kazem H A and Chaichan M 2018 Humidity impact on photovoltaic cells performance: a review *Int. J. Rec. Eng. Res. Dev.* **3** 27–37

[14] Chaichan M 2012 Effect of humidity on the PV performance in Oman *Asian Trans. Eng.* **2** 29–32

[15] Lin L and Ravindra N 2020 Temperature dependence of CIGS and perovskite solar cell performance: an overview *SN Appl. Sci.* **2** 1361

[16] Lee Y and Tay A A O 2013 Stress analysis of silicon wafer-based photovoltaic modules under IEC 61215 mechanical load test *Energy Procedia* **33** 265–71

[17] Johnston S and Silverman T 2015 *Photoluminescence and Electroluminescence Outdoor Module Imaging* (Golden, CO: National Renewable Energy Laboratory, 2015 PV Module Reliability Workshop) https://nrel.gov/docs/fy15osti/64438.pdf (accessed 17 December 2024)

[18] Dietrich S, Sander M, Pander M and Ebert M 2012 Interdependency of mechanical failure rate of encapsulated solar cells and module design parameters *Proc. of SPIE—The Int. Society for Optical Engineering* 8472 84720P

[19] Pingel S, Zemen Y-B, Frank O, Geipel T and Berghold J 2009 Mechanical stability of solar cells within solar panels *Proc. 24th European Photovoltaic Energy Conf.*

[20] Bender W, Waytuck D, Wang S and Reed D A 2018 In situ measurement of wind pressure loadings on pedestal style rooftop photovoltaic panels *Eng. Struct.* **163** 281–93

[21] Grand View Research I 2022 Concentrated Solar Power Market Size, Share and Trends Analysis Report by Application (Utility, EOR, Desalination), by Technology, by Region, and Segment Forecasts, 2020–2027 (https://grandviewresearch.com/industry-analysis/concentrated-solar-power-csp-market).

[22] Assmus M and Köhl M 2012 Experimental investigation of the mechanical behavior of photovoltaic modules at defined inflow conditions *J. Photon. Energy* **2** 022002

[23] Laha S K *et al* 2021 Analysis of mechanical stress and structural deformation on a solar photovoltaic panel through various wind loads *Microsyst. Technol.* **27** 3465–74

[24] Singh P and Ravindra N 2012 Temperature dependence of solar cell performance—an analysis *Sol. Energy Mater. Sol. Cells* **101** 36–45

[25] Singh P, Singh S N, Lal M and Husain M 2008 Temperature dependence of I–V characteristics and performance parameters of silicon solar cell *Sol. Energy Mater. Sol. Cells* **92** 1611–6

[26] Almehisni R and Al Naimat F 2018 Heat transfer influence of solar panel on spacecraft *2018 Advances in Science and Engineering Technology Int. Conferences (ASET)*

[27] Barde H 2024 Castles in the Sky? A skeptic's take on beaming power to earth from space *IEEE Spectr.* **61** 22–9

[28] Landis G A and Bailey S 2002 Photovoltaic power for future NASA missions *Paper AIAA-2002–0718, AIAA 40th Aerospace Sciences Meeting (Reno, Nevada, January 14–17)*

[29] Landis G A, Merritt D, Raffaelle R P and Scheimen D 2005 High-temperature solar cell development *NASA/CP-2005–213431* NASA 241–7

[30] Landis G A 1994 Review of solar celltemperature coefficients for space *Proc. XIII Space Photovoltaic Research and Technology Conf. (June 1994)* (NASACP-3278, NASA Lewis Research Center) pp 385–400

[31] Silverman T J *et al* 2013 Outdoor performance of a thin film gallium arsenide photovoltaic module, NREL/CP-5200–57902 *Presented at the 39th IEEE Photovoltaic Specialists Conf. (Tampa, Florida, June 16–21)*

[32] Verduci R *et al* 2022 Solar energy in space applications: review and technology perspectives *Adv. Energy Mater.* **12** 2200125

[33] Cotal H *et al* 2009 III–V multijunction solar cells for concentrating photovoltaics *Energy Environ. Sci.* **2** 174–92

[34] Du Y *et al* 2016 Evaluation of photovoltaic panel temperature in realistic scenarios *Energy Convers. Manage.* **108** 60–7

[35] IBM 2014 'Sunflower' solar concentrator produces energy and hot water https://edn.com/ibm-sunflower-solar-concentrator-produces-energy-and-hot-water/ (accessed 6 July 2024)

[36] Van Kessel T G, Martin Y C, Sandstrom R L and Guha S 2008 Extending photovoltaic operation beyond 2000 suns using a liquid metal thermal interface with passive cooling *2008 33rd IEEE Photovoltaic Specialists Conf. (San Diego, CA)*

[37] Meng Q *et al* 2021 Effect of temperature on the performance of perovskite solar cells *J. Mater. Sci., Mater. Electron.* **32** 12784–92

[38] Wu S, Li C, Lien S Y and Gao P 2024 Temperature matters: enhancing performance and stability of perovskite solar cells through advanced annealing methods *Chemistry* **6** 207–36

[39] Yu X *et al* 2020 Investigation on low-temperature annealing process of solution-processed TiO_2 electron transport layer for flexible perovskite solar cell *Materials (Basel)* **13** 1031

[40] Oyewole D O *et al* 2021 Annealing effects on interdiffusion in layered FA-rich perovskite solar cells *AIP Adv.* **11** 065327

[41] Fukuda K, Yu K and Someya T 2020 The future of flexible organic solar cells *Adv. Energy Mater.* **10** 2000765

[42] Li X *et al* 2021 Review and perspective of materials for flexible solar cells *Mater. Rep.: Energy* **1** 100001

[43] Castro-Hermosa S, Dagar J, Marsella A and Brown T 2017 Perovskite solar cells on paper and the role of substrates and electrodes on performance *IEEE Electron Device Lett.* **38** 1278–81

[44] Wang C *et al* 2017 Compositional and morphological engineering of mixed cation perovskite films for highly efficient planar and flexible solar cells with reduced hysteresis *Nano Energy* **35** 223–32

[45] Zhong M *et al* 2019 Highly efficient flexible MAPbI 3 solar cells with a fullerene derivative-modified SnO_2 layer as the electron transport layer *J. Mater. Chem.* A **7** 6659–64

[46] Heo J H *et al* 2017 Highly flexible, high-performance perovskite solar cells with adhesion promoted AuCl3-doped graphene electrodes *J. Mater. Chem.* A **5** 21146–52

[47] Deng W *et al* 2020 Anti-solvent free fabrication of FA-based perovskite at low temperature towards to high performance flexible perovskite solar cells *Nano Energy* **70** 104505

[48] Lin L, Bora B, Prasad B, Sastry OS, Mondal S and Ravindra N M 2023 Influence of outdoor conditions on PV module performance – an overview *Mater. Sci. Eng. Int. J.* **7** 88–101

Chapter 10

Approaches to hotspot reduction and shade loss minimization

Shadow-induced performance loss in solar modules is a critical consideration in its application as power sources, particularly in satellites, although it is also very important for terrestrial applications. Hotspots are areas of elevated temperature localized within specific areas of a photovoltaic (PV) module that lead to substantial increases in the temperature of solar cells and consequently decreasing their efficiency and overall performance. This chapter explores strategies for minimizing hotspots and shade-induced losses in photovoltaic systems. Various approaches that are being practiced in the PV industry are discussed to address these challenges, aiming to enhance system reliability, efficiency and overall performance.

10.1 Shaded cell loss and hotspot related challenges

In addition to the cell performance loss from weather-related conditions, as discussed in the previous chapter, it is also critical to investigate the performance and stability of solar modules, especially due to shadow-induced challenges [1]. Shadow loss effect in solar modules is an important consideration in its applications as power sources in satellites. In real-world conditions, several solar cells, among the entire PV module, are shaded from fallen leaves, non-uniform illumination, dust accumulation, animal activity, buildings, antennas, trees and other surrounding obstacles, leading to non-uniform power generation, or when there is intrinsic difference between each solar cell in the panel or microcracks in the cell etc [1]. The amount of power loss is a function of the size and shape of the shadow, the geometrical and electrical layout of the cells in the string, and the way the shadow falls across the particular solar cell module [1, 2]. A PV module is a combination of:

doi:10.1088/978-0-7503-5994-8ch10

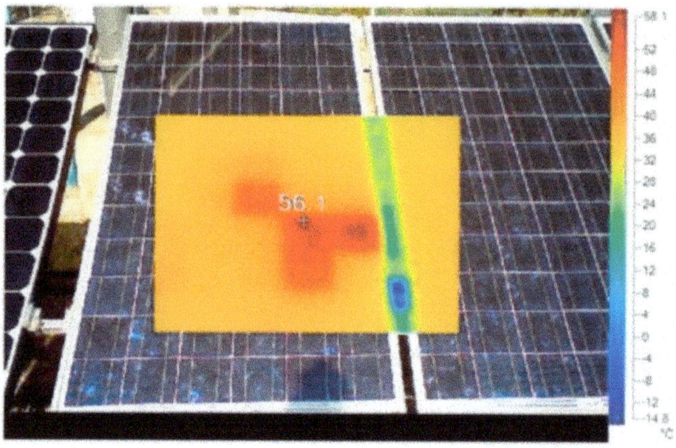

Figure 10.1. Example of a PV module that is affected by multiple hotspots. Reprinted from [4], copyright (2019), with permission from Elsevier .

(a) a series-connected string of cells for enhanced voltage, and (b) a parallel-connected string of cells for enhanced current and hence maximizing power. However, in crystalline/wafer-based solar modules, cells are connected in series, limiting thereby the output current of the module while minimizing Joule losses in associated cables and power converters. All the cells in the series- connected string must conduct the same amount of current. The PV module system, connected in series, carries the same current even when some of the solar cells are shaded; therefore, the shaded cells behave like reverse-bias diodes.

Hotspots are areas of elevated temperature localized within specific areas of a PV module that lead to substantial increases in the temperature of solar cells, consequently leading to decreasing efficiency [3]. Figure 10.1 illustrates a PV module in operation with four cells experiencing multiple hotspots with an intense localized peak temperature of 56.1 °C [4]. Hotspots may occur in a PV module when the solar cells are mismatched or with defects or microcracks, or especially when the cells are partially shaded [5]. Moreover, hotspot-induced heating occurs when there is one or several shaded cells among a large number of cells in the string, which causes reverse bias across the shaded cell. This phenomenon contributes to large power dissipation in the shaded cells. When a solar cell reaches the breakdown voltage, the cell is damaged permanently. In general, crystalline silicon solar cells have a breakdown voltage in the range of −10 to −30V.

IEC 61215 defines and verifies the hotspot tolerance of a solar module by a hotspot endurance test. Hotspot temperature of opaque PV modules is higher than semitransparent cells by 2 °C–3 °C under standard test conditions, which decreases with an increment in the numbers and areas of hotspots [6]. Zhang *et al* [7] have investigated the reasons for degradation and failure of PV modules of about 200 MWp that had been running in the United States for 1–3 years. Figure 10.2 shows a total of 115 defective PV modules that were observed, of which 22% were due to cell hotspots [7].

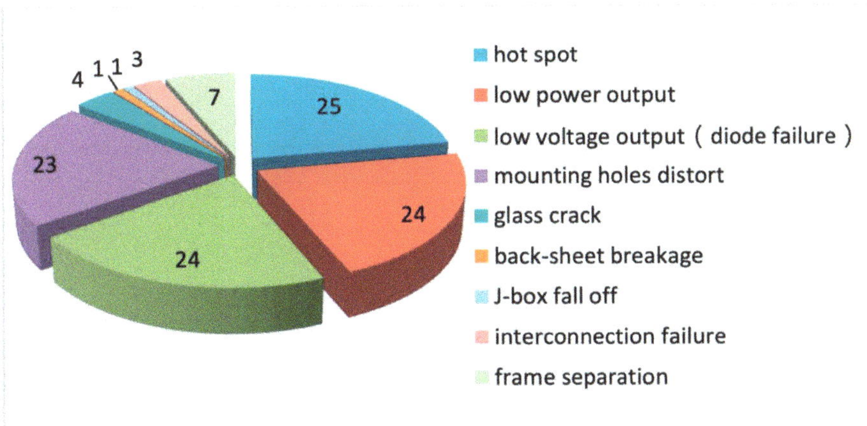

Figure 10.2. The short-term failure distribution of solar modules in the US. Reprinted from [5], copyright (2017) with permission from Elsevier.

The solar cells have a breakdown voltage where the current in the cell leads to reverse bias. Silicon cells generally break down in reverse bias by avalanche breakdown (magnitude > 15 V), whereas CIGS and CdTe exhibit breakdown voltage of magnitude < 10 V and a decrease in breakdown voltage under illumination [8]. Perovskite solar cells (PSCs) have much lower breakdown (reverse bias) than crystalline silicon and inorganic thin-film solar cells, wherein reverse bias occurs between −1 and −4 V, as shown in figure 10.3 [8]. By utilizing experimental methods, Yang *et al* [9] proposed that (reverse) current should be less than 1.0 A for monocrystalline-Si solar cells which corresponds to a reverse bias of less than −12 V. Bowring *et al* [8] reported the mechanism of reverse bias degradation for PSCs in the dark (long time shading), in which current flows in reverse bias most likely due to tunneling mediated by mobile ions. Both the series resistance, R_s, and breakdown voltage increase and V_{oc} decrease under this condition. However, even this degradation in efficiency is recoverable; it takes a much longer time to recover after each shading event. As a result, the study proposed two options to address the shading degradation problems; one approach involves designing thin cells that break down at lower voltage to reduce power dissipation. Another approach is to design cells with larger breakdown voltage with bypass diodes.

10.2 Minimization and prevention of shaded cell and hotspot loss

Investigation of the mitigating strategies for the hotspots are of importance in the PV module. Recent state-of-the-art technologies, that are focused on the investigation of the effect of shade loss and hotspot related problems, are categorized into the following types:

1. Partial shading, which revisits and estimates the effect of shading on the PV module;

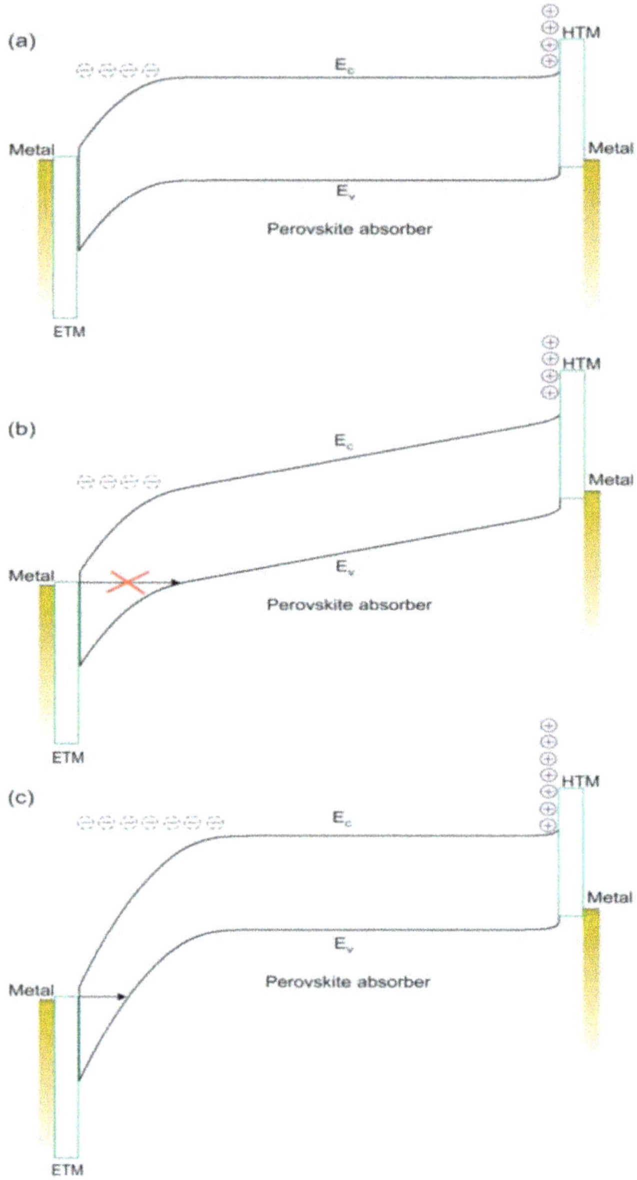

Figure 10.3. Band diagrams of the PSC (a) in dark equilibrium at 0 V (b) right after applying a reverse bias (−1 V), and (c) after ions have migrated to the contacts to equilibrate with the applied reverse bias. Reuse from [8] John Wiley & Sons. Copyright 2017 WILEY-VCH Verlag GmbH & Co. KGaA, Weinheim.

2. Field tests, which analyze the performance of PV module under real conditions in a specific period;
3. Bypass diode topology, which investigates the impact of diode arrangements within the solar cell module;

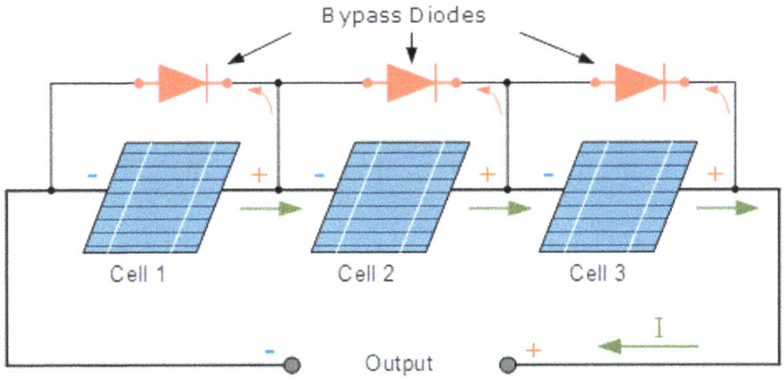

Figure 10.4. An illustration of bypass diode protection device connected in parallel with each of the cells. Reproduced with permission from [13] Alternative Energy Tutorials, copyright (2010–2024) all rights reserved.

4. Artificial intelligence (AI), which models and predicts the performance of solar module under specific conditions and new mitigation technologies, searches for new technologies to mitigate or solve the shading problem.

PV hotspots can be easily detected using infrared inspection, which has become a common practice in current PV applications in the industry [10]. Advancing beyond visual techniques, recent research has explored the application of machine learning (ML) algorithms capable of detecting hotspots through the analysis of electrical performance metrics, including output voltage, current, and dynamic series resistance [11]. Recent advancements in PV system technology persist in tackling the ongoing issue of hotspots, which notably impact both module reliability and output performance.

In order to prevent the hotspot failure, a passive bypass diode protection device is mounted on the solar cell module. The passive technique refers to the connection of a PV panel with the bypass diode in parallel. The use of bypass diodes across PV strings is a standard practice that is required in crystalline silicon PV panels [12]. This is illustrated in figure 10.4 [13].

Under the condition of shading, the cell in the middle in figure 10.5 [13] behaves as a load and forward biases the bypass diode that is connected in parallel resulting in letting the current from the two good cells to flow; this is illustrated in figure 10.5 [13].

Differential power processing (DPP) architectures are another type of power converters, that are being increasingly used in PV systems to lower the mismatch effect [14, 15]. DPP converters solely manage the mismatched power within PV modules, resulting in low power losses compared to alternative architectures. Bypass diode is the primary method to address the shade loss problem, providing the bypass path for the current flow; thus, the generated power will not be affected by the shaded cell. However, the activated bypass diode causes the inevitable voltage drop and consumes the generated power. Thus, it affects the maximum power delivered by the solar cell modules [16]. Mostly, the bypass diodes should have forward voltage as low as possible (e.g., Schottky diode). However, a lower forward voltage means a higher

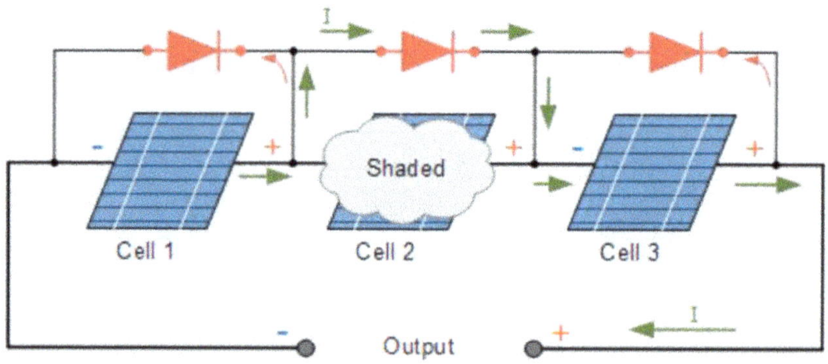

Figure 10.5. An illustration of bypass diode protection device connected in parallel with each of the cells. The shaded cell in the center becomes a load and switches on the parallel-connected diode permitting current flow from the two good diodes that are not in shadow. Reproduced with permission from [13] Alternative Energy Tutorials, copyright (2010–2024) all rights reserved.

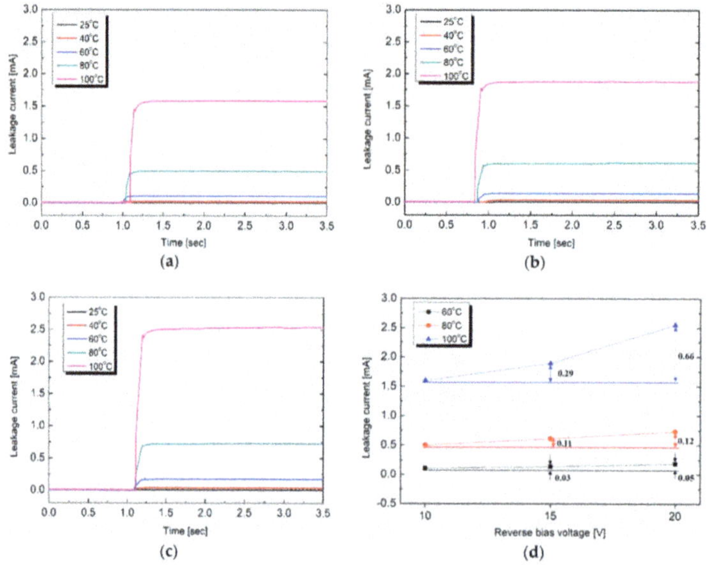

Figure 10.6. Leakage current of a bypass diode at various temperatures and reverse bias voltages: (a) 10 V, (b) 15 V, and (c) 20 V. (d) The leakage current of the bypass diode as a function of reverse bias voltage with different surrounding temperatures. (Reuse with permission open access) [18] CC BY 3.0.

leakage current, raising the risk of thermal runaway [17]. Thus, the bypass diodes must conduct when one cell is shaded, and the shaded cell must be under its breakdown voltage [17]. In 2018, Shin *et al* [18] studied the effect of operating conditions on the photovoltaic parameters of installed bypass diodes on the back side of PV modules in a closed junction box. This is shown in figure 10.6 [18]. This result illustrates the ability to monitor heat dissipation and leakage current under high temperature operation of solar panels when designing a bypass diode junction box.

The configuration of bypass diodes on the PV modules forming part of the array has an important influence on the possibility of hotspot presence [19]. In order to determine the shading pattern of PV modules, experimental investigations are needed on a given array size under different shaded conditions. In 2017, Bana *et al* [20] investigated the effect of uniform shading on m-Si PV module to obtain the output performance; they analyzed the results under real operating conditions for possible shading scenarios. Prevention is better than cure; thus, it is necessary to establish a flexible inspection methodology, follow up with cleaning procedures and utilize data science technologies to reduce the risk of fire under various weather conditions. In 2018, Kaid *et al* [21] proposed a surveillance approach based on the failure diagnosis of PV modules using an adaptive neuro-fuzzy inference approach, which allows early defect identification of shading cells. In 2019, Niazi *et al* [22] using ML approaches based on Naïve Bayes classification on the thermal images of PV modules, effectively and rapidly detected, evaluated and categorized failure conditions for PV modules. This ML-based monitoring method achieved an accuracy of 94.10% with low computational cost.

In proposing various approaches to address hotspot related problems in solar panels, active techniques have attracted more attention in the form of smart bypass diodes such as power metal–oxide–semiconductor field-effect transistors (MOSFETs) connected in series with PV panels. Such methods, proposed by Daliento *et al* [23] and bipolar junction transistors (BJTs), instead of bypass diodes, studied by Dhimish *et al* [24] can be adopted to achieve bypassing solar cells in case of mismatch-related problems. Furthermore, a divider circuit, consisting of resistors in series, designed not only to reduce the reverse voltage, power loss and temperature of the cell under hotspot conditions, but also increase the power output of the PV panels at a low cost, [25] is feasible to mitigate such issues in solar panels. Dhimish *et al* [24] designed a two hotspot mitigation technology which is capable of enhancing the power output of PV modules under hotspot and shading conditions. A comparison between the proposed bypass diode and the conventional bypass diode, when a monocrystalline silicon module is connected in series with a solar panel in a 1 kW string, is illustrated in figure 10.7 [23]. The proposed technology adopts the integration of MOSFETs with the affected PV module. This has been observed to increase the power output by 1.44 and 3.97 W by two hotspot mitigation techniques, respectively. Dhimish *et al* [26] proposed a current limiter circuit to overcome partial shading and hotspot related loss conditions by decreasing the temperature of the panel from 21.3 °C to 16.4 °C. In this way, the output power as well as the lifetime of PV modules can be enhanced, which in turn contributes to the improved performance of the entire PV system [27].

10.3 Real-world implementations

Besides the use of passive and active bypass diodes, the real-world implementation of addressing hotspots in solar cell panels typically involves several strategies such as optimized position of the solar panel, thermal management, monitoring system and high quality of solar panels. These approaches are all aimed at preventing excessive

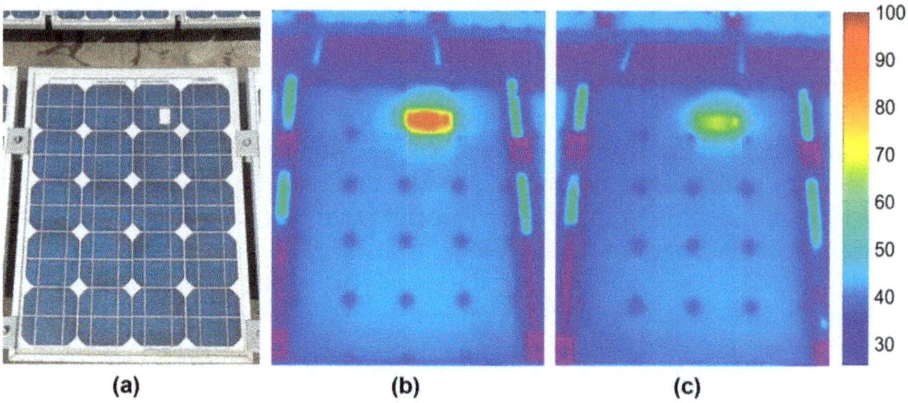

Figure 10.7. (a). Image of a monocrystalline silicon solar panel under test (higher shunt resistance), is shaded with a tape of an area optimized to maximize power dissipation. Thermographic images of (b). the conventional bypass diode and (c). the proposed bypass diode. Reprinted from [23], copyright (2016) with permission from Elsevier.

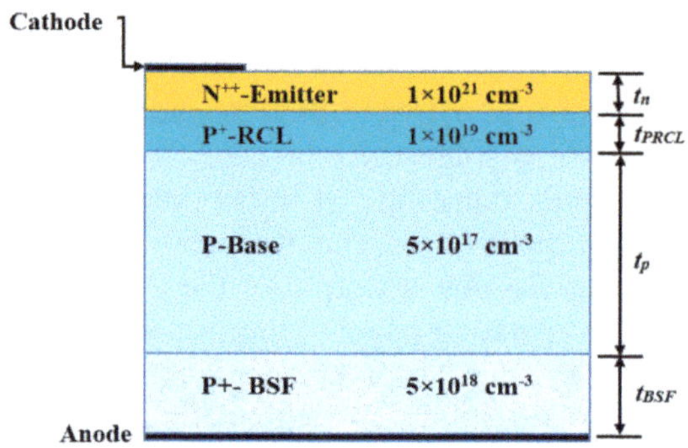

Figure 10.8. Illustration of the proposed self-protected thin film silicon solar cells including a reverse conducting layer; RCL: reverse conducting layer [29].

heating in localized areas, which can lead to reduced efficiency, performance and potential irreparable damage to the panels.

Although maximizing the utilization of both passive and active solar strategies can lead to a more sustainable framework for use in urban development, urban expansion and increasing population density in numerous countries are contributing to rapid growth of cities [28]. Saif *et al* [29] introduced a novel design of a self-protected thin-film crystalline silicon (c-Si) solar cell using TCAD simulation. This novel design proposes the use of a regular solar cell during forward bias while behaving as a bypass diode during reverse bias. The cell structure can be improved with doping concentration and thicknesses of layers, further enhancing the efficiency. The structure of the proposed design is illustrated in figure 10.8 [29].

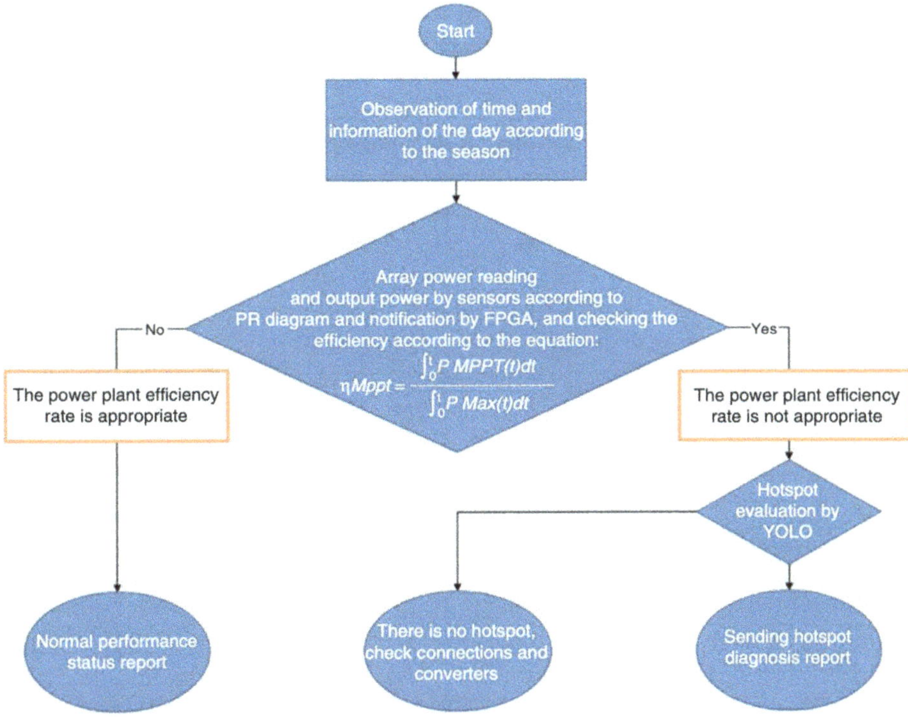

Figure 10.9. Execution algorithm of the error tracking system. Reproduced with permission from [31], copyright 2023, The Author(s). Published by Oxford University Press on behalf of National Institute of Clean-and-Low-Carbon Energy.

In order to identify hotspots early, allowing for timely intervention from the damage to the panel, economic and effective continuous monitoring systems are in urgent need. Pruthviraj *et al* [30] proposed, for the first time in India's solar farm operations, a drone-mounted thermal camera that was deployed to capture images of the solar field. Its purpose was to inspect solar panels for faulty cells and generate an orthomosaic image covering the entire area. A complete mapping of the hotspots with an accuracy of up to 97% was obtained through the analysis of image visualization algorithms such as RGB, NIR, and Thermal Index yield. Yazdani *et al* [31] simulated the solar panel performance by utilizing a smart monitoring error tracking system for the detection of hotspots in real time, as shown in figure 10.9. Besides, this simulation result was validated through the coordination with meteorological data and comparison with technical data from a local solar power plant. Chockalingam *et al* [32] introduced a non-invasive method for detecting and quantifying hotspots in PV systems, which utilized temperature sensors and micro-controllers to effectively identify and isolate hot regions, demonstrating high accuracy in hotspot detection. Cardinale–Villalobos *et al* [33] by using an innovative approach, combined AI and deep learning methods for the detection of hotspots; the team utilized infrared thermography (IRT) for diagnosing PV installations in smart cities. The system overcomes the limitation of IRT under certain irradiance conditions and achieves high sensitivity and accuracy, offering potential cost savings for PV installation management [33].

Besides the above-mentioned strategies, the quality assurance is critical for ensuring the use of high quality materials and manufacturing processes of the wafer-to-cell to module. This helps to minimize the probability of occurrence of defects, dislocations, microcracks and materials-related properties that may lead to hotspots. Gressler *et al* [34] have investigated the hotspot production in emerging photovoltaic systems (EPVs) such as in organic solar cells, dye-sensitized solar cells, PSCs, and quantum dots solar cells through life cycle analysis (LCA). LCA can identify materials and manufacturing processes that contribute most to the environmental impact of the overall product. The results show that EPV fabrication may lead to lower energy demand and shorter energy payback time compared to conventional PV technologies. The cut-cell patterns (for example, half-cell, one-third cell, already in practice in crystalline silicon solar cells/modules) can be potential alternative for conventional full cell patterns in terms of enhanced stability, reduced hotspot effects, etc [35]. Akram *et al* [35] have investigated the thermo-mechanical behavior of cut and full cell modules during manufacturing and hotspot formation. Their results show that cut-cell patterns offer improved stability and reduced hotspot effects over full-cells. Xu *et al* [36] conducted experiments and finite element analysis (FEA) to determine the hotspot temperature in high-wattage solar modules of varying designs. These designs encompassed different cell sizes (156.75, 166, 182, and 210 mm), cell numbers per bypass diode, and cell shapes (full cell, half-cell, and one-third-cells). The use of cut-cell technique has been justified as a way to enhance efficiency by reducing the size of the cell, which reduces the resistance of the cell while enhancing the fill factor and the power output [37]. Another reason for the use of this technique is associated with the increasing size of the silicon wafers (figure 10.10). While the current increases with increase in wafer size/cell area, it also increases the electrical losses. Cutting cells into half-cells and one-third-cells has been found to be a way to reduce current-related losses while, at the same time, it helps to reduce shading losses, that are associated with the metal electrodes of the cells. In recent years, rectangular monocrystalline silicon wafers have been introduced in the silicon photovoltaics industry [38]. For example, the current size of 182 R is 183.75 × 182 mm; 185 × 182 mm. 186 × 182 mm etc. Based on Trina Solar's 210R modules, in mid-July, last year, nine companies in China agreed on standardizing the dimensions of rectangular silicon modules. A summary of the standard specifications of these rectangular modules is presented in table 10.1 [39].

Das *et al* have performed detailed investigations into shading mitigation techniques in photovoltaics. The classification of various shading mitigation techniques is summarized in figure 10.11 [40]. As can be seen in this figure, while there are many approaches to shading mitigation techniques in photovoltaics, the bypass diode method is the most commonly deployed technique in the commercial sector of the industry.

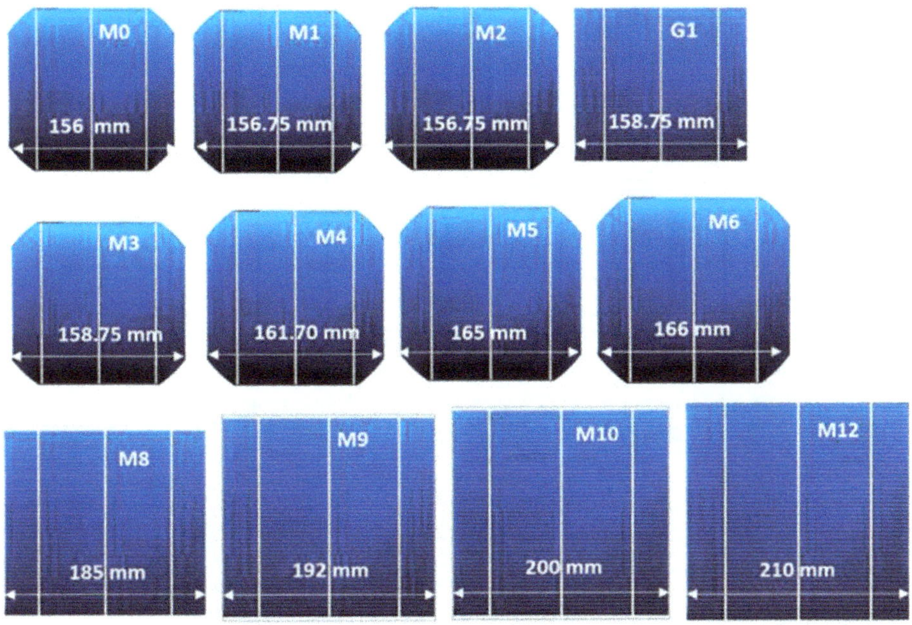

Figure 10.10. Illustration of the evolution of silicon wafer size in the manufacture of monocrystalline silicon solar cells [41].

Table 10.1. Standard module specifications based on trina solar's 210R products [39].

Cell type	Module type	Number of cells	Module lenth (mm)	Module width (mm)	Mounting hole distance in long side (mm)
Half-cell	Single-glass framed/ dual-glass framed	110	238 ± 2	1096 ± 2	400/1400 ± 1
		120	217 ± 2	1303 ± 2	400/1400 ± 1
		132	238 ± 2	1303 ± 2	400/1400 ± 1

10.4 Conclusion

Various approaches to hotspot reduction and shade loss minimization, which are practiced in the photovoltaics industry, have been briefly summarized in this chapter. The use of half-cells and one-third-cells is illustrated particularly in the context of the evolution in wafer size of monocrystalline silicon modules. The use of bypass diodes to mitigate shadow loss is anticipated to continue as the industry standard in photovoltaics.

This chapter was reproduced with permission from [1].

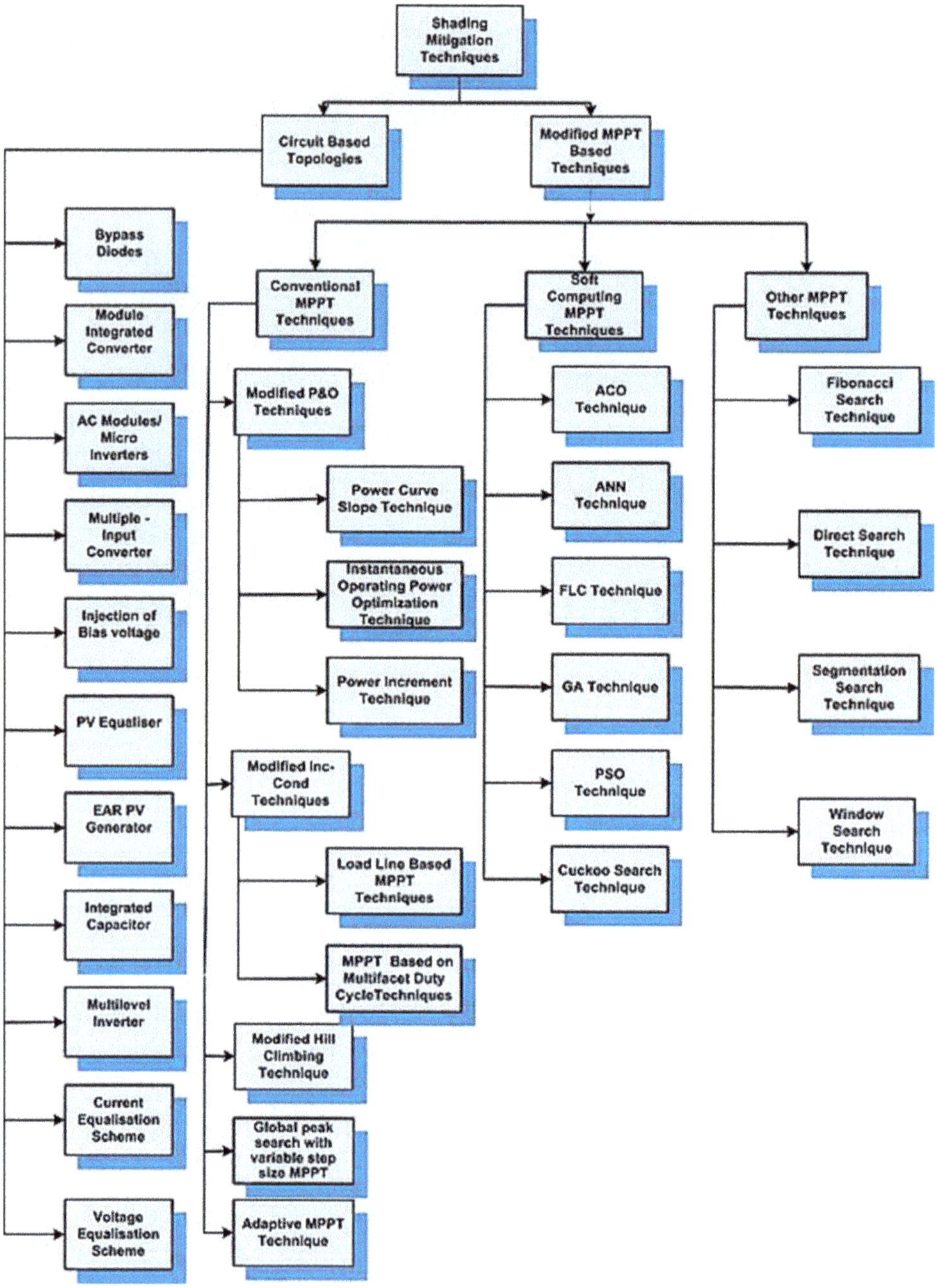

Figure. 10.11. Classification of various shading mitigation techniques. Reprinted from [40], copyright (2017) with permission from Elsevier.

References

[1] Lin L *et al* 2023 Influence of outdoor conditions on PV module performance—an overview *Mater. Sci. Eng* **7** 88–101

[2] Sullivan R M 1965 *Shadow Effects on Series-Parallel Array of Solar Cells* (Greenbelt, MD: NASA, NASA Goddard Space Flight Center)

[3] Dhimish M, Theristis M and d'Alessandro V 2024 Photovoltaic hotspots: a mitigation technique and its thermal cycle *Optik* **300** 171627

[4] Livera A, Theristis M, Makrides G and Georghiou G 2019 Recent advances in failure diagnosis techniques based on performance data analysis for grid-connected photovoltaic systems *Renew. Energy* **133** 126–43

[5] Deng S *et al* 2017 Research on hot spot risk for high-efficiency solar module *Energy Procedia* **130** 77–86

[6] Wu Z, Hu Y, Wen J X, Zhou F and Ye X 2020 A review for solar panel fire accident prevention in large-scale PV applications *IEEE Access* **8** 132466–80

[7] Zhang Z *et al* 2016 Study of bypass diode reliability under non-uniform irradiance distribution on pv module surface (not published)

[8] Bowring A R, Bertoluzzi L, O'Regan B C and McGehee M D 2018 Reverse bias behavior of Halide perovskite solar cells *Adv. Energy Mater.* **8** 1702365

[9] Yang H, Wang H and Wang M 2012 Investigation of the relationship between reverse current of crystalline silicon solar cells and conduction of bypass diode *Int. J. Photoenergy* **2012** 357218

[10] Dhimish M, Holmes V, Mather P and Sibley M 2018 Novel hot spot mitigation technique to enhance photovoltaic solar panels output power performance *Sol. Energy Mater. Sol. Cells* **179** 72–9

[11] Manno D *et al* 2021 Deep learning strategies for automatic fault diagnosis in photovoltaic systems by thermographic images *Energy Convers. Manage.* **241** 1–17

[12] Dhimish M, Holmes V, Mehrdadi B and Dales M 2018 Comparing Mamdani Sugeno fuzzy logic and RBF ANN network for PV fault detection *Renew. Energy* **117** 257–74

[13] https://alternative-energy-tutorials.com/photovoltaics/bypass-diode.html (accessed 17 August 2024)

[14] Shenoy P S and Krein P T 2012 Differential power processing for DC systems *IEEE Trans. Power Electron.* **28** 1795–806

[15] Kim K A, Shenoy P S, Krein P T and Chapman P L 2012 *38th IEEE Photovoltaic Specialists Conf., PVSC*

[16] Teo J C, Tan R, Mok V H, Ramachandaramurthy V K and Tan C K 2017 Effects of bypass diode configurations to the maximum power of photovoltaic module *Int. J. Smart Grid Clean Energy* **6** 225–32

[17] Vieira R G, de Araújo F M U, Dhimish M and Guerra M I S 2020 A comprehensive review on bypass diode application on photovoltaic modules *Energies* **13** 2472

[18] Shin W G, Ko S W, Song H J, Ju Y C, Hwang H M and Kang G H 2018 Origin of bypass diode fault in c-si photovoltaic modules: leakage current under high surrounding temperature *Energies* **11** 2416

[19] Silvestre S, Boronat A and Chouder A 2009 Study of bypass diodes configuration on PV modules *Appl. Energy* **86** 1632–40

[20] Bana S and Saini R P 2017 Experimental investigation on power output of different photovoltaic array configurations under uniform and partial shading scenarios *Energy* **127** 438–53

[21] Kaid I E *et al* 2018 Photovoltaic system failure diagnosis based on adaptive neuro fuzzy inference approach: South Algeria solar power plant *J. Clean. Prod.* **204** 169–82

[22] Niazi K A K, Akhtar W, Khan H A, Yang Y and Athar S 2019 Hotspot diagnosis for solar photovoltaic modules using a Naive Bayes classifier *Sol. Energy* **190** 34–43

[23] Daliento S, Di Napoli F, Guerriero P and D'Alessandro V 2016 A modified bypass circuit for improved hot spot reliability of solar panels subject to partial shading *Sol. Energy* **134** 211–8

[24] Dhimish M, Holmes V, Mehrdadi B, Dales M and Mather P 2018 PV output power enhancement using two mitigation techniques for hot spots and partially shaded solar cells *Electr. Power Syst. Res.* **158** 15–25

[25] Tang S, Xing Y, Chen L, Song X and Yao F 2021 Review and a novel strategy for mitigating hot spot of PV panels *Sol. Energy* **214** 51–61

[26] Dhimish M and Badran G 2020 Current limiter circuit to avoid photovoltaic mismatch conditions including hot-spots and shading *Renew. Energy* **145** 2201–16

[27] Niazi K A K, Yang Y and Sera D 2019 Review of mismatch mitigation techniques for PV modules *IET Renew. Power Gener.* **13** 2035–50

[28] Nations U 2022 *World Cities Report 2022: Envisaging the Future of Cities* (https://unhabitat.org/sites/default/files/2022/06/wcr_2022.pdf)

[29] Saif O M *et al* 2023 Design and optimization of a self-protected thin film c-si solar cell against reverse bias *Materials* **16** 2511

[30] Pruthviraj U, Kashyap Y, Baxevanaki E and Kosmopoulos P 2023 Solar photovoltaic hotspot inspection using unmanned aerial vehicle thermal images at a solar field in South India *Remote Sens.* **15** 1914

[31] Yazdani H, Radmehr M and Ghorbani A 2023 Smart component monitoring system increases the efficiency of photovoltaic plants *Clean Energy* **7** 303–12

[32] Chockalingam A, Naveen S, Sanjay S, Nanthakumar J and Praveenkumar V 2023 *9th Int. Conf. on Electrical Energy Systems (ICEES)* pp 371–6

[33] Cardinale-Villalobos L *et al* 2023 IoT system based on artificial intelligence for hot spot detection in photovoltaic modules for a wide range of irradiances *Sensors* **23** 6749

[34] Gressler S, Part F, Scherhaufer S, Obersteiner G and Huber-Humer M 2022 Advanced materials for emerging photovoltaic systems—environmental hotspots in the production and end-of-life phase of organic, dye-sensitized, perovskite, and quantum dots solar cells *Sustain. Mater. Technol.* **34** e00501

[35] Waqar Akram M *et al* 2020 Study of manufacturing and hotspot formation in cut cell and full cell PV modules *Sol. Energy* **203** 247–59

[36] Xu T, Deng S, Zhang G and Zhang Z 2021 Research on hot spot risk of high wattage solar modules *Sol. Energy* **230** 583–90

[37] *New Trend: Why Choosing 1/3 Cut Solar Cell?* https://maysunsolar.eu/blog/new-trend-why-choosing–1-3-cut-solar-cell (accessed 18 August 2024)

[38] *New Trend in PV Cells: Rectangular Silicon Wafers (182R and 210 R), Maysun Solar* https://linkedin.com/pulse/new-trend-pv-cells-rectangular-silicon-wafers-182r/(accessed 18 August 2024)

[39] *Trina Solar Leads Industry Standardization with 'Golden Size' Modules for All Settings* https://trinasolar.com/en-glb/resources/newsroom/matrina-solar-leads-industry-standardization-%E2%80%98golden-size%E2%80%99-modules-all-settings (accessed 18 August 2024)

[40] Das S K *et al* 2017 Shading mitigation techniques: state-of-the-art in photovoltaic applications *Renew. Sustain. Energy Rev.* **78** 369–90

[41] Novergy 2020 Significance of increasing size of mono-crystalline wafers in modules https://www.novergysolar.com/significance-increasing-size-mono-crystalline-wafers-modules/

N M Ravindra, Leqi Lin and Priyanka Singh

Chapter 11

Computational methods

The illuminated current–voltage (I–V) characteristics of solar cells provide information regarding the parameters such as the maximum power output (P_{max}), short-circuit current (I_{sc}), open-circuit voltage (V_{oc}), fill factor (FF), and efficiency (η). The diode parameters, which include photogenerated current (I_{ph}), ideality factor (n), reverse saturation current (I_o), series resistance (R_s), and shunt resistance (R_{sh}), affect the solar cell parameters. This chapter addresses analytical and numerical simulation techniques that are available in the literature to compute diode parameters for single and double-diode models. The computation of few other significant parameters e.g., spectral response, quantum efficiency, minority carrier lifetime and junction depth are discussed briefly.

11.1 Introduction

Solar photovoltaic (PV) technology enables the conversion of sunlight into electricity, offering a simple and useful way to sustain the growing energy demand [1]. The PV device is a p-n junction solar cell, which is described by its current–voltage (I–V) characteristics. The performance parameters, viz. I_{sc}, V_{oc}, FF and η are determined from the illuminated I–V characteristics of a solar cell. Diode parameters of solar cell, i.e., I_{ph}, n, I_o, R_s and R_{sh} control the performance of a PV cell at incident intensity of illumination (P_{in}) and operating temperature (T). Computation of performance parameters (I_{sc}, V_{oc}, FF, and η), along with diode parameters (I_{ph}, n, I_o, R_s and R_{sh}), is crucial to evaluate the behaviour and efficiency of a PV cell. Single-diode model (SDM) and double-diode model (DDM) are most frequently used to explain the I–V characteristics of solar cells through Shockley equation by incorporating R_s and R_{sh} to account for the current-dependent and voltage-dependent loss mechanisms [2, 3]. Since SDM ignores the recombination process, sometimes it fails to adequately model the physical processes for some PV technologies [4, 5] and DDM is more viable for these technologies. Additionally, DDM is more accurate and offers the required understanding of the moderately complex physics of

doi:10.1088/978-0-7503-5994-8ch11

solar cells [6, 7]. Despite the existence of more intricate models, such as the three-diode model, their application is constrained by their computational complexity [2]. Both SDM and DDM are characterized to be implicit, nonlinear, and multivariable. As a result, using the *I–V* data, an accurate solution cannot be determined. Consequently, a number of analytical techniques have been proposed in the literature to extract the five diode parameters (I_{ph}, n, I_o, R_s and R_{sh}) that are associated with SDM [8–10], whereas seven parameters (I_{ph}, n_1, n_2, I_{o1}, I_{o2}, R_s and R_{sh}) are extracted with DDM. Both analytical and numerical approaches can be implemented to solve SDM and DDM-based equations; some of these methods are based on electrical circuits [3, 8], partial differential equations or semiconductor equations [11], artificial intelligence (AI) techniques [12, 13], and empirical approaches by curve fitting methods [14].

This chapter explores the computation of performance parameters (I_{sc}, V_{oc}, FF, and η) as well as diode parameters (I_{ph}, n, I_o, R_s and R_{sh}). Furthermore, an overview of the analytical and simulation techniques for extracting diode parameters for single- and double-diode model is presented. In addition, certain other important solar cell metrics, such as spectral response (SR), quantum efficiency (QE), minority carrier lifetime (τ) and junction depth, are briefly addressed.

11.2 Illuminated current–voltage (*I–V*) characteristics

I–V measurements are important for solar cell characterization. The resulting graphs from the *I–V* measurements can be used to extract performance parameters (I_{sc}, V_{oc}, FF, and η) along with diode parameters (I_{ph}, n, I_o, R_s and R_{sh}), which provide insights into device performance. While dark *I–V* characteristics are more sensitive than illuminated *I–V* characteristics for determining the diode parameters [15], parameter extraction is typically performed using the illumination case in order to examine the PV cell performance under illumination conditions [4, 16]. The schematic of *I–V* measurement methodology is illustrated in figure 11.1(a); it is typically performed by illuminating the front surface of the solar cell with a simulated light which corresponds to AM 1.5 solar spectrum equivalent to an intensity of 100 mWcm^{-2}. The operating temperature of the device is kept constant at 25 °C (298 K) with the use of an appropriate temperature controller as shown in the figure 11.1(a). *I–V* characteristics can be measured via a source meter. Both the light intensity and the temperature can be maintained constant or could be varied to measure *I–V* characteristics for various temperatures and intensities if required. The intensity of the incident light needs to be calibrated with the help of a standard reference solar cell of known I_{sc} under AM1.5 condition (100 mWcm^{-2} and $T = 25$ °C).

Figure 11.1(b) shows a cross-sectional image of a solar cell. Emitter and base regions are very important and commonly used terms to illustrate the working of a p-n junction solar cell. These terms have different names for different solar cell technologies. When a solar cell is illuminated, the light enters the emitter first. In order to keep the depletion region close to the incident sunlight for strong absorption, the emitter is often narrow, while the base is typically thick enough to absorb the majority of the light.

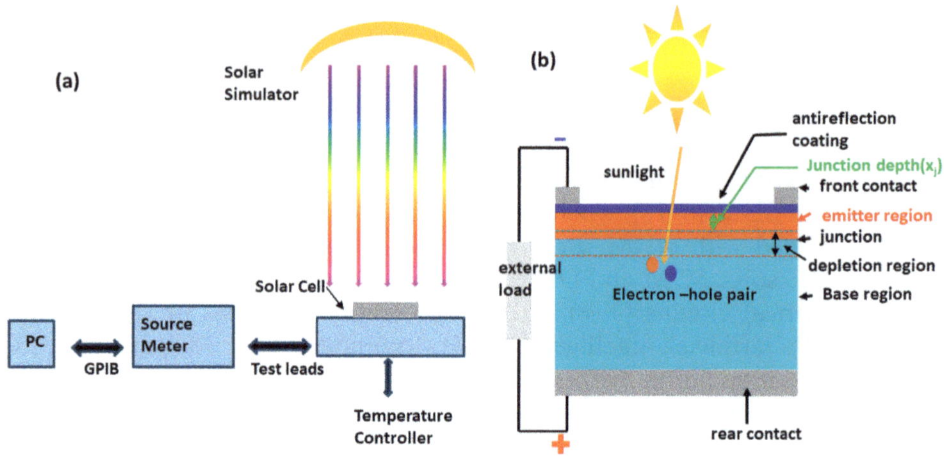

Figure 11.1. (a) A schematic representation of the experimental setup for the I–V measurements of solar cells under simulated AM1.5 corresponding to an intensity of 100 mWcm^{-2} (b) Cross-sectional image of a solar cell showing emitter, depletion and base region.

Both SDM and DDM have been extensively applied in the assessment of the electrical properties of solar cells and modules [17]. A number of publications [3, 8, 18, 19] have examined the benefits and limitations of both models. A comparison of the two models has revealed a trade-off between computation time and accuracy [20]. Nevertheless, because of the simplicity, SDM is still preferred for various PV solar technologies [21, 22] under different environmental conditions [23, 24]. Section 11.2.1 will elaborate on the analytical methods that are employed to extract the performance parameters as well as diode parameters of solar cells.

11.2.1 Single-diode model (SDM)

The SDM of a solar cell, with R_s and R_{sh} based on Shockley model, is depicted in figure 11.2(a) [25, 26]; it can be theoretically modelled under illumination using the following equations [27, 28],

$$I = -I_{ph} + I_d + \frac{V_j}{R_{sh}} \tag{11.1}$$

where, I_{ph} is the photogenerated current, I_d is current through the diode and V_j is the voltage developed across the p-n junction; the third term on the right hand side describes the leakage current through shunt resistance R_{sh}. Both I_d and V_j are defined as follows:

$$I_d = I_o\left\{\exp\left(\frac{V_j}{nV_t}\right) - 1\right\} \tag{11.2}$$

$$V_j = V - IR_s \tag{11.3}$$

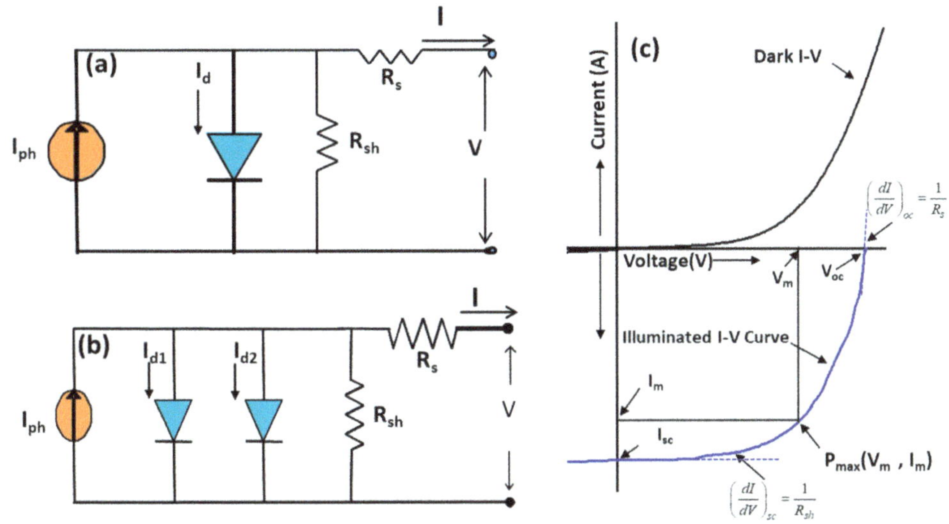

Figure 11.2. The equivalent circuit diagram of a solar cell described by (a) SDM, and (b) DDM having series (R_s) and shunt resistances (R_{sh}), I is the device current, V is the voltage, (c) dark and illuminated I–V curves of a solar cell to denote the parameters V_{oc}, I_{sc}, maximum voltage (V_m), maximum current (I_m) and maximum power output ($P_{\max}$) and slope dI/dV at open-circuit and short-circuit conditions.

Substituting the values of I_d and V_j from equation (11.2) and equation (11.3), respectively, in equation (11.1),

$$I = -I_{ph} + I_o\left\{\exp\left(\frac{V_j}{nV_t}\right) - 1\right\} + \frac{V_j}{R_{sh}} \tag{11.4}$$

In equations (11.3) and (11.4), I and V are output current and voltage, respectively (as shown in figure 11.2(a); $V_t = kT/q$ is thermal voltage; $q = 1.602\,166\,34 \times 10^{-19}$C is the electron charge; $k = 1.380\,649 \times 10^{-23} \cdot \mathrm{m^2 s^{-2} \cdot kgK^{-1}}$ is Boltzmann's constant, and T is operating temperature; $V_t = 0.025\,68$ V at room temperature ($T \sim 298$ K); I_o is reverse saturation current. An SDM has five unknown diode parameters (I_{ph}, n, I_o, R_s and R_{sh}) for a given set of operating conditions (i.e., solar irradiance and cell temperature) [28]. In order to determine the correct values of these parameters, several analytical representations of diode parameters employ different assumptions to simplify the calculations [29, 30]. For instance, it is assumed that R_{sh} is infinite [27]. I_{ph} is assumed to be equal to I_{sc} by neglecting the value of diode current compared to I_{ph} [31]. Various approximations for reducing the number of parameters from five to one have been reported in review papers [3, 18]. Several approximations and assumptions, which are utilized to solve the equations describing the behaviour of solar cells, are listed in table 11.1 [32].

11.2.2 Double-diode model (DDM)

In SDM, it has been assumed that there is no recombination loss in the space charge (or depletion) region. Additionally, in the case of SDM, for the majority of silicon

Table 11.1. Common assumptions in analytical techniques. Reproduced from [33] CC BY 4.0

No.	Assumption	Implication
1	$\exp\left(\frac{V_{OC}}{nV_t}\right) > > \exp\left(\frac{I_{SC}R_S}{nV_t}\right)$	Ignore $\exp\left(\frac{I_{SC}R_S}{nV_t}\right)$
2.	$\frac{I_O}{nV_t}\exp\left(\frac{V_{OC}}{nV_t}\right) > > \frac{1}{R_{sh}}$	Ignore $\frac{1}{R_{sh}}$
3.	$\frac{I_O}{nV_t}\exp\left(\frac{I_{SC}R_S}{nV_t}\right) < < \frac{1}{R_{sh}}$	$R_{sh} \approx R_{sh} - R_s$
4.	$R_s < < R_{sh}$	$\frac{R_S}{R_{sh}} \approx 0$ and $R_s + R_{sh} \approx R_{sh}$
5.	$I_{ph} + I_o > > I_o \exp\left(\frac{I_{SC}R_S}{nV_t}\right) + \frac{I_{SC}R_S}{R_{sh}}$	$I_{ph} + I_o = I_{sc}$
6.	$I_o(\exp\left(\frac{I_{SC}R_S}{nV_t}\right) - 1) < < I_{ph} - \frac{I_{SC}R_S}{R_{sh}}$	Neglect $I_o(\exp\left(\frac{I_{SC}R_S}{nV_t}\right) - 1)$
7.	$\exp\left(\frac{I_{SC}R_S}{nV_t}\right) < < 1$	$I_o(\exp\left(\frac{I_{SC}R_S}{nV_t}\right) - 1) \approx I_o$
8.	$\exp\left(\frac{V_m + I_m R_S}{nV_t}\right) > > 1$	$I = -I_{ph} + I_o \exp\left(\frac{V - IR_S}{nV_t}\right) + \frac{V - IR_S}{R_{sh}}$
9.	$\frac{V_{OC}}{R_{sh}} < < I_{sc}$	Neglect $\frac{V_{OC}}{R_{sh}}$
10.	$\frac{V_m}{R_{sh}} < < I_{sc}$	Neglect $\frac{V_m}{R_{sh}}$
11.	$I_{ph} > > I_o \exp\left(\frac{I_{SC}R_S}{nV_t} - 1\right) + \frac{I_{SC}R_S}{R_{sh}}$	$I_{ph} \approx I_{sc}$

(Si) solar cells, the assumption, $R_{sh} \gg R_s$ is valid; however, this is not true for thin-film and organic solar cells. A more accurate model for these PV cells is the DDM. Figure 11.2(b) depicts the two-diode model of a solar cell [34, 35]. Two additional parameters, saturation current (I_{o2}) and ideality factor (n_2), are included in DDM to describe charge recombination in the space charge region for generation and recombination currents. This recombination current in the depletion region is added to equation (11.1) and thus, in many cases, the I–V characteristics of the PV cells are better described by DDM as given by equation (11.5) along with equation (11.6).

$$I = -I_{ph} + I_{d1} + I_{d2} + \frac{V_j}{R_{sh}} \tag{11.5}$$

Equation (11.5) can be written as:

$$I = -I_{ph} + I_{o1}\left\{\exp\left(\frac{V_j}{n_1 V_t}\right) - 1\right\} + I_{o2}\left\{\exp\left(\frac{V_j}{n_2 V_t}\right) - 1\right\} + \frac{V_j}{R_{sh}} \tag{11.6}$$

where I_{o1}, n_1 are the reverse saturation current and ideality factor due to the recombination in quasi-neutral (or bulk) regions and I_{o2}, n_2 are due to the recombination in the space charge (or the depletion) region of the cell.

While DDM provides improved PV cell profiling, extracting the seven unknown parameters (I_{ph}, n_1, n_2, I_{o1}, I_{o2} R_s and R_{sh}) requires complex computational process [36]. Nevertheless, better methods for calculating unknown parameters using seven [37], six [38], five [39], and four parameters [40] for DDM have also been explored.

These techniques make use of numerical algorithms, such as the Newton–Raphson method or particle swarm optimization, which need a significant amount of data and a number of mathematical operations to evaluate the unknown parameters [41].

11.3 Dark current–voltage (*I–V*) characteristics

The dark current–voltage (I_D–V) characteristics of a PV cell having finite R_s and R_{sh} can be described by substituting $I_{ph} = 0$ in equation (11.4) and in equation (11.6) for SDM and DDM, respectively, and can be formulated as:

$$I_D = I_o\left\{\exp\left(\frac{V_j}{nV_t}\right) - 1\right\} + \left(\frac{V_j}{R_{sh}}\right) \tag{11.7a}$$

$$I_D = I_{o1}\left\{\exp\left(\frac{V_j}{n_1 V_t}\right) - 1\right\} + I_{o2}\left\{\exp\left(\frac{V_j}{n_2 V_t}\right) - 1\right\} + \left(\frac{V_j}{R_{sh}}\right) \tag{11.7b}$$

11.4 Solar cell performance parameters

The dark and illuminated *I–V* characteristics are shown in figure 11.2(c) for practical analysis. The dark *I–V* curve is shifted down by a photogenerated current, I_{ph} resulting in the illuminated *I–V* curve. The parameters I_{sc}, V_{oc}, maximum voltage (V_m) and maximum current (I_m), corresponding to the maximum power point (*MPP*) and maximum power output ($P_{\max}$), are mentioned on the illuminated *I–V* curve. This section discusses the extraction of the performance parameters using equations (11.3) and (11.4).

The open-circuit voltage is defined as the voltage when terminals are open in the solar cell. Substituting the total current, $I = 0$ and voltage, $V = V_{oc}$ in equations (11.3) and (11.4), the relation between V_{oc}, n, I_o and I_{ph} for SDM, assuming $R_{sh} \to \infty$,

$$V_{oc} = nV_t \ln\left(\frac{I_{ph}}{I_o} + 1\right) \tag{11.8}$$

Equation (11.8) represents V_{oc} of a solar cell, which is determined by n, I_{ph}, and I_o. A low I_o is essential for achieving higher V_{oc} values.

The fill factor is defined as the ratio of $P_{\max}$ at *MPP* to the product V_{oc} and I_{sc} ($\approx I_{ph}$) and its value is always less than unity and can be expressed as,

$$FF = \frac{P_{\max}}{V_{oc}\, I_{sc}} \tag{11.9}$$

The condition for *MPP* and expression for $P_{\max}$ is given by,

$$\frac{dP}{dV} = 0 \Rightarrow \frac{d(IV)}{dV} = 0 \tag{11.10}$$

$$P_{\max} = I_m V_m \tag{11.11}$$

The output power per unit area of the cell can be determined from the illuminated I–V characteristics by evaluating the area of the rectangle formed between the operating point (MPP) and the two axes (as shown in figure 11.2(c)). The operating point depends on the load resistance.

The efficiency of a solar cell is defined as the ratio of $P_{\max}$ per unit area to P_{in} and is represented as

$$\eta = \frac{P_{\max}}{P_{in}\text{Area}} \tag{11.12}$$

Solving equation (11.9) and equation (11.12) in terms of the V_{oc}, J_{sc} and FF, the efficiency of the cell can be represented as,

$$\eta = \frac{V_{oc} J_{sc} FF}{P_{in}} \tag{11.13}$$

where, J_{sc} is the short-circuit current density (current per unit area). The current density of the cell depends on the incident solar spectral irradiance and is given by,

$$J_{sc} = q \int_{h\nu = E_g}^{\infty} \frac{dN_{ph}}{dh\nu} d(h\nu) \tag{11.14}$$

Here, N_{ph} is the incident photon flux. In a practical solar cell, the value of J_{sc} may be limited by reflection losses, ohmic losses (R_s and R_{sh}), shadowing losses (front metal coverage) and recombination losses. Efficiency of a solar cell is derived from its performance parameters V_{oc}, J_{sc} and FF, using equation (11.13).

11.5 Solar cell diode parameters

Solar cell diode parameters affect the performance of solar cells [42, 43]. The increase in R_s and decrease in R_{sh} severely degrades the FF. I_{sc} and V_{oc} are decreased by high R_s and low R_{sh} values, respectively. Lower values of R_{sh} results in leakage and increases I_o, which in turn leads to a decrease in V_{oc}, FF, and ultimately η [42, 44]. Additionally, both n and I_o reveal the charge recombination that takes place at the surface, bulk and in depletion layer. Furthermore, because of the inverse relationships between I_o and V_{oc} in equation (11.8) and n and FF in equation (11.9), both I_o and n adversely affect V_{oc} and FF, respectively. Therefore, the computation of these diode parameters is crucial to improve the performance of PV cells. Singh and Ravindra have [44] calculated, I_o, n, R_s and R_{sh} for silicon solar cells using SDM and DDM. In the literature, various approaches for parameter extraction, such as analytical, numerical, hybrid, and optimization techniques have been introduced. The diode parameters (I_{ph}, R_s, R_{sh}, n, and I_o) can be estimated using any of these approaches [32, 45, 46]. Numerous parameter extraction techniques with various levels of accuracy and complexity have been proposed [4, 8, 19, 47]. These techniques are often divided into two categories: numerical and analytical. This chapter mainly emphasizes analytical and numerical simulation techniques.

The analytical techniques are quick because they derive a set of equations that characterize cell behaviour and may be solved without the need for iteration through simplifications and empirical information [47–49]. Additionally, accuracy of the analytical models is influenced by photovoltaic technology [50]. Usually, numerical approaches create a set of equations that can be solved by iterative or numerical algorithms [4, 51]. The subsequent section will go over common analytical methodologies and procedures for obtaining diode parameters; numerical or simulation techniques will be discussed later.

11.5.1 Analytical techniques for parameter extraction

Analytical approaches generally apply some approximations in deriving the standard equations from equations (11.3) and (11.4) by employing short-circuit (*sc*), open-circuit (*oc*) and *MPP* conditions on the *I–V* characteristics. Equation (11.4) can be evaluated by applying the conditions, *sc*, *oc* and *MPP*.

At short-circuit condition ($I = -I_{sc}$ and $V = 0$), equation (11.4) can be expressed as:

$$sc(0, -I_{sc}):\ -I_{sc} = -I_{ph} + I_o\left\{\exp\left(\frac{I_{sc}R_s}{nV_t}\right) - 1\right\} + \frac{I_{sc}R_s}{R_{sh}} \tag{11.15a}$$

Equation (11.15a) can be rearranged in terms of n, I_o, R_s, and R_{sh} to determine I_{ph} as:

$$I_{ph} = I_{sc}\left(1 + \frac{R_s}{R_{sh}}\right) + I_o\left\{\exp\left(\frac{I_{sc}R_s}{nV_t}\right) - 1\right\} \tag{11.15b}$$

Similarly, at an open-circuit condition ($V = V_{oc}$, $I = 0$), equation (11.4) can be written as:

$$oc(V_{oc}, 0):\ 0 = -I_{ph} + I_o\left\{\exp\left(\frac{V_{oc}}{nV_t}\right) - 1\right\} + \left(\frac{V_{oc}}{R_{sh}}\right) \tag{11.16a}$$

Equation (11.16a) can be rewritten to determine I_{ph} in terms of V_{oc}, n, I_o, and R_{sh},

$$I_{ph} = I_o\left\{\exp\left(\frac{V_{oc}}{nV_t}\right) - 1\right\} + \left(\frac{V_{oc}}{R_{sh}}\right) \tag{11.16b}$$

Equation (11.4) at *MPP* ($I = -I_m$, $V = V_m$) can be expressed as:

$$MPP(V_m, -I_m):\ -I_m = -I_{ph} + I_o\left\{\exp\left(\frac{V_m + I_mR_s}{nV_t}\right) - 1\right\} + \left(\frac{V_m + I_mR_s}{R_{sh}}\right) \tag{11.17a}$$

Equation (11.17a) can be used to express I_{ph} in terms of V_m, I_m, n, I_o, R_s, and R_{sh} as:

$$I_{ph} = I_m + I_o\left\{\exp\left(\frac{V_m + I_mR_s}{nV_t}\right) - 1\right\} + \left(\frac{V_m + I_mR_s}{R_{sh}}\right) \tag{11.17b}$$

Diode parameter extraction can be categorized into three cases, which are commonly addressed in the literature. Case I considers all five parameters (I_{ph}, n, I_o, R_s and R_{sh}); Case II incorporates four parameters (I_{ph}, n, I_o and R_s) while ignoring R_{sh}; and Case III takes into account three parameters (I_{ph}, n and I_o), ignoring both R_s and R_{sh} [52]. The above equations are used to determine the diode parameters for Case I, Case II and Case III.

Case I

This is known as the five-parameter model for PV solar cell/module and delivers the best results as it considers the power losses caused due to manufacturing defects by parasitic resistances (R_s and R_{sh}) [53, 54]. This model makes use of the parameters (V_{oc}, I_{sc}, V_m, I_m) as well as the slopes at the *sc* and *oc* points as shown on the *I–V* curve in figure 11.2(c) [53]. The ideality factor is computed as the primary parameter, and the remaining parameters are then computed using this value of n. The extraction of parameters has been restricted to standard temperature ($T = 298$ K) in this chapter.

11.5.1.1 Photocurrent (I_{ph})

The photogenerated current in the PV cell is linearly proportional to the incident illumination and also changes with temperature. Equations (11.15b), (11.16b), and (11.17b) can be used to compute I_{ph} for Case I, based on known parameters (I_{sc}, V_{oc}, V_m, I_m, I_o, n, R_s, R_{sh}). Any of the equations among equations (11.15b), (11.16b), and (11.17b) can be selected in accordance with known parameters.

11.5.1.2 Ideality factor (n) and reverse saturation current (I_o)

Equations (11.16b) and (11.17b) can be expressed logarithmically as,

$$\ln\left(I_{ph} + I_o - \frac{V_{oc}}{R_{sh}}\right) - \ln I_o = \frac{V_{oc}}{nV_t} \tag{11.18}$$

$$\ln\left\{I_{ph} + I_o - I_m - \left(\frac{V_m + I_m R_s}{R_{sh}}\right)\right\} - \ln I_o = \frac{V_m + I_m R_s}{nV_t} \tag{11.19}$$

Subtracting equation (11.18) from equation (11.19) yields,

$$\ln\left\{\frac{I_{ph} + I_o - \frac{V_{oc}}{R_{sh}}}{I_{ph} + I_o - I_m - \left(\frac{V_m + I_m R_s}{R_{sh}}\right)}\right\} = \frac{V_{oc} - V_m - I_m R_s}{nV_t} \tag{11.20}$$

This can be written as:

$$n = \frac{V_{oc} - V_m - I_m R_s}{V_t\left[\ln\left(I_{ph} + I_o - \frac{V_{oc}}{R_{sh}}\right)/\left(I_{ph} + I_o - I_m - \left(\frac{V_m + I_m R_s}{R_{sh}}\right)\right)\right]} \tag{11.21a}$$

Equation (11.21a) represents the ideality factor of the solar cell in terms of parameters (V_{oc}, V_m and I_m) and diode parameters (I_{ph}, I_o, R_s, R_{sh}); applying the approximations 1, 2, 3, 4 listed in table 11.1 and $\frac{I_o}{nV_t}\exp\left(\frac{I_{sc}R_s}{nV_t}\right) < < \frac{1}{R_{sh}-R_s}$ equation (11.21a) for Case I can be can be written as:

$$(n)_{\text{I}} = \frac{(V_m + I_m R_s - V_{oc})}{V_t\left[\ln\left(I_{sc} - \frac{V_m}{R_{sh}} - I_m\right) - \ln\left(I_{sc} - \frac{V_{oc}}{R_{sh}}\right) + \frac{I_m}{I_{sc} - (V_{oc}/R_{sh})}\right]} \tag{11.21b}$$

The ideality factor n is the most frequently utilized parameter for various solar cell technologies and ranges from 1 to 2 in Si solar cells. The value of n close to 1 represents ideal junctions, while n approaching 2 is associated with solar cell deterioration, non-uniformities or recombination centres, and R_{sh} effects [55]. This factor is more than 2 in organic solar cells [25], whereas in dye-sensitized solar cells, it is between 2 and 3 [56]. Ideality factor in the vicinity of 2 have been reported in perovskite solar cells as a result of carrier recombination and trap-assisted recombination in dark conditions [57, 58].

Three equations for saturation current computation can be obtained by eliminating I_{ph} from equations (11.15b), (11.16b), and (11.17b) for Case I. The first equation for I_o can be obtained from the *sc* (equation (11.15b)) and *oc* (equation (11.16b)) conditions as in [52]:

$$(I_{o_1})_{sc,\,oc} = \frac{I_{sc} + \frac{I_{sc}R_s}{R_{sh}} - \frac{V_{oc}}{R_{sh}}}{\left\{\exp\left(\frac{V_{oc}}{nV_t}\right) - \exp\left(\frac{I_{sc}R_s}{nV_t}\right)\right\}} \tag{11.22a}$$

Similarly, eliminating I_{ph} from equations (11.16b) and (11.17b), calculating I_o at *oc* and *MPP* conditions:

$$(I_{o_1})_{oc,\,MPP} = \frac{I_m + \frac{I_m R_s}{R_{sh}} + \frac{V_m}{R_{sh}} - \frac{V_{oc}}{R_{sh}}}{\exp\left(\frac{V_{oc}}{nV_t}\right) - \exp\left(\frac{V_m + I_m R_s}{nV_t}\right)} \tag{11.22b}$$

Similarly, calculating I_o from equations (11.15b) and (11.17b) by eliminating I_{ph} at *sc* and *MPP* conditions:

$$(I_{o_1})_{sc,\,MPP} = \frac{I_m + \frac{I_m R_s}{R_{sh}} + \frac{V_m}{R_{sh}} - \frac{I_{sc}R_s}{R_{sh}} - I_{sc}}{\exp\left(\frac{I_{sc}R_s}{nV_t}\right) - \exp\left(\frac{V_m + I_m R_s}{nV_t}\right)} \tag{11.22c}$$

Equations (11.22a), (11.22b), and (11.22c) can be used to compute I_o at a given temperature T. However, these equations are dependent on three unknown diode parameters: R_s, R_{sh}, and n. Depending on which parameters are estimated first, different approaches yield different I_o values. For instance, if n, R_s, and R_{sh} are computed first, I_o can be derived using equations (11.22a) and (11.22c). It has been discovered that equation (11.22b) is inappropriate for the determination of I_o as it produces negative values for I_o [59]. In order to determine n and I_o, numerous alternative methods have been explored in the literature [14, 60]. In this chapter, we will discuss a simple variable intensity method [61] for the simultaneous determination of n and I_o and is applicable to both SDM and DDM. In this method, the intensity of radiation incident on the PV cell is varied, and the corresponding V_{oc} and I_{sc} are measured. The setup for the cell measurement under illumination is shown in figure 11.1(a).

Under open-circuit condition equation (11.3) reduces to

$$V_j = V_{oc} \tag{11.23}$$

Equation (11.23) suggests that the entire photovoltage is produced across the p-n junction. This is frequently the case with solar cells meant to operate in normal sunlight or, more appropriately, in low injection level conditions. However, it may not be applicable for high-intensity illumination, when a cell may operate at high injection level conditions. Using the *sc* and *oc* conditions and applying the approximations $(R_s/R_{sh}) < < 1$, $I_{ph} = I_{sc}$, I_d can be written by using equation (11.16b) as:

$$I_d|_{oc} = \left(I_{sc} - \frac{V_{oc}}{R_{sh}}\right) = I_o \exp\left(\frac{V_{oc}}{nV_t}\right) \tag{11.24a}$$

$I_d|_{oc}$ is the diode current corresponding to $V_j = V_{oc}$.

Logarithmically, equation (11.24a) can be represented as:

$$\ln I_d = \ln I_o + \frac{1}{nV_t}(V_{oc}) \tag{11.24b}$$

n and I_o are computed from the ln (I_d) versus V_{oc} curve, which is illustrated in figure 11.3(a) [44]. The intercept on the ln (I_d) axis yields the value of I_o, whereas the slope $\frac{d(\ln I_d)}{dV_{oc}}$ of the curve which is equal to $\frac{1}{nV_t}$ yields the value of n. The graph of ln (I_d) versus V_{oc} for three Si solar cells cell#1, cell#2 and cell#3 is shown in figure 11.3

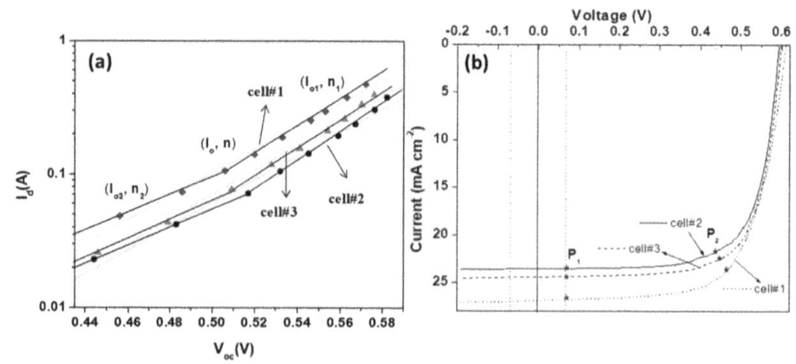

Figure 11.3. (a) ln (I_d) versus V_{oc} curve for silicon solar cells (cell#1, cell#2 and cell#3) obtained from V_{oc}–I_{sc} characteristics at room temperature (25 °C), (b) I–V curves without bias (4th quadrant) and with reverse bias (3rd quadrant) at room temperature (25 °C) under a simulated AM1.5 solar irradiance of 100 mW cm^{-2} intensity for the cells (cell#1, cell#2 and cell#3). Reproduced with permission from [44].

(a) [44, 62]. The corresponding illuminated I–V characteristics of the cells; cell#1, cell#2 and cell#3, measured at room temperature (25 °C), under a simulated AM 1.5 solar irradiance of 100 mWcm^{-2} intensity, are shown in figure 11.3(b). The details of the PV cells; cell#1, cell#2 and cell#3 are published in the paper [44]. In figure 11.3 (a), the dotted line represents the single exponential fitting and determines the value of I_o and n.

In order to determine the diode parameters (I_{o1}, n_1, I_{o2}, n_2) for DDM, applying the same assumptions as in equation (11.24a), equation (11.6) can be written as:

$$I_d = \left(I_{sc} - \frac{V_{oc}}{R_{sh}}\right) = \left\{I_{o1}\exp\left(\frac{V_{oc}}{n_1 V_t}\right) + I_{o2}\exp\left(\frac{V_{oc}}{n_2 V}\right)\right\} \tag{11.25a}$$

and

$$\ln I_d = (\ln I_{o1} + \ln I_{o2}) + \left(\frac{1}{n_1 V_t} + \frac{1}{n_2 V_t}\right)V_{oc} \tag{11.25b}$$

According to equation (11.25b), the graph of $\ln I_d$ versus V_{oc} can be used to obtain n_1 and n_2 from two slopes ($1/n_1 V_t$), ($1/n_2 V_t$), and I_{o1} and I_{o2} from two intercepts $\ln I_{o1}$, $\ln I_{o2}$. Solid lines represent the double exponential fitting of data in figure 11.3(a) for cell#1, cell#2 and cell#3.

The value of n can be computed using equation (11.21b) which requires the measured parameters (I_{sc}, V_{oc}, V_m and I_m) and diode parameters (R_s and R_{sh}) of a PV cell. However, equations (11.24b) and (11.25b) can be used to compute both n and I_o simultaneously for SDM and DDM, requiring V_{oc} and I_{sc} data at different intensities for a PV cell. The benefit of this method is that it is independent of R_s and thus more reliable than other methods that are available in the literature [61].

11.5.1.3 Series resistance (R_s*) and shunt resistance (*R_{sh}*)*

The series and shunt resistances can be evaluated by analyzing the slopes at three points: *oc*, *sc*, and *MPP*. The equations for R_s and R_{sh} are derived from the I–V characteristics [63, 64] via differentiation of equation (11.4) with respect to V:

$$\left(\frac{dI}{dV}\right) = \frac{I_o}{nV_t}\left(1 - R_s\frac{dI}{dV}\right)\exp\left(\frac{V_j}{nV_t}\right) + \frac{1}{R_{sh}}\left(1 - R_s\frac{dI}{dV}\right) \tag{11.26}$$

As demonstrated in [45], the derivative at *sc* point, using equation (11.26), can be expressed as:

$$\left(\frac{dI}{dV}\right)_{sc} = \frac{1}{R_{sh}} \tag{11.27}$$

At the open-circuit point, the derivative, using equation (11.26), can be approximated as:

$$\left(\frac{dI}{dV}\right)_{oc} = \frac{1}{R_s} \tag{11.28}$$

The slopes $(dI/dV)_{sc}$ and $(dI/dV)_{oc}$ are depicted in figure 11.2(c).

At MPP, the derivative of $P = IV$ with respect to V can be determined [28, 63, 65] by substituting $dP/dV = 0$ for MPP condition, yielding the following expression:

$$\left(\frac{dP}{dV}\right)_{MPP} = \frac{I_m}{V_m} = \left(\frac{\frac{1}{R_{sh}} + \frac{I_o}{nV_t}\exp\left(\frac{V_m + I_m R_s}{nV_t}\right)}{1 + \frac{R_s}{R_{sh}} + \frac{I_o R_s}{nV_t}\exp\left(\frac{V_m + I_m R_s}{nV_t}\right)}\right) \tag{11.29}$$

By rearranging the equation (11.29), R_{sh} in terms of R_s, V_m, I_m, n and I_o, can be derived for Case I as:

$$(R_{sh})_{\mathrm{I}} = \left(\frac{V_m - I_m R_s}{I_m + \frac{I_o}{nV_t}(V_m - I_m R_s)\exp\left(\frac{V_m + I_m R_s}{nV_t}\right)}\right) \tag{11.30}$$

Equations (11.27) and (11.28) can be used to determine R_{sh} and R_s. Equation (11.30) represents a relationship between R_{sh} and R_s. In general, R_s includes the material resistance, the resistances of the front and rear metallic contacts, and the contact resistances of the metallic contacts to the front and back surfaces [66]. R_{sh} is a parallel high conductive path across the p-n junction or at the cell edges, and it is related to the leakage current across the surfaces involving pin-holes, grain boundaries, and charge recombination processes [67]. Several approaches are available in the literature for measuring R_s and R_{sh} of a PV cell [43, 68–71]. It is important to mention that in both laboratory and commercial solar cells, FF is limited not only by high R_s values but also by low R_{sh} [72] values. Furthermore, resistive losses in commercial solar cells become larger as substrate size increases [43]. In our previous work, Priyanka *et al* [71], a novel method for determining R_{sh} and R_s of a solar cell has been developed utilizing illuminated I–V characteristics extending from the fourth to the third quadrant, as shown in figure 11.3(b). It does not assume that R_{sh} is infinitely large and also allows the calculation of R_s as a function of V_j using practical values of R_{sh}. The method is described in detail in the paper [71]. The methodology for determining R_s and R_{sh}, using SDM and DDM, is presented in this chapter, based on the papers [44, 71].

The values of R_s corresponding to the MPP (P_2 as shown in figure 11.3(b)) and near to the SC point (P_1 as shown in figure 11.3(b)) were determined from the I–V characteristics of the cells in the third and fourth quadrant using the relation [71]:

$$R_s = \frac{1}{I_f}\left(nV_t \ln\left(\frac{I_r P - (V_r + V_f + I_f P)}{I_o(P - R_s)}\right) - V_f\right) \tag{11.31}$$

where, $P = R_s + R_{sh}$.

The slope of the V_r–I_r curve in the third quadrant (as illustrated in figure 11.3(b)) computes $(R_s + R_{sh})^{-1}$, and R_{sh} may be assumed equal to P because $R_s << R_{sh}$. In equation (11.31), V_f, I_f represent voltage and current in the fourth quadrant, and V_r, I_r indicate voltage and current in the third quadrant of the I–V characteristics of a PV cell. The above method was used to calculate the R_s and R_{sh} values for cells:

Table 11.2. Diode parameters (I_o, n, I_{o1}, n_1, I_{o2}, n_2, R_s and R_{sh}) using SDM and DDM. R_s and R_{sh} values are determined at P_1 (near *SC* point) and P_2 (at *MPP*) for three cells (cells #1, #2, and #3). Reproduced with permission from [44].

	Diode parameters												
	Single-diode model (SDM)					Double-diode model (DDM)							
Cell	$I_o \times (10^{-6}A)$	n		R_s	R_{sh}	$I_{o1} \times (10^{-7}A)$	n_1	$I_{o2} \times (10^{-5}A)$	n_2		R_s	R_{sh}
cell#1	2.4	1.94	P_1	0.372	276.06	1.7	1.55	3.9	2.52	P_1	0.233	276.13
			P_2	0.038	276.47					P_2	0.034	276.47
cell#2	4.3	1.96	P_1	0.484	452.41	8.3	1.70	2.3	2.52	P_1	0.448	452.52
			P_2	0.094	452.75					P_2	0.084	452.74
cell#3	1.7	1.86	P_1	0.359	332.98	0.57	1.44	1.2	2.27	P_1	0.271	333.04
			P_2	0.071	333.37					P_2	0.051	333.36

Table 11.3. Performance parameters (V_{oc}, J_{sc}, *FF*, and η), for three cells (cells #1, cell#2, and cell#3) at room temperature (25 °C) under simulated AM1.5 Sun irradiation of 100 mWcm^{-2}. Reproduced with permission from [44].

	Performance parameters			
Cell	V_{oc} (V)	J_{sc} (mAcm^{-2})	FF	$\eta(\%)$
cell#1	0.609	26.78	0.68	11.1
cell#2	0.593	23.61	0.71	9.9
cell#3	0.599	24.45	0.72	10.4

cell#1, cell#2 and cell#3. In order to determine R_s for SDM and DDM, the relevant diode parameters I_o, n, and I_{o1}, n_1, I_{o2}, n_2 are computed using equations (11.24b) and (11.25b) as described previously. The diode parameters I_o, n, and I_{o1}, n_1, I_{o2}, n_2 along with R_s and R_{sh} for cell#1, cell#2 and cell#3 are listed in table 11.2. The performance parameters; V_{oc}, J_{sc}, *FF* and efficiency for the solar cells—cell#1, cell#2 and cell#3 are listed in table 11.3. Since both points P_1 and P_2 lie at $V_j < 0.5\ V$, I_{o2}, n_2 were used to determine R_s and R_{sh} for DDM. Using I_{o2} and n_2, lower R_s values were obtained as compared to R_s evaluated using I_o, n.

Case II

In Case II, R_{sh} is ignored; it is also known as the four parameters (I_{ph}, n, I_o and R_s) model [73, 74]. This model provides good accuracy, comparable to that of the DDM. The method does not require the full *I–V* curve; instead, it is based on the parameters (V_{oc}, I_{sc}, V_m and I_m) [75, 76]. The elimination of R_{sh} using approximation 5 in table 11.1, equations (11.15b), (11.16b), and (11.21b) can be reformulated for Case II as:

$$I_{ph} = I_{sc} \tag{11.32}$$

$$(n)_{\mathrm{II}} = \frac{(2V_m - V_{oc})}{V_t\left\{\frac{\mathrm{Im}}{I_{sc} - \mathrm{Im}} - \mathrm{In}\left(1\frac{\mathrm{Im}}{I_{sc}}\right)\right\}} \tag{11.33}$$

I_o for Case II can be determined from equations (11.22a), (11.22b) and (11.22c) and can be written as:

$$(I_{o_{\mathrm{II}}})_{sc,oc} = \frac{I_{sc}}{\left\{\exp\left(\frac{V_{OC}}{nV_t}\right) - \exp\left(\frac{I_{SC}R_S}{nV_t}\right)\right\}} \tag{11.34a}$$

$$(I_{o_{\mathrm{II}}})_{oc,MPP} = \frac{I_m}{\exp\left(\frac{V_{OC}}{nV_t}\right) - \exp\left(\frac{V_m + I_mR_S}{nV_t}\right)} \tag{11.34b}$$

$$(I_{o_{\mathrm{II}}})_{sc,MPP} = \frac{I_m - I_{sc}}{\exp\left(\frac{I_{SC}R_S}{nV_t}\right) - \exp\left(\frac{V_m + I_mR_S}{nV_t}\right)} \tag{11.34c}$$

The value of R_s is given by equation (11.29) as:

$$(R_s)_{\mathrm{II}} = \left(\frac{nV_t \ln\left(1 - \frac{I_m}{I_{sc}}\right) + V_{oc} - V_m}{I_m}\right) \tag{11.35}$$

Case III

This is the simplest representation of a solar cell where $R_s = 0$ and $R_{sh} \rightarrow \infty$; it decreases the number of five unknown parameters to three (I_{ph}, n, and I_o). The three-parameter model can further be reduced to two parameters for $n = 1$ which represents an ideal solar cell. Case III demonstrates SDM with no resistance [77–79]. In this case, the behavior of a PV cell is described by substituting $R_s = 0$ and $R_{sh} \rightarrow \infty$ in equations (11.15b), (11.16b), and (11.21), and is represented for Case III by the following equations [78]:

$$I_{ph} = I_{sc} \tag{11.36}$$

$$(n)_{\mathrm{III}} = \frac{V_m - V_{oc}}{V_t \ln\left(1 - \frac{I_m}{I_{sc}}\right)} \tag{11.37}$$

Equations (11.22a), (11.22b), and (11.22c) can be used to compute I_o for Case III as:

$$(I_{o_{\mathrm{III}}})_{sc,oc} = \frac{I_{sc}}{\left\{\exp\left(\frac{V_{OC}}{nV_t}\right) - 1\right\}} \tag{11.38a}$$

Table 11.4. List of solar cell models for parameter extraction using single and double-diode models. Reproduced from [32] CC BY 4.0.

Diode model	Extraction method	Extracted parameters	References
Single-diode model (SDM)	Analytical	I_{ph}, n, I_o, R_s, R_{sh}	[31, 45, 64]
	Numerical	I_{ph}, n, I_o, R_s, R_{sh}	[80, 81]
	Hybrid	I_{ph}, n, I_o, R_s, R_{sh}	[82]
	Optimization	I_{ph}, n, I_o, R_s, R_{sh}	[83–87]
	Numerical	n, R_s, R_{sh}	[88]
	Analytical	R_s, R_{sh}	[71, 89]
Double-diode model (DDM)	Numerical	I_{ph}, n_1, n_2, I_{o1}, I_{o2} R_s, R_{sh}	[37]
	Hybrid	I_{ph}, n_1, n_2, I_{o1}, I_{o2} R_s, R_{sh}	[6]
	Numerical	I_{ph}, I_{o1}, I_{o2} R_s, R_{sh}	[90]
	Optimization	I_{ph}, n_1, n_2, I_{o1}, I_{o2} R_s, R_{sh}	[83–85, 87, 91–94]
	Analytical	R_s, R_{sh}	[44]

$$(I_{o_{\mathrm{III}}})_{oc,MPP} = \frac{I_m}{\exp\left(\frac{V_{oc}}{nV_t}\right) - \exp\left(\frac{V_m}{nV_t}\right)} \tag{11.38b}$$

$$(I_{o_{\mathrm{III}}})_{sc,MPP} = \frac{I_{sc} - I_m}{\left\{\exp\left(\frac{V_m}{nV_t}\right) - 1\right\}} \tag{11.38c}$$

Nonetheless, the simplicity of the model is achieved at the expense of decreased accuracy, particularly at low irradiance due to the effect of the PV resistances. A recent study [52] evaluated the accuracy and complexity of these analytical techniques (Case I, Case II and Case III) for parameter extraction. The findings suggest that analytical procedures can produce results comparable to those obtained by numerical techniques [52]. Table 11.4 lists some of the available methods for diode parameter extraction using SDM and DDM.

11.5.2 Simulation techniques for parameter extraction

While analytical equations are easier to solve manually and provide valuable insight into cell functioning, they become more complex to solve when additional cell operating conditions are incorporated. Numerous researchers from across the world are engaged in the development of diverse simulation/numerical models that are intended for PV cells. Simulation techniques involve the development of mathematical models that describe the behaviour of solar cells under various operating conditions. These models typically comprise equations representing carrier generation, transport, and collection processes, as well as factors that describe material properties and device geometry.

PV solar cell/module modelling may be broadly classified into two types: electronic component-based modeling [95] and mathematical-based modelling [96–98]. A mathematics-based model may precisely characterize PV cells/modules utilizing complex mathematical algorithms. Computer-aided mathematical software, such as Matlab, has been used for decades to design and simulate mathematical models of solar cells [16]. The Matlab/SIMULINK model not only helps to anticipate the behaviour of any PV cell under different physical and environmental situations, but it can also be considered as a smart tool for extracting diode parameters (I_{ph}, n, I_o, R_s and R_{sh}). In contrast, electric/electronics-based simulation for PV cell is performed by software packages such as PSPICE [95]. However, it is unable to reflect all information of the surrounding/s (cell temperature, solar irradiance, etc). This may impact the accuracy of the simulation findings. Thus, modelling based on both mathematical and electronic components is significant, and a significant amount of research is being done in both areas. Additionally, simulation techniques significantly minimize experimental costs and manufacturing time by ensuring the most practical design of the solar cell under evaluation. In the mid-1980s, several solar simulators were developed for photovoltaic applications [99, 100]. Currently, a plethora of commercial solver packages have been designed particularly for modelling and simulating solar cells [101]. A variety of one-dimensional modelling tools, such as PC-1D, SCAPS-1D, and AMPS-1D, and multi-dimensional modelling tools, such as Silvaco ATLAS, COMSOL, and Sentaurus ASPIN3 are widely known to the photovoltaic community in current solar technology [102]. Table 11.5 lists the availability of these tools, including their operating system requirements and applications. This comparison analysis identifies the optimal tool for solar cell simulation [103]. PC1D is the most extensively utilized tool among commercially available solar cell modelling packages [99] for understanding device physics. PC1D is capable of simulating solar cells made of Si, germanium, GaAs, a-Si, InP, etc. PC1D enables the variation of parameters such as bulk doping levels, temperature, carrier lifetime, back surface field, and doping concentration in the emitter to visualize the solar cell performance. PC1D offers graphical representation of performance metrics, such as the internal quantum efficiency (*IQE*) and external quantum efficiency (*EQE*) of a PV cell. SCAPS-1D [104] software is utilized for electrical simulation and facilitates the characterization of a solar cell as a sequence of layers with varied optoelectrical properties and defect states in each layer and interface; it has been implemented in various cell technologies [105–108]. The most effective simulation tool for thin-film hydrogenated a-Si solar cells (a-Si:H) is ASA [109]. It is suitable for analyzing homo and heterojunctions, as well as multi-junctions such as perovskite silicon solar cells. Another widely used simulation tool is AMPS-1D, which is used to assess the characteristics and performance of polycrystalline (pc-Si) [110] a-Si:H [111], CZTS [102] and CIGS [112] solar cells. A modified version of AMPS, known as wxAMPS, is also used to simulate amorphous silicon p-i-n tandem cells [113], CZTS [114] PV cells and CZTS/CTS tandem solar cells [115]. Commercial packages for organic thin-film solar cells include SETFOS [116] and GPVDM [117] simulators. AFORS-HET is a free simulator package specifically designed for heterojunction solar cells and other optoelectronic devices [118]. The ASPIN [119] 2D simulator works well with CIGS cells and aSi:H/μc-Si:H tandem cells that are based on p-i-n junctions and

Table 11.5. List of solar cell simulators. Reproduced from [135] CC BY 4.0.

S/N	Simulators	Dimension	Availability	System requirement	Capability
1.	Simulink (MATLAB)	NA	Subscription-based	Compatible with Windows, Linux and Mac	Capable of modeling most PV cells
2.	PSPICE	NA	Free of charge	Compatible with Windows	Electrical simulation
3	PC1D	1D	Free of charge	All windows versions up to Windows 7	Capable of simulating most solar cells but perfect for (C-Si) solar cell.
4	ASA	1D	Subscription-based	Windows 7 or 8 Operating systems	Thin-film cells: a-Si:H, μc-Si:H and a-Si:H/μc-Si:H tandem cell
5	AMPS	1D	Free of charge	MacOS 10.9 and Windows 7 or Upgrade	C-Si, a-Si:H, CIGS, CZTS cells
6	wxAMPS	1D	Free of charge	All Windows versions up to Windows 7, Vista	Having all feature of AMPS including tandem solar cell.
7	SCAPS	1D	Free of charge	Compatible with Windows and Linux	Thin-film Solar cell: CIGS, CdTe, GaAs, C-Si and a-Si:H cell
8	SETFOS	1D	Subscription-based	Compatible with Windows and Linux (×86 and ARM)	Organic, perovskites, quantum-dots, perovskite/silicon tandem solar cells
9	Gpvdm	1D	Subscription-based	Windows Vista/7/8 Operating Systems	Perovskite, Polymer, C-Si, a-Si:H and CIGS cells
10	AFORS-HET	1D	Free of charge	Windows XP/Vista/7/8/8.1/Linux	a-Si:H/c-Sicell
11	ASPIN	2D	Free of charge	Windows 7/8.1/10	CIGS, a-Si:H/C-Si.a-Si:H p-i-n heterojunctioncell
12	PECSIM	N/A	Free of charge	Windows 7/8.1/10	Perovskite, dye-sensitized solar cells (DSCs)
13	ADEPT	1D	Free of charge	Windows 10, Windows Server 2016, Processor: Intel Core i7 or equivalen	C-Si, GaAs, AlGaAs/GaAs tandem, CIGS, CdTe and thin film a-Si:H solar cell.

14	TCAD	2D, 3D	Subscription-based	Windows 7, 8, 8.1 and 10 (64-bit), Linux 6 and 7	CMOS, power, memory, image sensors, solar cells
15	ATLAS	2D, 3D	Subscription-based	Windows and Linux platforms (Version not mentioned)	Organic solar cells, Tandem solar cells, Photodetectors
16	MSCS	1D	Free of charge	Compatible with Windows and Linux. Web version is very responsive.	III–V high efficiency Multi-junction cell
17	Quokka 2 (MATLAB)	1D, 2D or 3D	Free of charge	Compatible with Windows, Linux and Mac	Silicon heterojunction solar cell
18	SIESTA (DFT)	1D, 2D	Free of charge	Compatible with Windows, Linux and Mac	Si, CdTe, organic polymers
19	Quantum ESPRESSO (DFT)	1D, 2D	Free of charge	Compatible with Linux	Si, CdTe, or organic polymers

makes it possible to simulate lateral transport, grain boundaries, and combinations of various materials which are not viable with 1D simulators [120]. Modern solar cell simulation software, PECSIM, is created for DSSCs (dye-sensitized solar cells) and perovskite cells [121]. Another popular simulation tool is ADEPT [122] which is suitable for the performance simulation of c-Si, GaAs, AlGaAs/GaAs, CIGS, CdTe, and a-Si:H solar cells [123]. TCAD software is an electrical simulation [124] tool; it enables researchers and industrialists to simulate complex solar cells designs. Modern TCAD software package such as Nextnano is helpful for simulating Group III-nitride solar cells and exploring future nanoscale structured devices [125]. ATLAS is another outstanding simulator that has been developed to assess the optoelectronic behaviour of 2D, 3D semiconductor devices as well as to simulate the performance of tandem cells [126]. MSCS-1D is a faster tool than other complex simulators and has been created to address multi-junction solar cells with extremely high and ultrahigh efficiencies [127]. Quokka 2 is a quick and free computer simulation tool that may be used to model solar cells in 2D or 3D [128]. Quokka provides a fast and effective numerical solution for 1D/2D/3D charge carrier transport in quasi-neutral silicon devices.

11.5.2.1 Density functional theory (DFT)

In addition to simulation techniques (described in the preceding section), molecular dynamics (MD) and computational quantum theories such as density functional theory (DFT) reliably anticipate the properties of various layers to estimate device efficiency and stability. DFT is accurate and commonly used to assess potential candidate materials for solar cell applications. DFT is important in understanding the electronic structure and transport-related properties of semiconducting materials such as Si, CdTe, or organic polymers used in solar cells [129, 130]. In addition, DFT offers insights into the optoelectronic features of polymer based solar cells, including charge mobility, V_{oc}, reorganization energies, bandgap, ionization potential, electron affinity, and highest occupied and lowest unoccupied molecular orbitals (HOMO and LUMO) [131] which are essential for comprehending how materials absorb and transfer light-generated charge carriers. DFT has recently been used to examine 2100 structures, to select best candidates for solar cell materials using the computed bandgap values [132]. For the purpose of DFT applications, a number of software packages are available. Siesta [133] and Quantum Espresso (QE) [134] are software for DFT, available at no cost to the user, and are commonly used to investigate solar cells.

The aforementioned solar cell simulators are well-known and widely used to simulate various types of solar cells. While many simulators share many basic components, there are some differences between them in terms of user access, speed, efficacy, and graphical user interface (GUI).

11.6 Spectral response (*SR*)

The spectral response (*SR*) measurement describes how the solar cell reacts to different wavelengths of light. It evaluates the spectral range across which the solar cell performs well; it can give information regarding the material properties and device design. SR of a solar cell is the ratio of J_{sc}, generated under monochromatic illumination of a certain wavelength (λ) to P_{in}, at the same wavelength.

SR can be expressed mathematically as:

$$SR(\lambda) = \left(\frac{J_{sc}}{P_{in}}\right)_{\lambda} \text{Amp/Watt} \tag{11.39}$$

J_{sc} of the experimental cell for a given irradiance level E_λ (Watt m^{-2} unit^{-1} wavelength), with its spectral distribution $E_\lambda\, d\lambda$, can be calculated by integrating the measured SR (λ) in the appropriate wavelength range (e.g. 400–1200 nm) by using the expression,

$$J_{sc} = q \int_{\lambda_0}^{\lambda_g} E_\lambda\, SR(\lambda)\, d\lambda \tag{11.40}$$

where, λ_0 is the least significant wavelength of the supplied solar cell spectrum, e.g., AM0 or AM1.5., λ_g is the cutoff wavelength of the radiation that can be absorbed in a solar cell and E_λ is the irradiance of the solar spectrum.

In general, SR is measured in the wavelength range 400–1200 nm in steps of 50 nm wavelength interval using interference filters by illuminating the emitter region with a monochromatic radiation from a tungsten halogen lamp as the illuminating source. A calibrated solar cell is used as a reference cell.

Figure 11.4 depicts SR of various types of PV modules for AM1.5 G spectrum (up to 1300 nm). This figure illustrates that SR varies by technology [136]. Figure 11.4 shows that SR is lower in the higher wavelength region because photons have less energy than the material bandgap threshold. Therefore, in narrow SR technologies

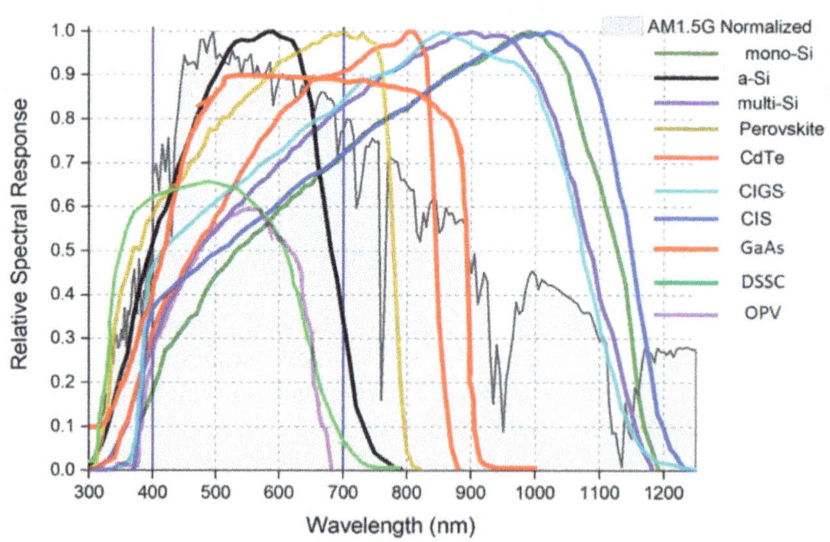

Figure 11.4. Spectral response characteristics of different solar module technologies. Reproduced with permission from [137] CC BY 3.0.

such as a-Si and CdTe, the impact of spectral variation on the output of PV devices is most noticeable.

11.7 Quantum efficiency (*QE*)

The quantum efficiency (*QE*) of a solar cell is defined as the ratio of the number of electrons in the external circuit produced by an incident photon of a given wavelength. Accordingly, it is possible to define two quantum efficiencies: internal quantum efficiency (*IQE*) and external quantum efficiency (*EQE*). The value of *EQE* can be computed by taking into account all of the incident photons. However, the value of *IQE*, takes into account photons that are not reflected. The *EQE* measurement evaluates the efficiency of a solar cell in converting the incident photons into electrical current as a function of wavelength. This technique yields information about *SR* of the solar cell and helps in the identification of factors influencing light absorption and carrier generation. Conceptually, the *SR* and *QE* are similar to each other.

Mathematically, *EQE* and *IQE* can be defined as:

$$EQE = \frac{hc}{q\lambda} SR(\lambda) \tag{11.41}$$

$$IQE = \frac{EQE}{(1 - R_\lambda)} \tag{11.42}$$

where, h is the Planck's constant, c is the velocity of light, and R_λ is the reflection coefficient. Figure 11.5 illustrates the graphical representations of *EQE* and *IQE*. Figure 11.5(a) shows *IQE*, *EQE*, and R_λ of a c-Si solar cell with λ and figure 11.5(b) indicates the *EQE* of PV cells based on various materials.

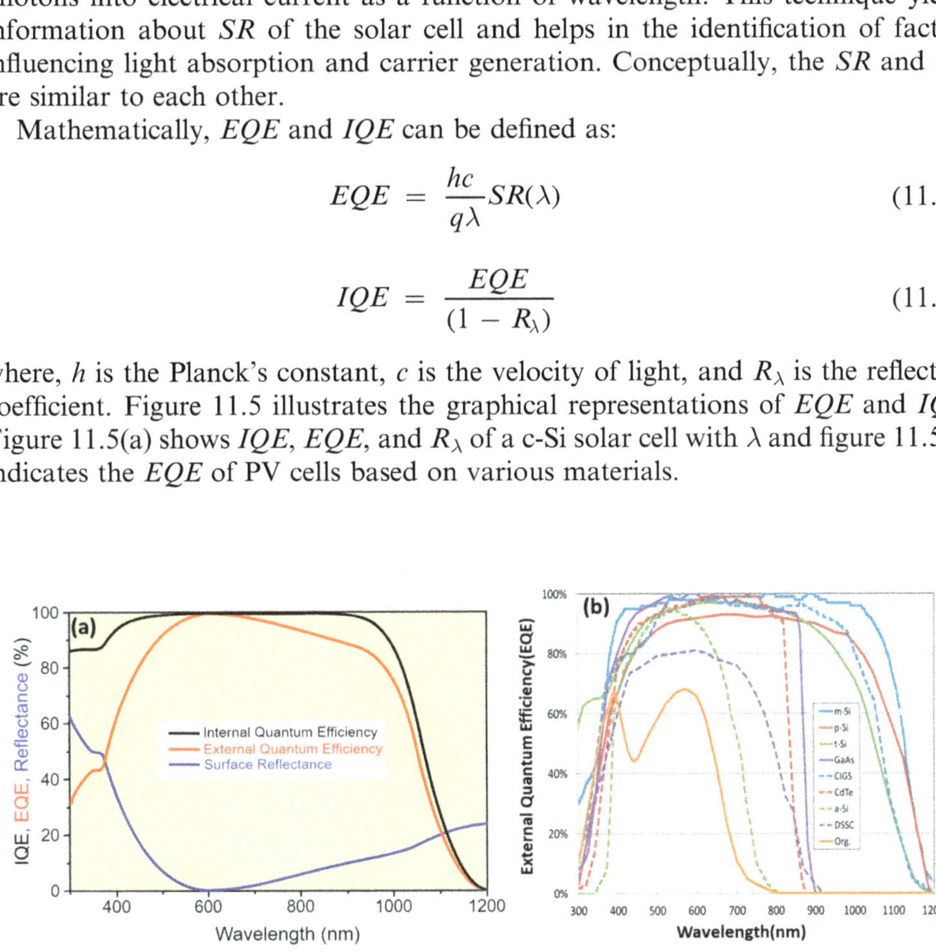

Figure 11.5. (a) *IQE*, *EQE*, and reflectance (R_λ) of a crystalline silicon solar cell with wavelength (λ) (adapted from Wikipedia) (b) *EQE* of photovoltaic solar cells based on various materials. Reprinted with permission from [138] CC BY 3.0.

11.8 Minority carrier lifetime (τ)

Minority carrier lifetime (τ) is an important parameter and its computation is crucial for understanding the PV cell performance and maximizing efficiency. The minority carrier lifetime is the average time a minority charge carrier (holes in n-type, electrons in p-type) spends in the semiconductor material before recombining with a majority charge carrier (holes in p-type, electrons in n-type). There are numerous methods for determining the minority carrier lifetime in solar cells. Generally, lifetime measurements are performed using transient or steady-state approaches. Transient approaches involve the measurement of the change in carrier concentration as a function of time after the PV cell is exposed to a light or a voltage pulse. In contrast, steady-state approaches examine the response of a PV cell to a continuous stimulation, such as a constant light source or a steady-state injection of carriers. A recent review paper covers almost all the techniques for lifetime measurement [139]. Common transient techniques include transient photocurrent (TPC), transient photovoltage (TPV) transient absorption spectroscopy (TAS), time-resolved electroluminescence (TREL), and time-resolved photoluminescence (TRPL), etc. Some steady-state approaches include photoluminescence (PL) and open-circuit voltage decay (OCVD). Some techniques, such as microwave photoconductance decay (MPCD), space-charge-limited current (SCLC) and electroluminescence (EL) decay can be used in both steady-state and transient techniques.

Nevertheless, some of the approaches are appropriate for inorganic materials, while others are for organic materials. TRPL is a common minority carrier lifetime measurement technique that is suitable for a wide range of solar cells, including both inorganic and organic materials. It is a fast, direct, and contactless method for performing the measurement of the excess minority-carrier concentration as a function of time. The transient decay of the PL signal determines the minority-carrier lifetime. Figure 11.6 depicts the TRPL schematic, and standard curve fitting method for measurements.

In the TRPL technique, a semiconductor sample is stimulated by a short pulse of light, usually a laser. This pulse produces an initial excess carrier density (electrons

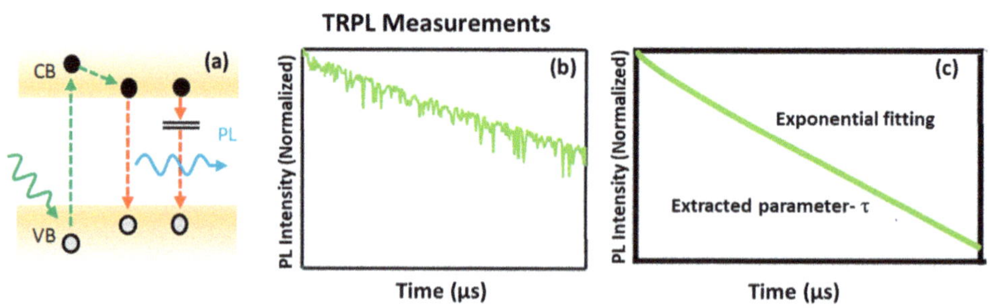

Figure 11.6. (a) TRPL signals generated from recombination pathways, (b) measured PL signal as a function of time using a sensitive detector (c) Conventional fitting method of TRPL readings (logarithmic scale). PL curves are commonly fitted by exponential function to derive lifetime τ (for $\beta = 1$) Reproduced and modified with permission from [141], copyright 2018 American Chemical Society.

or holes) in the sample. After the excitation pulse, the sample emits PL as the excess carriers recombine with the majority carriers as shown in figure 11.6(a). This PL signal is typically measured (depicted in figure 11.6(b)) as a function of time using a sensitive detector. The PL signal decays exponentially over time as the excess carriers recombine and can be represented by the following equation (11.43):

$$I(t) = I_0 \exp[(-t/\tau)^{\beta}] \tag{11.43}$$

where, $I(t)$ is PL intensity at time t and I_0 is the initial PL intensity at $t = 0$. τ is a time constant and represents the minority carrier lifetime and β $(0 \leqslant \beta \leqslant 1)$ is stretching index. The stretched exponential function is well known to describe the PL decay and transport properties of disordered systems [140]. The stretching index is a measure of the degree of disorder in the material.

For $\beta = 1$, equation (11.43) reduces to;

$$I(t) = I_0 \exp(-t/\tau) \tag{11.44a}$$

In logarithmic scale, equation (11.44a) can be written as:

$$\ln I(t) = \ln I_0 - t/\tau \tag{11.44b}$$

The slope of the curve ln $I(t)$ versus t can be used to calculate τ, according to above equation (11.44b). TRPL measurements can be used to investigate τ at different temperatures or excitation conditions. Generally, the decay of PL intensity is fitted by an exponential function according to equation (11.44b), as shown in figure 11.6(c).

11.9 Junction depth (x_j)

The junction depth (x_j), as depicted in the schematic representation of a solar cell in figure 11.1(b), is the depth at which the p-n junction is situated beneath the surface of the semiconductor material in a PV cell and is typically measured in nanometers (nm) or micrometers (μm). It is an important parameter which affects the efficiency and performance of the solar cell, as the majority of photogenerated carriers are separated and collected at this depth. It is preferable to have a shallow junction depth for efficient collection of photogenerated carriers and to minimize carrier recombination losses. However, x_j must be sufficiently deep to provide adequate light absorption and effective use of the semiconductor material. The optimal x_j involves a trade-off between maximizing carrier collection efficiency and ensuring sufficient light absorption. The junction depth is commonly controlled via diffusion processes during cell fabrication, by modifying the diffusion parameters and doping profile.

The sheet resistance (R_{sheet}) of the emitter region (n^+) in an n^+-p junction solar cell is related to x_j and is an important parameter that can influence the recombination losses at the front side of the solar cell. R_{sheet} (Ω/Ł) of the emitter layer is measured by the four point probe method [142]. In a study [143], the impact of R_{sheet} on the efficiency of a solar cell has been evaluated using PC1D simulation by fixing the emitter dopant concentration at 1×10^{20} cm^{-3}. The change in possible x_j is observed as shown in figure 11.7.

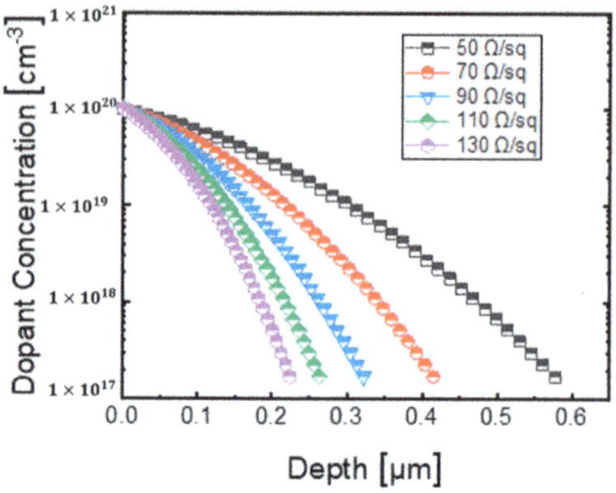

Figure 11.7. Variation in R_{sheet} when the emitter dopant concentration is fixed at 1×10^{20} cm^{-3}. Reproduced with permission from [143] CC BY 4.0.

It can be seen from the figure 11.7 that a lower R_{sheet} indicates a deeper junction. In order to achieve a R_{sheet} of 50 Ω sq^{-1}, the junction depth should be 0.59 μm. The shallow junction depth of 0.21 μm was obtained for a higher R_{sheet} of 130 Ω sq^{-1}. The high R_{sheet} at the emitter is favourable for low saturation current density at the front.

Different types of solar cells may necessitate different methodologies to determine x_j due to variation in material properties, fabrication methods, and device configurations. Though the measurement of R_{sheet} using four probe method provides x_j, surface analysis techniques are found to be suitable for measurement of x_j. Some of the techniques such as scanning electron microscopy (SEM), secondary ion mass spectrometry (SIMS), spectroscopic ellipsometry (SE), electrochemical capacitance-voltage (ECV) profiling are found to be appropriate for c-Si solar cells. However, techniques such as atomic force microscopy (AFM), PL imaging, *x*-ray diffraction (XRD) are found to be suitable for thin-film solar cells (e.g., CdTe, CIGS). x-ray photoelectron spectroscopy (XPS), TRPL, Kelvin probe force microscopy (KPFM) and ultraviolet photoelectron spectroscopy (UPS) are some of the techniques that are useful for characterization of perovskite solar cells. Conductive atomic force microscopy (C-AFM), Raman spectroscopy, grazing-incidence x-ray diffraction (GIXRD) are used for emerging solar cell technologies (e.g., organic photovoltaics, quantum dot solar cells).

By employing the aforementioned analytical and simulation techniques, researchers can comprehensively characterize solar cells, identify performance-limiting factors, and develop strategies for improving device efficiency, reliability, and durability.

11.10 Conclusions

Computational tools such as analytical and numerical simulation procedures, available in literature, are discussed in this chapter, to extract the parameters: I_{sc}, V_{oc}, FF and η and diode parameters i.e., I_{ph}, n, I_o, R_s and R_{sh}. Single- and double-diode models for parameter extraction are also described. The effect of diode parameters on the performance of solar cells is discussed. A few other significant solar cell parameters such as spectral response, quantum efficiency, minority carrier lifetime and junction depth and their computation are briefly discussed.

References

[1] Parida B, Iniyan S and Goic R 2011 A review of solar photovoltaic technologies *Renew. Sustain. Energy Rev.* **15** 1625–36

[2] Sarkar M N I 2016 Effect of various model parameters on solar photovoltaic cell simulation: a SPICE analysis *Renew.: Wind, Water, Solar* **3** 13

[3] Humada A M *et al* 2016 Solar cell parameters extraction based on single and double-diode models: a review *Renew. Sustain. Energy Rev.* **56** 494–509

[4] Villalva M G, Gazoli J R and Filho E R 2009 Comprehensive approach to modeling and simulation of photovoltaic arrays *IEEE Trans. Power Electron.* **24** 1198–208

[5] Gontean A *et al* 2017 A novel high accuracy PV cell model including self heating and parameter variation *Energies* **11** 36

[6] Yahya-Khotbehsara A and Shahhoseini A 2018 A fast modeling of the double-diode model for PV modules using combined analytical and numerical approach *Sol. Energy* **162** 403–9

[7] Chin V J, Salam Z and Ishaque K 2016 An accurate modelling of the two-diode model of PV module using a hybrid solution based on differential evolution *Energy Convers. Manage.* **124** 42–50

[8] Cotfas D T, Cotfas P A and Kaplanis S 2013 Methods to determine the dc parameters of solar cells: a critical review *Renew. Sustain. Energy Rev.* **28** 588–96

[9] Bashahu M and Nkundabakura P 2007 Review and tests of methods for the determination of the solar cell junction ideality factors *Sol. Energy* **81** 856–63

[10] Wang G *et al* 2017 An iterative approach for modeling photovoltaic modules without implicit equations *Appl. Energy* **202** 189–98

[11] Wong J 2013 Griddler: intelligent computer aided design of complex solar cell metallization patterns *2013 IEEE 39th Photovoltaic Specialists Conf. (PVSC)* (Piscataway, NJ: IEEE)

[12] Mellit A and Kalogirou S A 2008 Artificial intelligence techniques for photovoltaic applications: a review *Prog. Energy Combust. Sci.* **34** 574–632

[13] Ciulla G *et al* 2014 A comparison of different one-diode models for the representation of I–V characteristic of a PV cell *Renew. Sustain. Energy Rev.* **32** 684–96

[14] Liao P *et al* 2018 A new method for fitting current–voltage curves of planar heterojunction perovskite solar cells *Nano-micro Lett.* **10** 1–8

[15] King D *et al* 1997 Dark current–voltage measurements on photovoltaic modules as a diagnostic or manufacturing tool *Conf. Record of the Twenty Sixth IEEE Photovoltaic Specialists Conf.-1997* (Piscataway, NJ: IEEE)

[16] Ishaque K and Salam Z 2011 A comprehensive MATLAB Simulink PV system simulator with partial shading capability based on two-diode model *Sol. Energy* **85** 2217–27

[17] Franzitta V, Orioli A and Gangi A D 2017 Assessment of the usability and accuracy of two-diode models for photovoltaic modules *Energies* **10** 564

[18] Chin V J, Salam Z and Ishaque K 2015 Cell modelling and model parameters estimation techniques for photovoltaic simulator application: a review *Appl. Energy* **154** 500–19

[19] Abbassi R *et al* 2018 Identification of unknown parameters of solar cell models: a comprehensive overview of available approaches *Renew. Sustain. Energy Rev.* **90** 453–74

[20] Shannan N M A A, Yahaya N Z and Singh B 2013 Single-diode model and two-diode model of PV modules: a comparison *2013 IEEE Int. Conf. on Control System, Computing and Engineering* (Piscataway, NJ: IEEE)

[21] Chegaar M, Nehaoua N and Bouhemadou A 2008 Organic and inorganic solar cells parameters evaluation from single I–V plot *Energy Convers. Manage.* **49** 1376–9

[22] Easwarakhanthan T *et al* 1986 Nonlinear minimization algorithm for determining the solar cell parameters with microcomputers *Int. J. Sol. Energy* **4** 1–12

[23] Boutana N *et al* 2017 An explicit IV model for photovoltaic module technologies *Energy Convers. Manage.* **138** 400–12

[24] Lim L H I *et al* 2015 A linear method to extract diode model parameters of solar panels from a single I–V curve *Renew. Energy* **76** 135–42

[25] Chegaar M, Azzouzi G and Mialhe P 2006 Simple parameter extraction method for illuminated solar cells *Solid-State Electron.* **50** 1234–7

[26] Ruschel C S, Gasparin F P and Krenzinger A 2021 Experimental analysis of the single diode model parameters dependence on irradiance and temperature *Sol. Energy* **217** 134–44

[27] Celik A N and Acikgoz N 2007 Modelling and experimental verification of the operating current of mono-crystalline photovoltaic modules using four-and five-parameter models *Appl. Energy* **84** 1–15

[28] Kennerud K L 1969 Analysis of performance degradation in CdS solar cells *IEEE Trans. Aerosp. Electron. Syst.* **AES-5** 912–7

[29] Shongwe S and Hanif M 2015 Comparative analysis of different single-diode PV modeling methods *IEEE J. Photovolt.* **5** 938–46

[30] Ortiz-Conde A *et al* 2014 A review of diode and solar cell equivalent circuit model lumped parameter extraction procedures *FU: Elec Energ* **27** 57–102

[31] Tivanov M *et al* 2005 Determination of solar cell parameters from its current–voltage and spectral characteristics *Sol. Energy Mater. Sol. Cells* **87** 457–65

[32] Zidan M N *et al* 2021 Organic solar cells parameters extraction and characterization techniques *Polymers* **13** 3224

[33] Velilla E *et al* 2018 Numerical analysis to determine reliable one-diode model parameters for perovskite solar cells *Energies* **11** 1963

[34] Sulyok G and Summhammer J 2018 Extraction of a photovoltaic cell's double-diode model parameters from data sheet values *Energy Sci Eng* **6** 424–36

[35] Dehghanzadeh A, Farahani G and Maboodi M 2017 A novel approximate explicit double-diode model of solar cells for use in simulation studies *Renew. Energy* **103** 468–77

[36] Hovinen A 1994 Fitting of the solar cell IV-curve to the two diode model *Phys. Scr.* **1994** 175

[37] Sandrolini L, Artioli M and Reggiani U 2010 Numerical method for the extraction of photovoltaic module double-diode model parameters through cluster analysis *Appl. Energy* **87** 442–51

[38] Garrido-Alzar C 1997 Algorithm for extraction of solar cell parameters from I–V curve using double exponential model *Renew. Energy* **10** 125–8

[39] Hejri M *et al* 2014 On the parameter extraction of a five-parameter double-diode model of photovoltaic cells and modules *IEEE J. Photovolt.* **4** 915–23

[40] Ishaque K, Salam Z and Taheri H 2011 Simple, fast and accurate two-diode model for photovoltaic modules *Sol. Energy Mater. Sol. Cells* **95** 586–94

[41] Gow J and Manning C 1996 Development of a model for photovoltaic arrays suitable for use in simulation studies of solar energy conversion systems *1996 Sixth Int. Conf. on Power Electronics and Variable Speed Drives (Conf. Publ. No. 429) (Nottingham)* 69–74

[42] McIntosh K and Honsberg C 2000 The influence of edge recombination on a solar cell's IV curve *16th European Photovoltaic Solar Energy Conf. 2000*

[43] Wolf M and Rauschenbach H 1963 Series resistance effects on solar cell measurements *Adv. Energy Convers.* **3** 455–79

[44] Singh P and Ravindra N 2012 Analysis of series and shunt resistance in silicon solar cells using single and double exponential models *Emerg. Mater. Res.* **1** 33–8

[45] Phang J, Chan D and Phillips J 1984 Accurate analytical method for the extraction of solar cell model parameters *Electron. Lett.* **20** 406–8

[46] Khan F *et al* 2013 Extraction of diode parameters of silicon solar cells under high illumination conditions *Energy Convers. Manage.* **76** 421–9

[47] Batzelis E 2019 Non-iterative methods for the extraction of the single-diode model parameters of photovoltaic modules: a review and comparative assessment *Energies* **12** 358

[48] Jain A, Sharma S and Kapoor A 2006 Solar cell array parameters using Lambert W-function *Sol. Energy Mater. Sol. Cells* **90** 25–31

[49] Song Z *et al* 2021 An effective method to accurately extract the parameters of single diode model of solar cells *Nanomaterials* **11** 2615

[50] Pindado S *et al* 2018 Assessment of explicit models for different photovoltaic technologies *Energies* **11** 1353

[51] De Soto W, Klein S A and Beckman W A 2006 Improvement and validation of a model for photovoltaic array performance *Sol. Energy* **80** 78–88

[52] Ibrahim H and Anani N 2017 Evaluation of analytical methods for parameter extraction of PV modules *Energy Procedia* **134** 69–78

[53] Van Dyk E and Meyer E L 2004 Analysis of the effect of parasitic resistances on the performance of photovoltaic modules *Renew. Energy* **29** 333–44

[54] Dieme N and Sane M 2016 Impact of parasitic resistances on the output power of a parallel vertical junction silicon solar cell *Energy Power Eng.* **8** 130–6

[55] Jain A and Kapoor A 2005 A new method to determine the diode ideality factor of real solar cell using Lambert W-function *Sol. Energy Mater. Sol. Cells* **85** 391–6

[56] Murayama M and Mori T 2006 Equivalent circuit analysis of dye-sensitized solar cell by using one-diode model: effect of carboxylic acid treatment of TiO_2 electrode *Jpn. J. Appl. Phys.* **45** 542

[57] Wetzelaer G-J A *et al* 2015 Trap-assisted non-radiative recombination in organic-inorganic perovskite solar cells *Adv. Mater.* **27** 1837–41

[58] Agarwal S *et al* 2014 On the uniqueness of ideality factor and voltage exponent of perovskite-based solar cells *J. Phys. Chem. Lett.* **5** 4115–21

[59] Ndegwa R *et al* 2020 A fast and accurate analytical method for parameter determination of a photovoltaic system based on manufacturer's data *J. Renew. Energy* **2020** 7580279

[60] Almora O *et al* 2018 Discerning recombination mechanisms and ideality factors through impedance analysis of high-efficiency perovskite solar cells *Nano Energy* **48** 63–72

[61] Arora N 1982 *Studies on Solar Grade Polycrystalline Silicon Solar Cell* (University of Delhi)

[62] Singh P 2009 Fabrication, characterization and other related studies for performance improvement of crystalline silicon solar cells *Physics (N.Y.)* (New Delhi: Jamia Millia Islamia and National Physical Laboratory)

[63] Sera D, Teodorescu R and Rodriguez P 2007 PV panel model based on datasheet values *2007 IEEE Int. Symp. on Industrial Electronics* (Piscataway, NJ: IEEE)
[64] Lo Brano V and Ciulla G 2013 An efficient analytical approach for obtaining a five parameters model of photovoltaic modules using only reference data *Appl. Energy* **111** 894–903
[65] El Achouby H *et al* 2018 New analytical approach for modelling effects of temperature and irradiance on physical parameters of photovoltaic solar module *Energy Convers. Manage.* **177** 258–71
[66] Li Y *et al* 2013 Evaluation of methods to extract parameters from current–voltage characteristics of solar cells *Sol. Energy* **90** 51–7
[67] Mialhe P *et al* 1986 The diode quality factor of solar cells under illumination *J. Phys. D: Appl. Phys.* **19** 483
[68] Agarwal S K *et al* 1981 A new method for the measurement of series resistance of solar cells *J. Phys. D: Appl. Phys.* **14** 1643
[69] Singh V N and Singh R P 1983 A method for the measurement of solar cell series resistance *J. Phys. D: Appl. Phys.* **16** 1823
[70] Sharma S *et al* 2000 Overcoming the problems in determination of solar cell series resistance and diode factor *J. Phys. D: Appl. Phys.* **23** 1256
[71] Priyanka M L and Singh S N 2007 A new method of determination of series and shunt resistances of silicon solar cells *Sol. Energy Mater. Sol. Cells* **91** 137–42
[72] Bowden S and Rohatgi A Rapid and accurate determination of series resistance and fill factor losses in industrial silicon solar cells *17th European Photovoltaic Solar Energy Conf. (Munich)*
[73] Khezzar R, Zereg M and Khezzar A 2014 Modeling improvement of the four parameter model for photovoltaic modules *Sol. Energy* **110** 452–62
[74] Aoun N and Bailek N 2019 Evaluation of mathematical methods to characterize the electrical parameters of photovoltaic modules *Energy Convers. Manage.* **193** 25–38
[75] Aldwane B 2014 Modeling, simulation and parameters estimation for photovoltaic module *2014 First Int. Conf. on Green Energy ICGE 2014 (Sfax, Tunisia)* (Piscataway, NJ: IEEE) 101–6
[76] Sera D, Teodorescu R and Rodriguez P 2008 Photovoltaic module diagnostics by series resistance monitoring and temperature and rated power estimation *2008 34th Annual Conf. of IEEE Industrial Electronics (Orlando, FL)* (Piscataway, NJ: IEEE) 2195–9
[77] Xiao W, Dunford W G and Capel A 2004 A novel modeling method for photovoltaic cells *2004 IEEE 35th Annual Power Electronics Specialists Conf. (IEEE Cat. No. 04CH37551)* (Piscataway, NJ: IEEE)
[78] Saloux E, Teyssedou A and Sorin M 2011 Explicit model of photovoltaic panels to determine voltages and currents at the maximum power point *Sol. Energy* **85** 713–22
[79] Mahmoud Y, Xiao W and Zeineldin H 2011 A simple approach to modeling and simulation of photovoltaic modules *IEEE Trans. Sustain. Energy* **3** 185–6
[80] Ghani F *et al* 2014 The numerical calculation of single-diode solar-cell modelling parameters *Renew. Energy* **72** 105–12
[81] Ayodele T, Ogunjuyigbe A and Ekoh E 2016 Evaluation of numerical algorithms used in extracting the parameters of a single-diode photovoltaic model *Sustain. Energy Technol. Assess.* **13** 51–9

[82] Hejri M *et al* 2016 An analytical-numerical approach for parameter determination of a five-parameter single-diode model of photovoltaic cells and modules *Int. J. Sustain. Energy* **35** 396–410

[83] Kumari P A and Geethanjali P 2017 Adaptive genetic algorithm based multi-objective optimization for photovoltaic cell design parameter extraction *Energy Procedia* **117** 432–41

[84] Luu T V and Nguyen N S 2020 Parameters extraction of solar cells using modified JAYA algorithm *Optik* **203** 164034

[85] Ye M, Wang X and Xu Y 2009 Parameter extraction of solar cells using particle swarm optimization *J. Appl. Phys.* **105** 094502

[86] Muhsen D H *et al* 2015 Extraction of photovoltaic module model's parameters using an improved hybrid differential evolution/electromagnetism-like algorithm *Sol. Energy* **119** 286–97

[87] Gao X *et al* 2018 Parameter extraction of solar cell models using improved shuffled complex evolution algorithm *Energy Convers. Manage.* **157** 460–79

[88] Arcipiani B 1985 Generalization of the area method for the determination of the parameters of a non-ideal solar cell *Rev. Phys. Appl.* **20** 269–72

[89] El-Adawi M and Al-Nuaim I 2001 A method to determine the solar cell series resistance from a single I–V. Characteristic curve considering its shunt resistance—new approach *Vacuum* **64** 33–6

[90] Enebish N *et al* 1993 Numerical analysis of solar cell current–voltage characteristics *Sol. Energy Mater. Sol. Cells* **29** 201–8

[91] Ben Messaoud R 2020 Extraction of uncertain parameters of double-diode model of a photovoltaic panel using ant lion optimization *SN Appl. Sci.* **2** 239

[92] AlRashidi M R, El-Naggar K and AlHajri M F 2013 Parameters estimation of double diode solar cell model *Int. J. Electr. Comput. Eng.* **7** 118–21

[93] Babu T S *et al* 2016 Parameter extraction of two diode solar PV model using Fireworks algorithm *Sol. Energy* **140** 265–76

[94] Jacob B *et al* 2015 Solar PV modelling and parameter extraction using artificial immune system *Energy Procedia* **75** 331–6

[95] Gow J A and Manning C D 1999 Development of a photovoltaic array model for use in power-electronics simulation studies *IEE Proc.—Electr. Power Appl.* **146** 193–200

[96] Jiang Y, Qahouq J and Batarseh I 2010 Improved solar PV cell matlab simulation model and comparison *2010 IEEE Int. Symp. on Circuits and Systems (ISCAS) (Paris)* (Piscataway, NJ: IEEE) 2770–3

[97] Patel H and Agarwal V 2008 MATLAB-based modeling to study the effects of partial shading on PV array characteristics *IEEE Trans. Energy Convers.* **23** 302–10

[98] Ramaprabha R and Mathur B L 2009 MATLAB based modelling to study the influence of shading on series connected SPVA *2009s Int. Conf. on Emerging Trends in Engineering and Technology*

[99] Basore P A, Rover D and Smith A 1996 PC-1D version 2: Enhanced numerical solar cell modelling *Conf. Record of the Twentieth IEEE Photovoltaic Specialists Conf.* (Piscataway, NJ: IEEE)

[100] Rover D, Basore P and Thorson G 1985 Solar cell modeling on personal computers *IEEE Photovoltaic Specialists Conf.* 18

[101] Liu Y *et al* 2018 Modeling multijunction solar cells by nonlocal tunneling and subcell analysis *IEEE J. Photovolt.* 1–7

[102] Haddout A, Raidou A and Fahoume M 2019 A review on the numerical modeling of CdS/CZTS-based solar cells *Appl. Phys.* A **125** 124

[103] Kowsar A *et al* 2020 Comparative study on solar cell simulators *2019 2nd International Conference on Innovation in Engineering and Technology (ICIET) (Dhaka)* (Piscataway, NJ: IEEE) 1–6

[104] Burgelman M, Nollet P and Degrave S 2000 Modelling polycrystalline semiconductor solar cells *Thin Solid Films* **361–362** 527–32

[105] Tulka T K *et al* 2022 Optimization of a high-performance lead-free cesium-based inorganic perovskite solar cell through numerical approach *Heliyon* **8** e11719

[106] Hazeghi F and Ghorashi S M B 2019 Simulation of perovskite solar cells by using CuSCN as an inorganic hole-transport material *Mater. Res. Express* **6** 095527

[107] Thakur N, Kumar P and Sharma P 2023 Simulation study of chalcogenide perovskite ($BaZrSe_3$) solar cell by SCAPS-1D *Mater. Today Proc.* (in press)

[108] Islam M T and Thakur A 2023 Design simulation of chalcogenide absorber-based heterojunction solar cell yielding manifold enhancement in efficiency *Phys. Status Solidi (A)* **220** 2300290

[109] Zeman M *et al* 1997 Computer modelling of current matching in a-Si: H/a-Si: H tandem solar cells on textured TCO substrates *Sol. Energy Mater. Sol. Cells* **46** 81–99

[110] Hernández-Como N and Morales-Acevedo A 2010 Simulation of hetero-junction silicon solar cells with AMPS-1D *Sol. Energy Mater. Sol. Cells* **94** 62–7

[111] Belfar A and Mostefaoui R 2011 Simulation of n1-p2 microcrystalline silicon tunnel junction with AMPS-1D in a-Si: H/μc-Si: H tandem solar cells *J. Appl. Sci.* **11** 2932–9

[112] Bouloufa A, Djessas K and Zegadi A 2007 Numerical simulation of $CuIn_xGa_{1-x}Se_2$ solar cells by AMPS-1D *Thin Solid Films* **515** 6285–7

[113] Liu Y, Sun Y and Rockett A 2012 A new simulation software of solar cells—wxAMPS *Sol. Energy Mater. Sol. Cells* **98** 124–8

[114] Yasar S *et al* 2016 Numerical thickness optimization study of CIGS based solar cells with wxAMPS *Optik* **127** 8827–35

[115] Olopade M A *et al* 2015 Modeling and simulation of CZTS/CTS tandem solar cell using wxAMPS software *2015 IEEE 42nd Photovoltaic Specialist Conf. (PVSC)*

[116] Neukom M T *et al* 2019 Consistent device simulation model describing perovskite solar cells in steady-state, transient, and frequency domain *ACS Appl. Mater. Interfaces* **11** 23320–8

[117] MacKenzie R C I *et al* 2011 Modeling nongeminate recombination in P3HT:PCBM solar cells *J. Phys. Chem.* C **115** 9806–13

[118] Froitzheim A *et al* 2003 AFORS-HET: A computer-program for the simulation of hetero-junction solar cells to be distributed for public use *Proc. 3rd World Conf. on Photovoltaic Energy Conversion, 2003 (Osaka)* (Piscataway, NJ: IEEE) 279–82

[119] Vukadinović M *et al* 2001 Numerical modelling of trap-assisted tunnelling mechanism in a-Si:H and μc-Si n/p structures and tandem solar cells *Sol. Energy Mater. Sol. Cells* **66** 361–7

[120] ASPIN3 2019 http://lpvo.fe.uni-lj.si/en/software/aspin3/ (Accessed 9 Sepember 2019)

[121] Wenger S *et al* 2011 Coupled optical and electronic modeling of dye-sensitized solar cells for steady-state parameter extraction *J. Phys. Chem.* C **115** 10218–29

[122] Gray J L 1991 Adept: a general purpose numerical device simulator for modeling solar cells in one-, two-, and three-dimensions *The Conf. Record of the Twenty-Second IEEE Photovoltaic Specialists Conf.-1991* (Piscataway, NJ: IEEE)

[123] Safa Sultana R *et al* 2017 Numerical modeling of a CdS/CdTe photovoltaic cell based on ZnTe BSF layer with optimum thickness of absorber layer *Cogent Eng.* **4** 1318459

[124] Warren E L *et al* 2018 Maximizing tandem solar cell power extraction using a three-terminal design *Sustain. Energy Fuels* **2** 1141–7

[125] Sarollahi M *et al* 2022 Modeling of temperature dependence of Λ-graded InGaN solar cells for both strained and relaxed features *Front. Mater.* **9** 1006071

[126] Kowsar A and Farhad S F U 2018 High efficiency four junction III–V bismide concentrator solar cell: design, theory, and simulation *Int. J. Renew. Energy Res. (IJRER)* **8** 1762–9

[127] Kowsar A *et al* 2020 A novel simulator of multijunction solar cells-MSCS-1D *Int. J. Renew. Energy Res.* **10** 1369–75

[128] Fell A 2013 A free and fast three-dimensional/two-dimensional solar cell simulator featuring conductive boundary and quasi-neutrality approximations *IEEE Trans. Electron Devices* **60** 733–8

[129] Vaschetto M E, Monkman A P and Springborg M 1999 First-principles studies of some conducting polymers: PPP, PPy, PPV, PPyV, and PANI *J. Mol. Struct. THEOCHEM* **468** 181–91

[130] Scharber M C *et al* 2010 Influence of the bridging atom on the performance of a low-bandgap bulk heterojunction solar cell *Adv. Mater.* **22** 367

[131] Numbury Surendra B 2021 Applications of current density functional theory (DFT) methods in polymer solar cells *Density Functional Theory* ed G-M Daniel (Rijeka: IntechOpen) ch 11

[132] Rasukkannu M, Velauthapillai D and Vajeeston P 2017 Computational modeling of novel bulk materials for the intermediate-band solar cells *ACS Omega* **2** 1454–62

[133] Sahu J *et al* 2022 Interface layer DFT study of first transition series XO oxides (X = Ti, Cr, Mn, Fe, Co, Ni, Cu and Zn) *Mater. Today Proc.* **59** 1831–8

[134] Idrissi S *et al* 2021 DFT and TDDFT studies of the new inorganic perovskite CsPbI3 for solar cell applications *Chem. Phys. Lett.* **766** 138347

[135] Kowsar A *et al* 2019 Comparative study on solar cell simulators *2019 2nd Int. Conf. on Innovation in Engineering and Technology (ICIET)*

[136] RReDC Glossary of Solar Radiation Resource Terms 2017 (Accessed 25 November 2017)

[137] Mohammad Aminul I *et al* 2021 Assessing the impact of spectral irradiance on the performance of different photovoltaic technologies *Solar Radiation* ed A Mohammadreza (Rijeka: IntechOpen) ch 6

[138] Minnaert B and Veelaert P 2014 A proposal for typical artificial light sources for the characterization of indoor photovoltaic applications *Energies* **7** 1500–16

[139] Idris H and Smaili G B H 2023 A review of minority carrier recombination lifetime measurements *Ijraset J. Res. Appl. Sci. Eng. Technol.* **11** 1351–63

[140] Sturman B, Podivilov E and Gorkunov M 2003 Origin of stretched exponential relaxation for hopping-transport models *Phys. Rev. Lett.* **91** 176602

[141] Baloch A A B *et al* 2018 Analysis of photocarrier dynamics at interfaces in perovskite solar cells by time-resolved photoluminescence *J. Phys. Chem.* C **122** 26805–15

[142] Smits F M 1958 Measurement of sheet resistivities with the four-point probe *Bell Syst. Tech. J.* **37** 711–8

[143] Subramanian M *et al* 2022 Optimization of effective doping concentration of emitter for ideal c-Si solar cell device with PC1D simulation *Crystals* **12** 244

IOP Publishing

Recent Advances in Solar Cells

N M Ravindra, Leqi Lin and Priyanka Singh

Chapter 12

Tandem solar cells

Presently, tandem technology is the most promising solar photovoltaic (PV) approach for efficient performance of solar cells. This chapter presents a review of the recent developments in various types of tandem solar cells (TSCs), their architecture, and methods of processing. The performance of TSCs is discussed in the context of various bandgap combinations and suitable materials. TSCs are classified as inorganic, organic and hybrid depending on the choice of material/s. Inorganic TSCs such as III–V/silicon-based tandem have achieved certified efficiency of 35.9% whereas hybrid tandem devices, based on perovskite (pk) such as pk/silicon, pk/CIGS, pk/organic, have achieved certified efficiency of 33.9%, 24.2% and 23.4%, respectively. Organic/organic TSCs have achieved efficiency of 20.2%.

12.1 Introduction

The world's energy demand is increasing every year due to the growth in population and the economy. Energy from fossil fuels is not sustainable because of its limited supply as well as the associated environmental effects. Hence, the need for the use of alternative energy sources such as geothermal, solar, wind, and biofuels has increased. Solar photovoltaics is a practical, clean, and sustainable energy source to meet energy demand and lower electricity costs. A PV device is a p-n junction solar cell which converts solar energy to electrical energy. Nonetheless, the power conversion efficiency (*PCE*) of single p-n junction solar cells is restricted by the Shockley–Queisser (S–Q) limit [1] and they can only convert a portion of the solar spectrum into electricity. Multijunction solar cells (MJSCs) are designed to overcome the limitations of single-p-n-junction solar cells and to achieve higher efficiencies. An MJSC consists of several individual semiconductor p-n junctions, stacked together (also called subcells), that are connected in series to attain high performance [2]. TSCs are a special case of MJSCs, typically consisting of two subcells of two different materials, having considerably different bandgaps (figure 12.1(a)). This enables them to capture a larger portion of the solar spectrum

doi:10.1088/978-0-7503-5994-8ch12

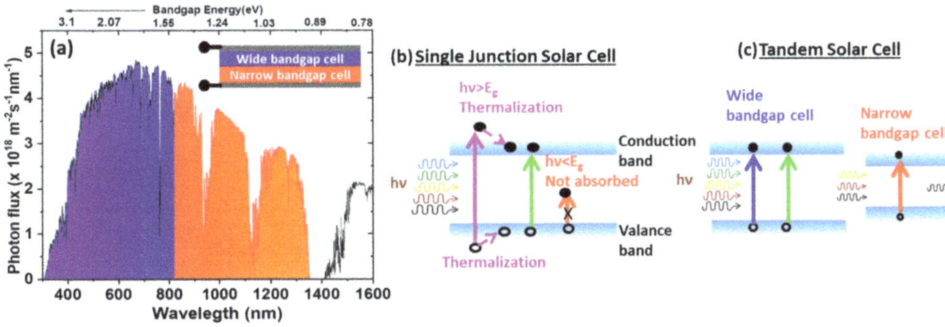

Figure 12.1. (a) Typical absorption spectrum range of subcells for TSCs: top cell (blue) and bottom cell (red). Solar spectra data taken from ASTM: ASTM G173–03 (AM 1.5G); (b) Schematic energy band diagram of single-junction solar cell; (c) TSC with two p-n junctions. Reproduced from [28]. Published under licence by IOP Publishing Ltd.

and increase the overall efficiency, as shown in figure 12.1(a). In single-p-n-junction solar cells, excess energy from photons with energies higher than the bandgap (E_g) is generally lost in the form of heat through thermalization processes as depicted through the energy band diagram in figure 12.1(b). TSCs can absorb these high-energy photons and utilize their energy more effectively, reducing thermalization losses and improving efficiency, as shown in figure 12.1(c).

Each p-n junction in a tandem cell is optimized to absorb a specific portion of the solar spectrum, allowing for more efficient utilization of sunlight. Currently, emerging TSCs, made up of two or more commercially feasible solar cells, are attracting significant interest due to their high efficiencies. MJSCs, based on III–V semiconductors, such as gallium arsenide (GaAs), indium phosphide (InP), and gallium indium phosphide (GaInP), are designed for various applications, including terrestrial photovoltaics, space missions, concentrated photovoltaics, and portable electronics [3–6]. TSCs require careful optical design to ensure that each subcell absorbs the appropriate component of the solar spectrum without significant losses due to reflection or absorption by other layers.

This chapter will address the processing and device structure of TSCs including their types and recent developments. The performance of TSCs using different bandgap combinations and appropriate materials is also discussed. Beginning with III–V MJSCs, inorganic and emerging hybrid tandem technologies as well as organic TSCs (OTSCs), will be explored.

12.2 Processing of tandem solar cells

TSCs can be fabricated via two different approaches—monolithic and mechanical stacking. The following section will describe the processing and device structure of monolithic and mechanically stacked TSCs.

12.2.1 Monolithic tandem solar cells

In monolithic TSCs, the individual subcells are grown on top of each other in a single continuous process, i.e., cells are directly integrated and share a common

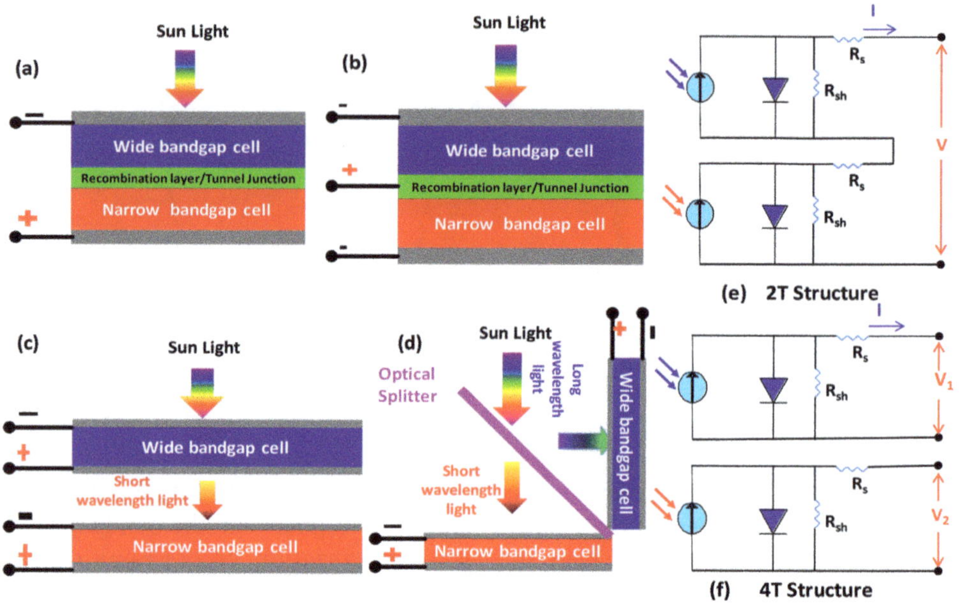

Figure 12.2. Structural representation of the four different types of TSCs (a) Two-terminal (2T) tandem device which usually has a monolithic structure, where the layer connecting the two cells works as a tunnel junction or acts as a recombination layer (b) three-terminal (3T) monolithic tandem device (c) four-terminal (4T) tandem device mechanically stacked and (d) 4T optical splitting tandem device. (e) Equivalent circuit diagram of 2T tandem cell and (f) equivalent circuit diagram of 4T tandem cell. In figure 12.2(e) and figure 12.2(f), R_{sh}, R_s, V and I represent the following: shunt resistance, series resistance, generated voltage and current, respectively. The top and bottom cells operate in the visible (400–800 nm) and infrared (800–1100 nm) ranges; they are indicated by blue and red arrows, respectively. Figure 12.2(e) and figure 12.2(f) are modified and reproduced with permission from [10] CC BY 3.0.

substrate (figure 12.2(a)). The layers in monolithic tandem cells are typically lattice-matched, i.e., their crystal structures are aligned to minimize defects and maximize performance. The upper wide-bandgap subcell in monolithic device architecture is constructed above the lower narrow bandgap subcell. Two subcells are connected using a transparent conductive adhesive or a tunnel junction. A monolithic TSC has high efficiency because of its continuous structure, which allows for efficient charge carrier transport between layers. However, fabrication of monolithic tandems often requires specialized equipment and processes to deposit and grow each layer with high precision.

12.2.2 Mechanically stacked tandem solar cells

Mechanical stacking involves separately fabricating each subcell and then physically stacking them on top of each other (figure 12.2(c)). This stacking can be achieved by using various techniques such as epitaxial liftoff, wafer bonding, or transfer printing. The layers in mechanical stacking tandems are made of different materials and may not necessarily be lattice-matched. One advantage of this kind of tandem is that it does not require interfacial tunneling. Mechanical stacking allows more flexibility in choosing materials, but it may also introduce interface defects that affect performance. Furthermore, mechanically stacked tandems have lower efficiency compared to monolithic tandems due to interface recombination and other losses at the interfaces between layers. Fabrication of mechanical stacking tandems can be more straightforward compared to monolithic tandems since each layer is optimized independently, but aligning the layers precisely can be challenging. In addition, mechanical stacking requires transparent conducting electrode (TCE) layers and thus imposes severe optical losses [7].

12.3 Device architecture of tandem solar cells

TSCs can be designed with various device architectures, including two-terminal (2T), three-terminal (3T) and four-terminal (4T) configurations. Each architecture has its own benefits and challenges in terms of fabrication complexity, efficiency, and stability. Figure 12.2 shows four distinct TSCs configurations.

12.3.1 Two-terminal (2T) tandem solar cells

The architecture of a 2T TSC is shown in figure 12.2(a), where the top and bottom cells are monolithically integrated on a single substrate. The benefit of a 2T tandem cell is its simplicity and ease of integration and can be coupled in parallel or series with minimal space between them (discussed later, as shown in figure 12.8) [8]. In addition, the role of tunnel junction interlayer is very important for achieving high efficiency as it provides high optical transmittance and low electrical resistance between the subcells. The equivalent circuit of a 2T TSC is shown in figure 12.2(e) under illumination. In figure 12.2(e), R_s is the series resistance, R_{sh} is the shunt resistance, I is the output current, and V is output voltage. As the subcells in a 2T tandem device are connected in series, current matching of the top and bottom cell is always required to ensure that the overall device current is not limited by the subcell with lower current. Three main obstacles need to be overcome in order to fabricate 2T devices: current matching, high performance recombination junctions, and performance losses under varying spectra [9]. It is anticipated that 2T devices would be advantageous for upcoming applications with the resolution of these issues.

12.3.2 Three-terminal (3T) tandem solar cells

In 3T designs, the two subcells can be connected by wiring, recombination layers or tunnel junctions [11–14]. As shown in figure 12.2(b), the third electrode can be formed at the contact between the top and bottom cells, and can extract additional generated power through the variation in the incident wavelength of the solar radiation. In addition, the third electrode may also be formed at the back side of the bottom cell in the case of interdigitated back contact bottom cell [15]. The third terminal allows more precise control over operating conditions of each subcell and may help mitigate current-matching issues that arise in 2T tandem cells. With individual electrical contacts for each subcell, 3T tandem cells can achieve higher efficiency and performance compared to 2T configurations. Schnabel *et al* produced a *PCE* of 27.3% for 3T tandem device having GaInP top cell [16], as shown in figure 12.4(c).

12.3.3 Four-terminal (4T) tandem solar cells

Two common structures exist for 4T configurations: mechanically stacked 4T tandem devices (figure 12.2(c)) and optical splitting 4T tandem devices (figure 12.2 (d)).

12.3.3.1 Mechanically stacked 4T tandem solar cells

A mechanically stacked 4T device is made up of independently fabricated subcells which yield the best possible performance (figure 12.2(c)). Four terminal connections are required as each subcell produces its own electricity. Despite optical connection, the two subcells are electrically separate from one another. Therefore, to enable some light (infrared photons) transmission to the bottom cell, bifacial TCEs are used on the top cell. Thin Ag films or nanowires have been employed for this purpose but more intricate structures (e.g., multilayer metallic/dielectric structures) might enhance their performance [17, 18]. Although efficiency can be maximized for a 4T tandem device, there may be an increase in the cost of module integration and connections. One easier method for an efficient device involves the use of a photon down conversion mechanism, which is simply done by covering a standard silicon-based solar cell with a thin film that absorbs blue light and re-emits photons at a longer wavelength. This allows red and infrared solar light to pass towards the rear cell, thereby increasing the amount of energy produced [19]. The equivalent circuit of a 4T TSC is shown in figure 12.2(f). 4T TSCs are not subject to current matching since the top and bottom cells are electrically separated. A 4T TSC offers greater flexibility and control over the performance of each subcell. However, considerable optical and electrical losses are unavoidable because of the intermediary resistance of electrodes and parasitic absorption. 4T configurations are typically used in advanced tandem solar cell designs, where precise optimization and control over each subcell is crucial for achieving maximum efficiency. An example of this kind of 4T tandem has achieved an efficiency of more than 30% for pk/Si [20] and 25.9% for pk/CIGS [21] structure (figure 12.4(c)).

12.3.3.2 Optical splitting 4T tandem solar cells

In an optical splitting 4T TSCs, a dichroic mirror is employed [22], as illustrated in figure 12.2(d), to split high energy photons toward the wide-bandgap cell and low energy photons toward the narrow bandgap cell. The benefit of this kind of design is that additional TCEs are not required. In addition, this design permits more freedom for the fabrication of an individual solar cell. However, the economic feasibility of this tandem arrangement is hindered by the higher cost of the optical splitter [9]. Uzu *et al* have demonstrated a *PCE* of 28.0% [22] and all-perovskite TSCs have attained a *PCE* of 23.26% [23].

12.4 Tandem solar cell performance and possible bandgap combinations

The bandgap combination is a necessary prerequisite for fabricating efficient TSCs successfully. When a solar cell is illuminated, only photons with energy greater than the bandgap of the semiconductor are absorbed and produce electron–hole pairs [24]. The wavelength of absorbed photons in the solar cell device depends on E_g and provides cut-off wavelength (λ_g), which is useful for carrier generation. It is given by following equation,

$$\lambda_g = \frac{1240}{E_g(\text{eV})}\,(\text{nm}) \tag{12.1}$$

The initial photon flux (N_{ph}) and the absorption coefficient (α_λ) of incident light determine the photogeneration of electron–hole pairs in the semiconductor [24]. As discussed in chapter 11, *PCE* or efficiency (η) of a solar cell can be expressed in terms of open-circuit voltage (V_{oc}), short-circuit current density (J_{sc}) and fill factor (FF),

$$\eta = \frac{V_{oc} J_{sc} FF}{P_{in}} \tag{12.2}$$

where, P_{in} is the intensity of incident light. V_{oc} and FF can be calculated from equations (11.8) and (11.9) described in chapter 11. In equation (12.2), J_{sc} depends on the incident solar spectral irradiance, and can be expressed by the following equation,

$$J_{sc} = q\int_{h\nu=E_g}^{\infty} \frac{dN_{ph}}{dh\nu} d(h\nu) \tag{12.3}$$

As discussed in chapter 11, ohmic losses (R_s and R_{sh}) and recombination losses could be the limiting factors for J_{sc}. The maximum achievable efficiency for a single p-n junction solar cell can be calculated using the relationship between the energy distribution of photons in the solar spectrum and E_g of the semiconductor material using equations (12.1–12.3). Theoretical S–Q (Shockley–Queisser) detailed-balance efficiency limit for a single-junction solar cell is shown in figure 12.3(a) as a function of bandgap (top solid line). The bottom grey lines indicate the S–Q efficiency limits corresponding to 75% and 50%. Solid and open symbols indicate the record efficiencies of solar cells made of various materials with respect to bandgap. The solid symbols display the data from July 2020, while the open symbols display the efficiency record from April 2016. It can be seen from figure 12.3(a), that at the maximum theoretical efficiency, the matching E_g of the material can be precisely chosen. For single-junction solar cells with a bandgap of 1.34 eV, the theoretical S–Q efficiency limit is 33.7% [25]. In order to maximize the efficiency of TSC, the current generated by each subcell must be well-matched. This requires precise tuning of the thickness and bandgap of each subcell to ensure that they contribute equally to the total current output. MJSCs, with up to three bandgaps, are commercially available, but they are used almost exclusively for space applications, especially for

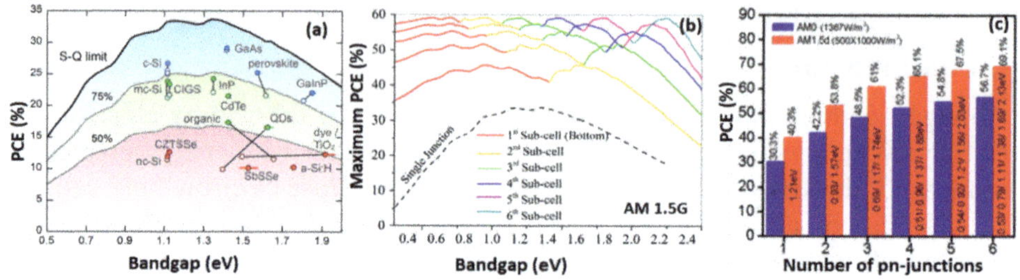

Figure 12.3. (a) Theoretical S–Q detailed-balance efficiency limit as a function of bandgap, for single-junction solar cells is shown (top solid line). The bottom grey lines indicate the corresponding 75% and 50% of S–Q efficiency limits. Record efficiencies of solar cells of different materials versus the bandgap is shown by open and solid symbols. The open symbols show the record efficiency reported in April 2016, the solid symbols show the record efficiency in July 2020. Reproduced with permission from [30] Copyright, 2020 American Chemical Society. (b) Maximum efficiency for a bandgap value in a MJSC stack for scenarios up to six bandgaps under the standard solar spectrum for AM1.5G is shown. Reproduced with permission from [31] Copyright 2016, Elsevier (c) Maximum theoretical efficiency for MJSCs related to the number of p-n junctions. The blue and red bars correspond to AM0 (1367 $W\,m^{-2}$) and AM1.5d (500 × 1000 $W\,m^{-2}$) solar spectrum, respectively. Reproduced with permission from [32] Copyright 2018, Elsevier.

satellite power. With a bandgap of 1.42 eV, the highest efficiency of GaAs that has been reported is 29.1%, whereas a perovskite solar cell has a *PCE* of 25.2% (figure 12.3(a)). The simulation results show that the maximum limiting *PCE* for TSCs, based on silicon, is 45% for two-junction (2 J) and 50% for three-junction (3 J) (figure 12.3(b)) [26, 27]. In 3 J devices with a silicon bottom cell, 2.01 and 1.5 eV are the ideal bandgap combinations for the top and middle cells. In figure 12.3(b), it can be seen that as the number of subcells in the TSCs increase, the maximum efficiency becomes less sensitive for lower subcells.

De Vos [28] has investigated thoroughly the fundamental (detailed balance) limit of tandem cell performance. The study reveals that stacking many subcells in series can increase the theoretical efficiencies above the S–Q limit, due to increase in the electrochemical potential of charge carrier extraction. As shown in table 12.1 [28], the maximum efficiency increases to 42% for a tandem cell made up of two subcells (2 J) with bandgaps of 1.9 and 1.0 eV, respectively, and to 49% for a tandem made up of three subcells (3 J) with bandgaps of 2.3, 1.4, and 0.8 eV, respectively. Under maximum light concentration, these efficiencies are 40% for a single cell, 55% for two cells, and 63% for three cells. The ideal efficiency of a stack with an infinite number of solar cells was predicted by this model. A tandem system of this kind could convert 86% of the concentrated sunlight and 68% of the non-concentrated sunlight. For tandem cells based on GaInP/GaInAs/GaInAs, under AM1.5G, experimental efficiencies as high as 33.8% have been achieved [29].

Theoretical *PCE* of 3 J tandem cell based on Si has the potential to surpass 50% [32]. The simulation studies show that, theoretically, GaInP/GaAs/Si has a maximum *PCE* of around 44%. However, world-record *PCE* of GaInP/GaAs/Si tandem was less than 36%, well below the theoretical limit [33]. Therefore, future advancements in this area

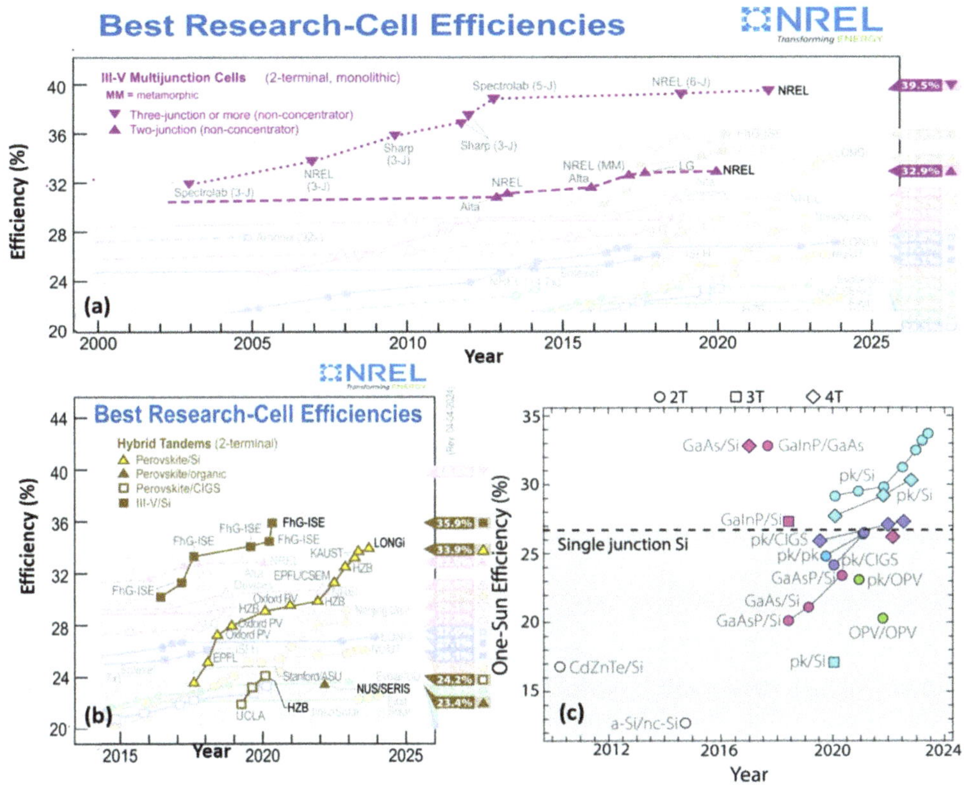

Figure 12.4. Developments in efficiency of tandem devices for, (a) 2T monolithic multijunction cells from III–V semiconductors for 2 J (▲), 3 J or more (▼) (b) 2T hybrid technologies; perovskite/silicon (▲), perovskite/organic (▲), perovskite/ CIGS (□) and III–V/silicon (■) and (c) 2T (O), 3T (□) and 4T (◇) based on III–V/Si, perovskite/silicon(pk/Si), pk/CIGS, pk/organic (pk/OPV),pk/pk,OPV/OPV. Figure 12.4(a) and figure 12.4(b) are provided by National Renewable Energy Laboratory (NREL) and is modified to show efficiencies for 2T multijunction and hybrid TSCs, respectively. Figure 12.4(c) is reproduced from [35] CC BY 4.0.

Table 12.1. The optimal bandgaps (E_{gi}) in unconcentrated sunlight for tandem structure with n stacked cells. Reproduced from [28]. Published under licence by IOP Publishing Ltd.

Number of subcells (n)	E_{g1} (eV)	E_{g2} (eV)	E_{g3} (eV)	E_{g4} (eV)	Efficiency η (%)
1	1.3	—	—	—	30
2	1.9	1.0	—	—	42
3	2.3	1.4	0.8	—	49
4	2.6	1.8	1.2	0.8	53

Table 12.2. Possible material combination for TSCs. Reproduced from [35]. CC BY 4.0.

Cell material	Approximate bandgap (E_g) (eV)	Top cell	Bottom cell	Wafer or film	Film configuration
Si	1.12		*x*	Wafer	—
CIGS (and related)	Range, typically 1.15		*x*	Film	Substrate
Perovskite narrow	Range, 1.2–1.4		*x*	Film	Superstrate
InSb	1.23		*x*	wafer	—
Organics	Range	*x*	*x*	Film	Super/ substrate
GaAs	1.42	*x*	*x*	Wafer/ film	Substrate
Cd(Se,Te)	Range, 1.4–1.5	*x*		Film	Superstrate
Perovskite wide	Range, 1.6–1.9	*x*		Film	Superstrate
$CuGaSe_2$	1.68	*x*		Film	Substrate
Amorphous Si (a-Si)	1.7	*x*		Film	Superstrate
GaInP	1.8–1.9	*x*		Film	Substrate

are essential. A *PCE* of 51.8% is achieved in the radiative limit by a 3 J solar cell with optimal bandgap energies of 1.90, 1.37, and 0.93 eV. Only aluminium-gallium-arsenide (AlGaAs) or aluminium-gallium-phosphide (AlGaInP) can have the necessary bandgap lattice suited to GaAs for the top junction. In a study, using theoretical ideal bandgap combination, the actual required subcell thicknesses and bandgaps have been calculated using an optical model [34]. In a recent report [35], potential absorber material combinations have been addressed in the bandgap range of 1.3–1.4 eV, with an emphasis on the best performing top and bottom cells. Table 12.2 lists the wide variety of potential materials for TSCs. The performance status of each cell type can be found in a recent publication in the form of an efficiency table [29].

12.5 Tandem solar cell classification

This category refers to record tandem cells with layers composed of two different materials. Some subcategories of hybrid tandems (perovskite/Si and perovskite/CIGS) were already included in the previous efficiency chart under 'Emerging PV', whereas other configurations (III–V/Si and Perovskite/organic) are new. All of these subcategories have been moved into the new hybrid tandems category—with the exception of perovskite/perovskite and organic/organic tandems, which are listed under 'Emerging PV'.

TSCs are fabricated by combinations of various materials, such as, perovskite/silicon (pk/Si) [36, 37], thin-film/Si [38] III–V/Si [39], pk/organic (pk/OPV) [40] and pk/CIGS. When both top and bottom subcells are made of the same material, they are described as: all-perovskite (pk/pk) [41], III–V/III–V [42], all-organic (OPV/OPV) [43] tandems. Each of the combinations offers unique advantages and challenges regarding processing, material compatibility, and efficiency potential.

TSCs are generally classified into two categories based on the material used in the cell: inorganic TSCs (ITSCs) have both the top and bottom subcells made of inorganic semiconductors, while OTSCs have both the top and bottom subcells made of organic components. Recently, due to the growing interest in TSCs, NREL [44] added a new 'Hybrid tandems' category to its efficiency chart, which shows the advancements in hybrid TSCs such as pk/Si, pk/CIGS, pk/organic, and III–V/Si (as shown in figure 12.4(b)). Nonetheless, all-perovskite (pk/pk) and all-organic OPV/OPV are listed under 'Emerging PV'. The following section presents an overview of ITSCs, OTSCs, and hybrid TSCs along with some examples and recent developments.

12.5.1 Inorganic tandem solar cells

The highly expensive ITSCs [45–47] are utilized in space applications and have an efficiency of up to 46%. ITSCs consist of subcells made from inorganic materials stacked on top of each other. Silicon is the most widely used semiconductor material in PV devices due to its abundance, stability [27] and well-established fabrication processes. Furthermore, Si solar cells are an excellent choice for the bottom subcells in TSCs because of their high V_{oc} of up to 0.75 V [48], high efficiency [49, 50], and narrower bandgap (~1.1 eV) which absorbs lower-energy photons efficiently [9]. Silicon TSCs have garnered attention from both academia and industry as a viable approach to increase the efficiency of Si PV technology. With continued advancements, silicon TSC holds great promise for further enhancing the performance and competitiveness of solar energy technologies.

12.5.1.1 III–V/silicon tandem solar cells

Early efforts focused on combining silicon with other materials such as GaAs or amorphous silicon (a-Si) to form tandem structures. However, these early tandem cells faced challenges relating to material compatibility, fabrication techniques, and performance optimization. Later on, the development of wafer bonding and epitaxial growth techniques enabled the integration of III–V compound semiconductors e.g., GaAs with Si. III–V/Si TSCs emerged as a promising approach, with III–V materials serving as the top cell to absorb high-energy photons, while Si served as the bottom cell to capture lower-energy photons. Research efforts focused on optimizing the design, material quality, and interface properties of III–V/Si tandems to achieve higher efficiencies. III–V compound semiconductors, such as GaAs, InP and GaInP, have direct bandgaps and high absorption coefficients, making them excellent candidates for top cells in TSCs. Cadmium telluride (CdTe) is a thin-film semiconductor material with a direct bandgap of around 1.5 eV, making it suitable for absorbing a broad range of wavelengths in the solar spectrum. CdTe-based tandem solar cells have been investigated typically using a wide-bandgap material as the top cell. CdTe is known for its low manufacturing costs and high efficiencies in single-junction configurations. Copper indium gallium selenide (CIGS) is another thin-film semiconductor material with a tunable bandgap (typically around 1.0 to 1.7 eV) and high absorption coefficient. CIGS-based tandem solar cells have been explored, leveraging their compatibility with solution-based and low-cost manufacturing techniques. These materials are chosen for their excellent optoelectronic properties, including high carrier mobilities, tunable bandgaps, and good stability. These inorganic materials have been combined in various configurations to create TSCs with optimized absorption of different portions of the solar spectrum. Figure 12.4 shows significant advancements in efficiency for tandem devices [35], including contributions from several universities and research centers. Figure 12.4(a) shows the evolution in efficiencies for 2T monolithic multijunction cells from III–V semiconductors for two, and more than two junctions [44]. Figure 12.4(b) shows the development in efficiencies for 2T hybrid tandem devices for pk/Si, pk/organic, pk/CIGS and III–V/Si [44]. Figure 12.4(c) displays the efficiency achievements for 2T, 3T and 4T TSCs based on III–V/Si, pk/Si, pk/pk, pk/CIGS, pk/OPV and OPV/OPV [35].

As mentioned in figure 12.4, the efficiency of III–V MJSCs has increased gradually due to enhancements in material quality and the addition of junctions to minimize thermalization losses and achieve the ideal bandgap combination [1, 51–53]. Lattice-matched material quality improvements produced record GaAs and GaInP single-junction solar cells [29, 53] and record GaInP/GaAs TSCs [54, 55]. Advanced techniques such as metamorphic (MM) epitaxy [56–58] have progressed to record efficiencies (figure 12.4(a)) of III–V MJSCs with three to six junctions converting up to 39.5% of the global spectrum [44, 59, 60].

PCE of more than 32% [61] has been obtained in TSCs for 2T GaInP/GaAs and 4T GaAS/Si (as shown in figure 12.4(a) and figure 12.4(c)). In addition, a 4T mechanically stacked 3 J structure GaInP/GaAs/Si tandem device has a *PCE* of 35.9% [33]. Although III–V/Si tandems have demonstrated high efficiencies, manufacturing is challenging due to the need for precise control over material properties, layer thicknesses, and interfaces. Advanced deposition techniques such as molecular beam epitaxy (MBE) or chemical vapor deposition (CVD) are often used to deposit high-quality semiconductor layers with excellent control over thickness and composition. Therefore, the high cost of producing III–V class semiconductors limits their use in space-based applications, high-concentration solar systems or spacecraft [27]. In a three-junction III–V MJSC, GaAs is generally used as a middle cell due to its excellent material quality, even though its bandgap is larger than ideal for the global spectrum. Recently, high-performance of thin GaInAs/GaAsP strain-balanced quantum well (QW)-based solar cells has been reported. These QWs are integrated into a 3 J inverted MM multijunction device composed of highly optimized GaInP top cell, a GaInAs/GaAsP QW middle cell, and a lattice-mismatched GaInAs bottom cell. Three-junction demonstrates *PCEs* of 39.5% and 34.2%, respectively, under AM1.5 G and AM0 space spectra (figure 12.4(a)). The global efficiency surpasses previous records for a six-junction device [44, 62].

PCE of 35.9% was demonstrated for a monolithic 2T device based on a III–V/Si 3 J under an AM1.5G spectrum by using a novel absorber material in the middle cell. The 2T device is as efficient as the record-breaking 4T device [33]. Figure 12.5(a) shows the device structure and PV performance, e.g. external quantum efficiency

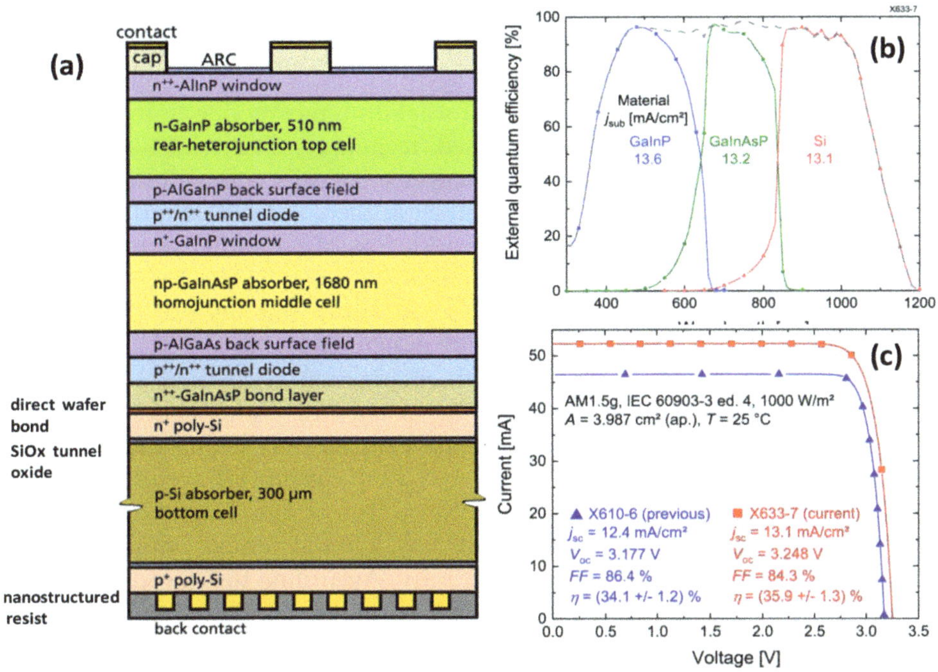

Figure 12.5. (a) Schematic of layer stack in the III–V/Si 3 J solar cell architecture, with a double layer of antireflection coating (ARC), GaInP-rear heterojunction top cell, a highly doped n-GaAs cap layer beneath the contacts, a silicon bottom cell with tunnel-oxide passivating contacts and a nanostructured diffractive rear-side grating for light path enhancement and a GaInAsP homojunction middle cell. (b) *EQE*s for subcells in the top (•), middle (•), bottom (▸), and overall total (- -) of the two-terminal record III–V//Si solar cell X633–7 (c) Current–voltage (*I–V*) characteristics of the current, two-terminal record III–V//Si solar cell X633–7 (■) compared to the previous champion device X610–6 (▲) under the AM1.5G spectrum. Reproduced with permission from [63] Copyright 2021, John Wiley & Sons).

(*EQE*) and illuminated *I–V* characteristics including performance parameters (J_{sc}, V_{oc}, *FF* and η) of 3 J (GaInP/GaAs/Si) solar cell.

Figure 12.5(b) displays the sum of the *EQE*s for the three subcells of a 3 J solar cell with an upright grown top cell structure. In the wavelength range of 470–910 nm, it is seen that the sum of the *EQE*s remains constant at a high level, ranging between 95% and 98%. The illuminated *I–V* characteristics of the current cell is compared with prior III–V//Si record device in figure 12.5(c).

12.5.1.2 Perovskite/silicon tandem solar cells

The emergence of metal halide perovskite materials as efficient absorbers sparked interest in pk/Si TSCs and are found to be a viable substitute for III–V semiconductors in tandem technology. Perovskite materials were combined with silicon to create efficient tandem structures (as can be seen in figure 12.4(b) and figure 12.4 (c). Perovskites have become prominent for both single-junction and TSCs due to their high absorption coefficients, tunable bandgaps, solution-processability [64], high V_{oc} and large charge carrier diffusion durations [65, 66]. The perovskite/Si TSC has silicon as the bottom cell which absorbs lower-energy photons efficiently, while the top perovskite cell absorbs higher-energy photons. The perovskite tandem has already demonstrated a high *PCE* and affordability, mostly due to the low cost of the materials and utilizing solution-based technique. Perovskites can be well matched with a-Si/Si, in silicon heterojunction solar cells when fabricated using a low temperature solution technique. Nonetheless, there are still important problems that need to be solved, including the higher V_{oc} loss of nonradiative recombination and the high sensitivity of perovskite to humidity [15]. Perovskite TSCs have several challenges during tandemization because of issues with stability and short lifespan.

The significant improvements in efficiency of perovskite based TSCs are shown in figure 12.4(b) and figure 12.4(c). The single-junction perovskite solar cell has increased its efficiency from 3.8% to 25.7% in only 12 years. This has expedited the performance development of TSCs based on perovskites, as can be seen in figure 12.4(b). A 2T monolithic pk/Si TSC with 13.7% steady efficiency for 1 cm^2 cell was achieved by Mailoa *et al* in 2015 [67]. Since then, a significant amount of work has been performed to increase overall efficiency, and the number of publications on pk/Si tandem devices has increased significantly. According to Oxford PV, certified efficiency of 29.5% is reported for 2T pk/Si TSC (figure 12.4(b) and figure 12.4(c)). Furthermore, pk/Si TSC has achieved efficiency of 33.9% by Longi [44, 68] (figure 12.4(b)) breaking its previous record of 33.7% by KAUST [44, 69].

Figure 12.6(a) represents a pk/Si monolithic tandem device fabricated with 1.70 eV wide-bandgap perovskite cell on a silicon cell. The perovskite top cell and the silicon bottom cell were joined via ITO as the recombination layer. The silicon cell has a smooth front surface and a textured back surface. The illuminated *J–V* characteristics of the device is shown in figure 12.6(b). The *PCE* is 22.4%, while a high V_{oc} of 1.83 V is attained; *PCE* is limited by lower J_{sc} of 16.4 mA cm^{-2} [70]. Figure 12.6(c) shows *EQE* and total absorbance *(1-R)* of the $Cs_{0.15}(FA_{0.83}MA_{0.17})_{0.85}Pb(I_{0.7}Br_{0.3})_3$/Si tandem device, where *R* is the reflectance.

The mechanically stacked 4T pk/Si TSC structure is depicted in figure 12.6(d) [71]. The efficiency of the 4T pk/Si TSCs is determined by measuring the silicon bottom cell filtered by the semitransparent perovskite top cell. The *J–V* characteristics for the subcells are presented in figure 12.6(e). The *PCE* of the original Si heterojunction solar cell is 23.4%, with performance parameters; V_{oc}—708 mV, J_{sc}—40.1 mA cm^{-2}, and *FF*—82.5%. The filtered silicon bottom cell has a *PCE* of 8.5%, and performance parameters; J_{sc}—14.5 mA cm^{-2}, V_{oc}—698 mV and *FF*—83.5%, and a substantial decrease in J_{sc} is caused, as the top perovskite cell absorbs the majority of high energy photons, only a small fraction of near-infrared photons are able to enter the filtered silicon bottom cell. *EQE* of the bottom silicon cell and the top perovskite cell is shown in figure 12.6(f). The bottom silicon cell and top perovskite cell have integrated photocurrent densities of 14.1 and 21.3 mA cm^{-2}, respectively, which are very close to *J–V* results [71]. A summary of the latest advancements in 2T and 4T pk/Si TSCs is provided in table 12.3 [10]. In addition, table 12.3 lists numerous works that used ITO as the recombination layer in the configuration of 2T devices. TSCs have been manufactured using perovskite films having a bandgap between 1.58 and 1.74 eV, as

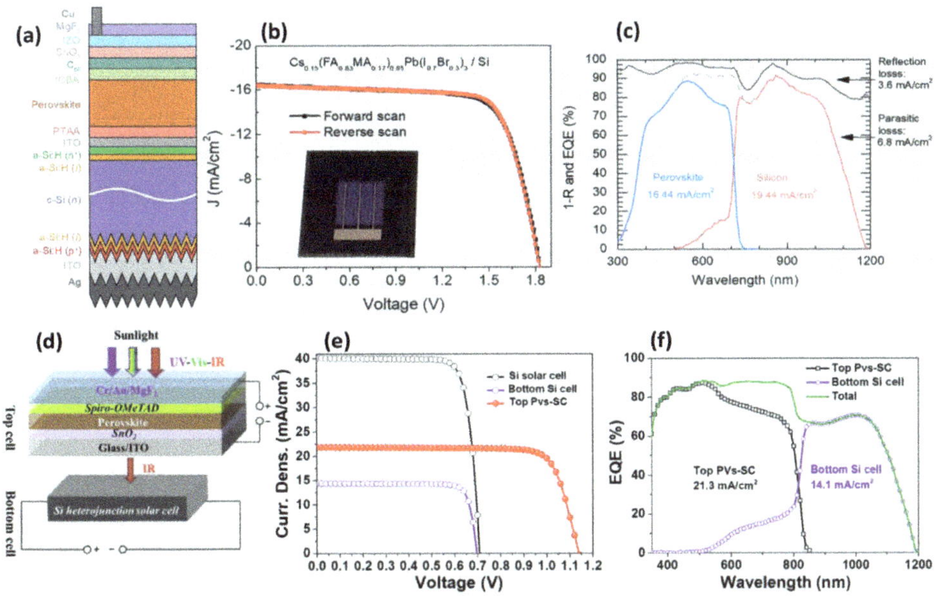

Figure 12.6. (a) Schematic structure of 2T monolithic pk/Si tandem cell with a 1.70-eV perovskite top cell both sides planar top subcell in p–i–n configuration and textured rear side of the bottom subcell (b) *J–V* characteristics of $Cs_{0.15}(FA_{0.83}MA_{0.17})_{0.85}Pb(I_{0.7}Br_{0.3})_3$/Si tandem cell having *PCE* 25.4% during forward and reverse scans, along with an inset picture of the device (c) *EQE* and total absorbance (*1–R*) of the $Cs_{0.15}(FA_{0.83}MA_{0.17})_{0.85}Pb(I_{0.7}Br_{0.3})_3$/Si tandem device, where *R* is the reflectance. Figure 12.6(a), figure 12.6(b) and figure 12.6(c) are reprinted with permission from [70] Copyright, 2019 Elsevier. (d) Mechanically stacked 4T perovskite/silicon (pk/Si) TSCs structure (e) Current density–voltage (*J–V*) characteristics and (f) corresponding *EQE*s for various solar cell types. Figure 12.6(d), figure 12.6(e) and figure 12.6(f) are reproduced with permission from [71] Copyright, 2021 Elsevier.

Table 12.3. Summary of the recent advancements in 2T and 4T perovskite/Si TSCs. Reproduced from [10] CC BY 3.0.

Tandem device	Perovskite composition	Perovskite bandgap (eV)	Interconnecting layer	Top electrode	Tandem PCE (%)	Year references
2T	$MAPbI_3$	1.58	n^{++} Si/p^{++} Si	Ag nanowire	13.7	2015 [67]
2T	$FAMAPbI_{3-x}Br_x$	1.62	Sputter ITO	Sputter ITO	18.0	2016 [72]
2T	$MAPbI_3$	1.58	Sputter IZO	Sputter IO:H/ITO	21.2	2016 [73]
2T	$CsRbFAMAPbI_{3-x}Br_x$	1.62	Interlayer-free	IZO	24.5	2018 [74]
2T	$CsFAMAPbI_{3-x}Br_x$	1.63	Sputter ITO	Sputter ITO	25.2	2019 [75]
2T	$CsFAMAPbI_{3-x}Br_x$	1.64	Sputter ITO	Sputter IZO	25.4	2019 [76]
2T	$CsFAMAPbI_{3-x}Br_x$	1.68	Sputter ITO	Sputter ITO	26.7	2020 [77]
2T	$CsFAMAPbI_{3-x}Br_x$	1.68	Sputter ITO	Sputter IZO	25.7	2020 [78]
2T	$CsFAMAPbI_{3-x}Br_x$	1.68	Sputter ITO	Sputter IZO	29.15	2020 [79]
4T	$MAPbI_3$	1.58	NA	Sputter ITO	13.4	2014 [80]
4T	$MAPbI_3$	1.58	NA	Ag nanowire	17.0	2015 [81]

4T	$FACsPbI_{3-x}Br_x$	1.74	NA	Sputter ITO	25.2	2016 [82]
4T	$RbFAMAPbI_{3-x}Br_x$	1.73	NA	Sputter ITO	26.6	2017 [83]
4T	$CsFAPbI_{3-x}Br_x$	1.77	NA	Sputter ITO	27.1	2018 [84]
4T	$CsFAMAPbI_{3-x}Br_x$	1.68	NA	Sputter IZO	28.2	2020 [85]
4T	$FACsPbI_3$	1.46	NA	Cr (1 nm)/Au (7 nm)	28.3	2020 [71]

shown in table 12.3. Furthermore, it is evident that tandem cells made of perovskite films, with bandgaps between 1.6 and 1.7 eV, are able to attain an overall efficiency of more than 25% [10].

12.5.1.3 Perovskite/CIGS tandem solar cells

This type of TSCs combines a top perovskite cell with a bottom cell made from CIGS thin-film technology. Perovskite provides efficient absorption of higher-energy photons, while CIGS absorbs lower-energy photons. pk/CIGS tandems offer the potential for high efficiencies and compatibility with scalable manufacturing processes. Although lower *PCE*s are being reported at the moment (as can be seen from figure 12.4(b) and figure 12.4(c)), pk/CIGS tandems have certain advantages over Si-based TSCs. The device can be made on flexible substrates [86] and has a substantially smaller carbon footprint per kWh produced, making these cells an effective, adaptable, lightweight, and sustainable alternative. pk/CIGS tandems with a combined radiation hardness of both subcells [87, 88] offers a high energy yield option for space applications. These benefits suggest that pk/CIGS devices will find use in a wide range of terrestrial and space applications in the future. There are relatively few papers on pk/CIGS solar cells [86–92] in comparison to the many on pk/Si TSCs. This is probably because of the lower *PCE*, smaller market share of CIGS, and difficult integration of the perovskite subcell on top of a (nano) rough CIGS cell surface. Nonetheless, in a recent work, a monolithic 2T pk/CIGS tandem device with certified record *PCE* of 24.2% [44, 93] has been reported (figure 12.4(b) and figure 12.4(c)).

Figure 12.7(a) shows a cross-section scanning electron microscope (SEM) micrograph of each layer of pk/CIGS tandem device. *PCE* of the fabricated device has been independently certified at Fraunhofer ISE. At a stabilized maximum power point, *PCE* of 24.2% is exhibited by pk/CIGS which is 1% higher than the previous record [92]. Figure 12.7(b) shows the certified I–V characteristics for the device active area of 1.04 cm^2 and has performance parameters; J_{sc} = 18.8 mA cm^{-2}, V_{oc} = 1.77 V, and FF = 71.2%. Measurements of *EQE* and absorbance ($1-R$) are used to analyse J_{sc} and optical performance of the manufactured device (figure 12.7(c)).

12.5.2 Organic tandem solar cells

OTSCs consist of two subcells stacked on top of each other, the top cell absorbs high energy photons while the bottom cell absorbs low energy photons. Organic solar

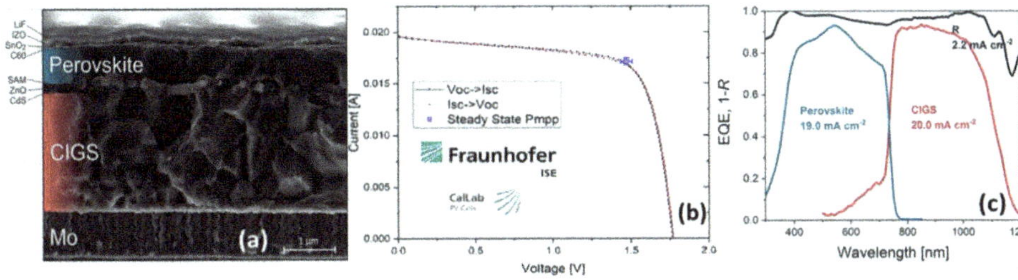

Figure 12.7. (a) Schematic diagram of a monolithic pk/CIGS TSC with all of the layers superimposed on a cross-sectional SEM image [93] (b) *I–V* characteristics of the pk/CIGS solar cell. Performance of the cell was certified at CalLab Fraunhofer ISE. The certified values are PCE_{MPP} = 24.2%, J_{sc} = 18.8 mA cm^{-2}, V_{oc} = 1.77 V, and *FF* = 71.2% for the device active area 1.04 cm^2. (c) *EQE* and *1–R* spectra for the pk/CIGS TSC. Photogenerated current densities, obtained from integration of the *EQE* spectra with AM1.5G spectrum. The perovskite bandgap is 1.68 eV, while CIGS has a bandgap of 1.1 eV, determined at the inflection point from the *EQE* spectra. Reproduced from [93] CC BY 4.0.

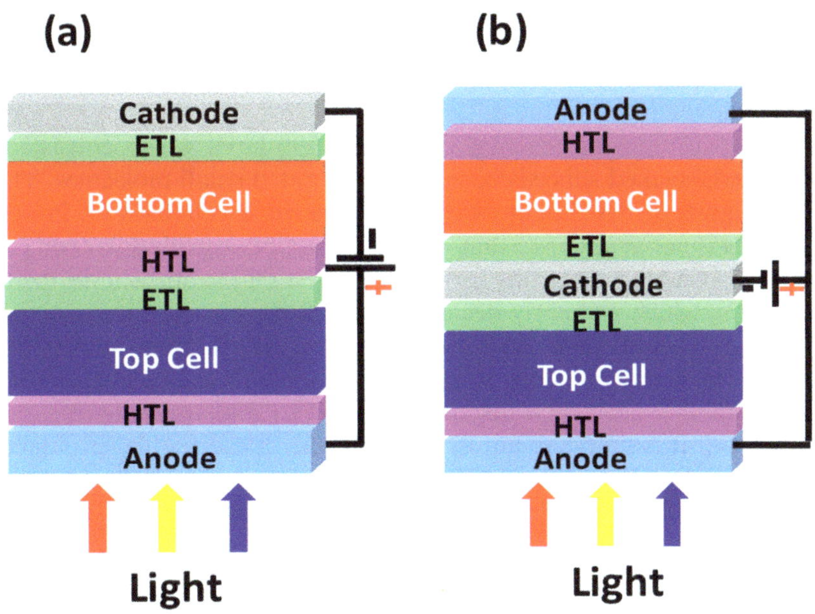

Figure 12.8. Schematic representation of OTSCs in (a) series connection (b) parallel connection.

cells have attracted significant interest because of their low weight, high mechanical elasticity, and adjustable semitransparency [94–96]. The absorption spectra of organic materials can be tuned by modifying their chemical structure or blending them with other materials. This allows one to design tandem cells with tailored absorption profiles optimized for specific solar spectrum conditions. Although it is most cost-effective and suitable, OTSC has achieved lower *PCE*s compared to other

TSCs [44]. Efficient charge generation and transport at the interfaces between the organic subcells are crucial for maximizing the performance of OTSCs. Efficiency of 20.6% for OTSCs has been made possible by lower voltage loss methodology [43]. OTSCs utilize a variety of organic semiconductor materials, including conjugated polymers and small-molecule organic compounds [97]. OTSCs may be connected in series or parallel, similar to other TSCs as shown in figure 12.8(a) and figure 12.8(b), respectively. V_{oc} of OTSC with series connections is equal to the sum of the V_{oc} of its two subcells while the J_{sc} value is limited by the minimum J_{sc} among both cells [98–101]. On the other hand, V_{oc} in parallel connection (figure 12.8(b)) is limited by the minimum V_{oc} between the top and bottom cells sets, and the projected J_{sc} is the sum of the short-circuit current for both cells. It should be noted that the subcell with lower J_{sc} controls the J_{sc} in the tandem cell. If the top and bottom cells have J_{sc} values of 16 and 18 mA cm^{-2}, respectively, then the resulting J_{sc} is 16 mA cm^{-2} assuming the same *FF* for each subcell [102]. If J_{sc} and *FF* differ between two subcells, the cell with a larger *FF* controls J_{sc} in the tandem cell [103]. J_{sc} is limited by the top cell; therefore, tandem cells must be engineered to achieve an identical J_{sc} from each subcell known as current matching [102]. It is important to note here that tandem devices are high V_{oc} and low J_{sc} devices that can be used to minimize power loss in a large-area PV module since power loss in a large-area electrode (follows Ohm's law, $P = (\text{current})^2 \times R_s$ where R_s is the series resistance in the cell) can be suppressed by low current in the device [102].

Hiramoto *et al* reported the first OTSC in 1990 [104] which consisted of two series-connected, stacked subcells based on evaporated small molecules. The V_{oc} of the tandem cell was 0.78 V, about twice that of the single cell (0.44 V). Presently, the *PCE* of OTSCs is not superior to that of single-junction organic solar cell [105]. The lack of narrow bandgap materials for rear cells is a major obstacle to the development of OTSCs, which primarily decides how well the near-infrared portion of the solar spectrum is used [106–108]. Furthermore, the efficient tandem device is dependent on the interlayer that exists between the subcells, which creates ohmic contact through the electron transport layer (ETL) and the hole-transport layer (HTL) [109, 110]. It is simple to adjust the thickness of each sublayer independently to achieve maximum photon-to-energy efficiency and a balanced absorbance in each wavelength range [111–113]. Li *et al* used tandem architecture to obtain *PCE* of 17.3% [114] by employing two different active layers: a narrow bandgap polymer in the bottom cell, PTB7-Th:O6T-4F:PC71BM, and a wide-bandgap polymer in the front cell, PBDB-T:F-M. Various developments in OTSCs have been reported in the literature [100, 115–118]. The hybrid TSC perovskite/organic (pk/OPV), as shown in figure 12.4(b) and figure 12.4(c), has achieved an efficiency of 24.2% [119].

Figure 12.9 represents a schematic of an OTSC with a *PCE* of 20.2%. In the device, electron beam evaporation method is used to create interconnecting layers [120]. Figure 12.9(a) shows a schematic of electron beam evaporation. PBDB-TF:GS-ISO are selected as the bulk heterojunction (BHJ) of the bottom subcell, and the PBDB-TF: BTP-eC9 is used as the BHJs in the top subcell. This is the first report of *PCE* of the tandem cell being more than 20.27% with the $TiO_{1.76}$/PEDOT:PSS interconnecting layer. A sharp, smooth, and dense TiO_x/PEDOT:PSS interface is achieved via electron

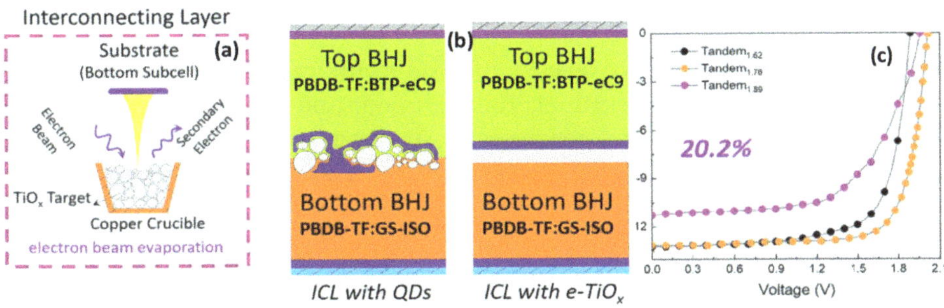

Figure 12.9. Performance of recent high efficiency OTSC (a) schematic diagram showing electron-beam evaporation (b) device architectures of the OTSCs (c) The current density–voltage (*J–V*) curves of OTSCs based on different depositions under AM 1.5G, 100 mW cm^{-2}. Reproduced with permission from [120] Copyright Elsevier 2022.

Table 12.4. List of some of the recent OTSCs. The performance parameters J_{sc}, V_{oc}, *FF* and *PCE* are mentioned including top and bottom cells modified from. Reprinted from [102], copyright (2020), with permission from Elsevier.

Top cell	Bottom cell	Performance parameters				Year (reference)
		J_{sc} mA cm^{-2}	V_{oc} (V)	*FF*	*PCE*	
PBDB-TF:BTP-eC9	PBDB-TF:GS-ISO	13.14	2.01	76.75	20.27%,	2022 [120]
PBDB-T:F-M(200)	PTB7-Th:O6T-4F: PC71BM (125)	14.35	1.64	73.37	17.3%	2018 [118]
PBDB-T-2F:TfIF-4FIC (150)	PTB7-Th:PCDTBT: IEICO-4F (90)	13.6	1.6	69	15%	2019 [121]
DTDCPB:C70(160)	PCE-10:BT-CIC(75)	12.6	1.59	59	15%	2018 [122]
PBDB-T:F-M (80)	PTB7-Th:NPBDT (100)	11.45	1.7	63	14.11%	2018 [116]
PBDB-T:ITCC-M (130)	PBDTTT-E-T:IEICO (120)	11.4	1.79	64	13.8%	2017 [100]
PSTzBI-EHp: PC71BM (150)	PBDTTT-E-T:IEICO (90)	10.88	1.72	68	12.9%	2018 [123]
P3TEA:FTTB-PDI4(100)	PTB7-Th:IEICS-4F (100)	9.83	1.72	65.5	10.8%	2018 [124]

beam evaporation [120]. Some of the recent developments in OTSCs, with top and bottom cells, include the following performance parameters; J_{sc}, V_{oc}, *FF* and *PCE*; they are listed in table 12.4 [102].

The aforementioned are just a few examples of the types of inorganic, hybrid and OTSCs being researched and developed. As can be seen, all of the above configurations have their benefits and challenges, and ongoing research aims to optimize efficiency, stability, and manufacturing processes to make TSCs commercially viable for widespread deployment in PV applications.

12.6 Conclusions

This chapter has provided an overview of TSC technology. The structure of TSCs such as two-terminal, three-terminal, four-terminal and the associated processing methods using monolithic and mechanical stacking is discussed briefly. Recent advancements in ITSCs such as perovskite/silicon, III–V/silicon, perovskite/CIGS, including OTSCs, are considered. The performance of TSC with number of p-n junctions and various bandgap combinations is discussed. ITSCs such as III–V/silicon tandem have achieved certified efficiency of 35.9%. Monolithic, perovskite/silicon and perovskite/CIGS tandem devices have record certified efficiencies of 33.9% and 24.2%, respectively. OTSCs have achieved efficiency of 20.2%, where electron beam evaporation is used to create the interconnecting layers.

References

[1] Shockley W and Queisser H J 1961 Detailed balance limit of efficiency of p-n junction solar cells *J. Appl. Phys.* **32** 510–9

[2] Green M A *et al* 2016 Solar cell efficiency tables (version 47) *Prog. Photovolt. Res. Appl.* **24** 3–11

[3] Fetzer C *et al* 2008 Production ready 30% efficient triple junction space solar cells *2008 33rd IEEE Photovoltaic Specialists Conf.* (Piscataway, NJ: IEEE)

[4] Chiu P T *et al* 2019 Qualification of 32% BOL and 28% EOL efficient XTE solar cells *2019 IEEE 46th Photovoltaic Specialists Conf. (PVSC)* (Piscataway, NJ: IEEE)

[5] Takamoto T, Washio H and Juso H 2014 Application of InGaP/GaAs/InGaAs triple junction solar cells to space use and concentrator photovoltaic *2014 IEEE 40th Photovoltaic Specialist Conference (PVSC)* (Piscataway, NJ: IEEE)

[6] Yamaguchi M *et al* 2005 Multi-junction III–V solar cells: current status and future potential *Sol. Energy* **79** 78–85

[7] Todorov T, Gunawan O and Guha S 2016 A road towards 25% efficiency and beyond: perovskite tandem solar cells *Mol. Syst. Des. Eng* **1** 370–6

[8] Leijtens T *et al* 2018 Opportunities and challenges for tandem solar cells using metal halide perovskite semiconductors *Nat. Energy* **3** 828–38

[9] Werner J, Niesen B and Ballif C 2018 Perovskite/silicon tandem solar cells: marriage of convenience or true love story ?—An overview *Adv. Mater. Interfaces* **5** 1700731

[10] Cheng Y and Ding L 2021 Perovskite/Si tandem solar cells: fundamentals, advances, challenges, and novel applications *SusMat* **1** 324–44

[11] Park I J *et al* 2019 A three-terminal monolithic perovskite/Si tandem solar cell characterization platform *Joule* **3** 807–18

[12] Gota F *et al* 2020 Energy yield advantages of three-terminal perovskite-silicon tandem photovoltaics *Joule* **4** 2387–403

[13] Tayagaki T *et al* 2019 Three-terminal tandem solar cells with a back-contact-type bottom cell bonded using conductive metal nanoparticle arrays *IEEE J. Photovolt.* **10** 358–62

[14] Tockhorn P *et al* 2020 Three-terminal perovskite/silicon tandem solar cells with top and interdigitated rear contacts *ACS Appl. Energy Mater.* **3** 1381–92

[15] Li X *et al* 2021 Silicon heterojunction-based tandem solar cells: past, status, and future prospects *Nanophotonics* **10** 2001–22

[16] Schnabel M *et al* 2020 Three-terminal III–V/Si tandem solar cells enabled by a transparent conductive adhesive *Sustain. Energy Fuels* **4** 549–58
[17] Guo F *et al* 2015 Fully printed organic tandem solar cells using solution-processed silver nanowires and opaque silver as charge collecting electrodes *Energy Environ. Sci.* **8** 1690–7
[18] Tyagi B *et al* 2022 High-performance, large-area semitransparent and tandem perovskite solar cells featuring highly scalable a-ITO/Ag mesh 3D top electrodes *Nano Energy* **95** 106978
[19] Lamanna E *et al* 2020 Mechanically stacked, two-terminal graphene-based perovskite/silicon tandem solar cell with efficiency over 26% *Joule* **4** 865–81
[20] Kim S *et al* 2021 Over 30% efficiency bifacial 4-terminal perovskite-heterojunction silicon tandem solar cells with spectral albedo *Sci. Rep.* **11** 15524
[21] Kim D H *et al* 2019 Bimolecular additives improve wide-band-gap perovskites for efficient tandem solar cells with CIGS *Joule* **3** 1734–45
[22] Uzu H *et al* 2015 High efficiency solar cells combining a perovskite and a silicon heterojunction solar cells via an optical splitting system *Appl. Phys. Lett.* **106** 013506
[23] Yao Y *et al* 2020 Highly efficient Sn–Pb perovskite solar cell and high-performance all-perovskite four-terminal tandem solar cell *Sol. RRL* **4** 1900396
[24] Sze S M 1981 *Physics of Semiconductor Devices* (NewYork: Wiley) ch 14
[25] Li H and Zhang W 2020 Perovskite tandem solar cells: from fundamentals to commercial deployment *Chem. Rev.* **120** 9835–950
[26] Almansouri I *et al* 2015 Supercharging silicon solar cell performance by means of multijunction concept *IEEE J. Photovolt.* **5** 968–76
[27] Yamaguchi M *et al* 2018 A review of recent progress in heterogeneous silicon tandem solar cells *J. Phys. D: Appl. Phys.* **51** 133002
[28] Vos A D 1980 Detailed balance limit of the efficiency of tandem solar cells *J. Phys. D: Appl. Phys.* **13** 839
[29] Green M *et al* 2021 Solar cell efficiency tables (version 57) *Prog. Photovolt. Res. Appl.* **29** 3–15
[30] Bruno *et al* 2020 Photovoltaics reaching for the Shockley–Queisser limit *ACS Energy Lett.* **5** 3029–33
[31] Bremner S P *et al* 2016 Optimum band gap combinations to make best use of new photovoltaic materials *Sol. Energy* **135** 750–7
[32] Philipps S P, Dimroth F and Bett A W 2018 High-efficiency III–V multijunction solar cells *McEvoy's Handbook of Photovoltaics* 3rd edn ed S A Kalogirou (New York: Academic) ch I-4-B pp 439–72
[33] Essig S *et al* 2017 Raising the one-sun conversion efficiency of III–V/Si solar cells to 32.8% for two junctions and 35.9% for three junctions *Nat. Energy* **2** 1–9
[34] Harbecke B 1986 Coherent and incoherent reflection and transmission of multilayer structures *Appl. Phys.* B **39** 165–70
[35] Alberi K *et al* 2024 A roadmap for tandem photovoltaics *Joule* **8** 658–92
[36] Aydin E *et al* 2023 Enhanced optoelectronic coupling for perovskite/silicon tandem solar cells *Nature* **623** 732–8
[37] Shi Y, Berry J J and Zhang F 2024 Perovskite/silicon tandem solar cells: insights and outlooks *ACS Energy Lett.* **9** 1305–30
[38] Okil M *et al* 2023 Investigation of polymer/Si thin film tandem solar cell using TCAD numerical simulation *Polymers* **15** 2049
[39] Yu S, Rabelo M and Yi J 2022 A brief review on III-V/Si tandem solar cells *Trans. Electr. Electron. Mater.* **23** 327–36

[40] Brinkmann K O *et al* 2022 Perovskite–organic tandem solar cells with indium oxide interconnect *Nature* **604** 280–6

[41] Boukortt N E *et al* 2023 All-perovskite tandem solar cells: from certified 25% and beyond *Energies* **16** 3519

[42] Shoji Y *et al* 2022 28.3% Efficient III–V tandem solar cells fabricated using a triple-chamber hydride vapor phase epitaxy system *Sol. RRL* **6** 2100948

[43] Wang J *et al* 2023 Tandem organic solar cells with 20.6% efficiency enabled by reduced voltage losses *Natl Sci. Rev.* **10** nwad085

[44] NREL 2024 Best research-cell efficiency chart by NREL https://www.nrel.gov/pv/cell-efficiency.html

[45] Fang J *et al* 2018 Amorphous silicon/crystal silicon heterojunction double-junction tandem solar cell with open-circuit voltage above 1.5 V and high short-circuit current density *Sol. Energy Mater. Sol. Cells* **185** 307–11

[46] Hajijafarassar A *et al* 2020 Monolithic thin-film chalcogenide–silicon tandem solar cells enabled by a diffusion barrier *Sol. Energy Mater. Sol. Cells* **207** 110334

[47] Amiri O *et al* 2017 Design and fabrication of a high performance inorganic tandem solar cell with 11.5% conversion efficiency *Electrochim. Acta* **252** 315–21

[48] Taguchi M *et al* 2013 24.7% record efficiency HIT solar cell on thin silicon wafer *IEEE J. Photovolt.* **4** 96–9

[49] Yoshikawa K *et al* 2017 Silicon heterojunction solar cell with interdigitated back contacts for a photoconversion efficiency over 26% *Nat. Energy* **2** 1–8

[50] Yoshikawa K *et al* 2017 Exceeding conversion efficiency of 26% by heterojunction interdigitated back contact solar cell with thin film Si technology *Sol. Energy Mater. Sol. Cells* **173** 37–42

[51] Bremner S, Levy M and Honsberg C B 2008 Analysis of tandem solar cell efficiencies under AM1. 5G spectrum using a rapid flux calculation method *Prog. Photovolt. Res. Appl.* **16** 225–33

[52] Brown A S and Green M A 2002 Detailed balance limit for the series constrained two terminal tandem solar cell *Physica* E **14** 96–100

[53] Kayes B M *et al* 2011 27.6% conversion efficiency, a new record for single-junction solar cells under 1 sun illumination *2011 37th IEEE Photovoltaic Specialists Conf.* (Piscataway, NJ: IEEE)

[54] Steiner M A *et al* 2021 High efficiency inverted GaAs and GaInP/GaAs solar cells with strain-balanced GaInAs/GaAsP quantum wells *Adv. Energy Mater.* **11** 2002874

[55] Geisz J F *et al* 2013 Enhanced external radiative efficiency for 20.8% efficient single-junction GaInP solar cells *Appl. Phys. Lett.* **103**

[56] Yamaguchi M and Amano C 1985 Efficiency calculations of thin-film GaAs solar cells on Si substrates *J. Appl. Phys.* **58** 3601–6

[57] France R M *et al* 2016 Metamorphic epitaxy for multijunction solar cells *MRS Bull.* **41** 202–9

[58] Fitzgerald E 1991 Dislocations in strained-layer epitaxy: theory, experiment, and applications *Mater. Sci. Rep.* **7** 87–142

[59] Geisz J F *et al* 2020 Six-junction III–V solar cells with 47.1% conversion efficiency under 143 Suns concentration *Nat. Energy* **5** 326–35

[60] France R M *et al* 2015 Design flexibility of ultrahigh efficiency four-junction inverted metamorphic solar cells *IEEE J. Photovolt.* **6** 578–83

[61] Cariou R *et al* 2018 III–V-on-silicon solar cells reaching 33% photoconversion efficiency in two-terminal configuration *Nat. Energy* **3** 326–33

[62] France R M *et al* 2022 Triple-junction solar cells with 39.5% terrestrial and 34.2% space efficiency enabled by thick quantum well superlattices *Joule* **6** 1121–35

[63] Schygulla P *et al* 2022 Two-terminal III–V//Si triple-junction solar cell with power conversion efficiency of 35.9 % at AM1.5g *Prog. Photovolt. Res. Appl.* **30** 869–79

[64] Unger E *et al* 2017 Roadmap and roadblocks for the band gap tunability of metal halide perovskites *J. Mater. Chem.* A **5** 11401–9

[65] Tress W 2017 Perovskite solar cells on the way to their radiative efficiency limit–insights into a success story of high open-circuit voltage and low recombination *Adv. Energy Mater.* **7** 1602358

[66] Stranks S D *et al* 2013 Electron-hole diffusion lengths exceeding 1 micrometer in an organometal trihalide perovskite absorber *Science* **342** 341–4

[67] Mailoa J P *et al* 2015 A 2-terminal perovskite/silicon multijunction solar cell enabled by a silicon tunnel junction *Appl. Phys. Lett.* **106** 121105

[68] Shaw V 2023 Longi claims 33.9% efficiency for perovskite-silicon tandem solar cell *PV Magazine—Photovoltaics Markets and Technology* **June 14** https://www.pv-magazine.com/2024/06/14/longi-claims-34-6-efficiency-for-perovskite-silicon-tandem-solar-cell/

[69] Bellini E *KAUST claims 33.7% efficiency for perovskite/silicon tandem solar cell.* May 30, 2023. https://pv-magazine.com/2023/05/30/kaust-claims-33-7-efficiency-for-perovskite-silicon-tandem-solar-cell/

[70] Chen B *et al* 2019 Grain engineering for perovskite/silicon monolithic tandem solar cells with efficiency of 25.4% *Joule* **3** 177–90

[71] Yang D *et al* 2021 28.3%-efficiency perovskite/silicon tandem solar cell by optimal transparent electrode for high efficient semitransparent top cell *Nano Energy* **84** 105934

[72] Albrecht S *et al* 2016 Monolithic perovskite/silicon-heterojunction tandem solar cells processed at low temperature *Energy Environ. Sci.* **9** 81–8

[73] Werner J *et al* 2016 Efficient monolithic perovskite/silicon tandem solar cell with cell area >1 cm^2 *J. Phys. Chem. Lett.* **7** 161–6

[74] Shen H *et al* 2018 In situ recombination junction between p-Si and TiO_2 enables high-efficiency monolithic perovskite/Si tandem cells *Sci. Adv.* **4** eaau9711

[75] Mazzarella L *et al* 2019 Infrared light management using a nanocrystalline silicon oxide interlayer in monolithic perovskite/silicon heterojunction tandem solar cells with efficiency above 25% *Adv. Energy Mater.* **9** 1803241

[76] Liu X *et al* 2019 20.7% highly reproducible inverted planar perovskite solar cells with enhanced fill factor and eliminated hysteresis *Energy Environ. Sci.* **12** 1622–33

[77] Kim D *et al* 2020 Efficient, stable silicon tandem cells enabled by anion-engineered wide-bandgap perovskites *Science* **368** 155–60

[78] Hou Y *et al* 2020 Efficient tandem solar cells with solution-processed perovskite on textured crystalline silicon *Science* **367** 1135–40

[79] Al-Ashouri A *et al* 2020 Monolithic perovskite/silicon tandem solar cell with > 29% efficiency by enhanced hole extraction *Science* **370** 1300–9

[80] Löper P *et al* 2015 Organic–inorganic halide perovskite/crystalline silicon four-terminal tandem solar cells *Phys. Chem. Chem. Phys.* **17** 1619–29

[81] Bailie C D *et al* 2015 Semi-transparent perovskite solar cells for tandems with silicon and CIGS *Energy Environ. Sci.* **8** 956–63

[82] McMeekin D P *et al* 2016 A mixed-cation lead mixed-halide perovskite absorber for tandem solar cells *Science* **351** 151–5
[83] Duong T *et al* 2017 Rubidium multication perovskite with optimized bandgap for perovskite-silicon tandem with over 26% efficiency *Adv. Energy Mater.* **7** 1700228
[84] Jaysankar M *et al* 2018 Minimizing voltage loss in wide-bandgap perovskites for tandem solar cells *ACS Energy Lett.* **4** 259–64
[85] Chen B *et al* 2020 Enhanced optical path and electron diffusion length enable high-efficiency perovskite tandems *Nat. Commun.* **11** 1257
[86] Fu F *et al* 2019 Flexible perovskite/Cu(In,Ga)Se_2 monolithic tandem solar cells arXiv:1907.10330 https://doi.org/10.48550/arXiv.1907.10330
[87] Lang F *et al* 2019 Efficient minority carrier detrapping mediating the radiation hardness of triple-cation perovskite solar cells under proton irradiation *Energy Environ. Sci.* **12** 1634–47
[88] Lang F *et al* 2020 Proton radiation hardness of perovskite tandem photovoltaics *Joule* **4** 1054–69
[89] Todorov T *et al* 2015 Monolithic perovskite-CIGS tandem solar cells via *in situ* band gap engineering *Adv. Energy Mater.* **5** 1500799
[90] Han Q *et al* 2018 High-performance perovskite/Cu(In,Ga)Se_2 monolithic tandem solar cells *Science* **361** 904–8
[91] Jošt M *et al* 2019 21.6%-Efficient monolithic perovskite/Cu(In,Ga)Se_2 tandem solar cells with thin conformal hole transport layers for integration on rough bottom cell surfaces *ACS Energy Lett.* **4** 583–90
[92] Al-Ashouri A *et al* 2019 Conformal monolayer contacts with lossless interfaces for perovskite single junction and monolithic tandem solar cells *Energy Environ. Sci.* **12** 3356–69
[93] Jošt M *et al* 2022 Perovskite/CIGS tandem solar cells: from certified 24.2% toward 30% and beyond *ACS Energy Lett.* **7** 1298–307
[94] Søndergaard R *et al* 2012 Roll-to-roll fabrication of polymer solar cells *Mater. Today* **15** 36–49
[95] Ameri T, Li N and Brabec C J 2013 Highly efficient organic tandem solar cells: a follow up review *Energy Environ. Sci.* **6** 2390–413
[96] Dong S *et al* 2020 Single-component non-halogen solvent-processed high-performance organic solar cell module with efficiency over 14% *Joule* **4** 2004–16
[97] Ameri T *et al* 2009 Organic tandem solar cells: a review *Energy Environ. Sci.* **2** 347–63
[98] Guo B *et al* 2016 A wide-bandgap conjugated polymer for highly efficient inverted single and tandem polymer solar cells *J. Mater. Chem.* A **4** 13251–8
[99] Guo B *et al* 2018 Exceeding 14% efficiency for solution-processed tandem organic solar cells combining fullerene-and nonfullerene-based subcells with complementary absorption *ACS Energy Lett.* **3** 2566–72
[100] Cui Y *et al* 2017 Fine-tuned photoactive and interconnection layers for achieving over 13% efficiency in a fullerene-free tandem organic solar cell *JACS* **139** 7302–9
[101] Jia Z *et al* 2021 High performance tandem organic solar cells via a strongly infrared-absorbing narrow bandgap acceptor *Nat. Commun.* **12** 178
[102] Salim M B, Nekovei R and Jeyakumar R 2020 Organic tandem solar cells with 18.6% efficiency *Sol. Energy* **198** 160–6
[103] Sista S *et al* 2011 Tandem polymer photovoltaic cells—current status, challenges and future outlook *Energy Environ. Sci.* **4** 1606–20
[104] Hiramoto M, Suezaki M and Yokoyama M 1990 Effect of thin gold interstitial-layer on the photovoltaic properties of tandem organic solar cell *Chem. Lett.* **19** 327–30

[105] Cui Y *et al* 2020 Single-junction organic photovoltaic cells with approaching 18% efficiency *Adv. Mater.* **32** 1908205

[106] Li N *et al* 2013 Towards 15% energy conversion efficiency: a systematic study of the solution-processed organic tandem solar cells based on commercially available materials *Energy Environ. Sci.* **6** 3407–13

[107] Dou L *et al* 2012 Systematic investigation of benzodithiophene-and diketopyrrolopyrrole-based low-bandgap polymers designed for single junction and tandem polymer solar cells *JACS* **134** 10071–9

[108] Chen F X *et al* 2018 Near-infrared electron acceptors with fluorinated regioisomeric backbone for highly efficient polymer solar cells *Adv. Mater.* **30** 1803769

[109] Yang F, Kang D-W and Kim Y-S 2017 An efficient and thermally stable interconnecting layer for tandem organic solar cells *Sol. Energy* **155** 552–60

[110] Lee S *et al* 2015 Polymer/small-molecule parallel tandem organic solar cells based on MoO_x–Ag–MoO_x intermediate electrodes *Sol. Energy Mater. Sol. Cells* **137** 34–43

[111] Simone G *et al* 2018 Near-infrared tandem organic photodiodes for future application in artificial retinal implants *Adv. Mater.* **30** 1804678

[112] Li M *et al* 2017 Solution-processed organic tandem solar cells with power conversion efficiencies > 12% *Nat. Photonics* **11** 85–90

[113] Liu W *et al* 2016 Nonfullerene tandem organic solar cells with high open-circuit voltage of 1.97 V *Adv. Mater.* **28** 9729–34

[114] Li G, Chang W-H and Yang Y 2017 Low-bandgap conjugated polymers enabling solution-processable tandem solar cells *Nat. Rev. Mater.* **2** 17043

[115] Jiang Z, Gholamkhass B and Servati P 2019 Effects of interlayer properties on the performance of tandem organic solar cells with low and high band gap polymers *J. Mater. Res.* **34** 2407–15

[116] Zhang Y *et al* 2018 Nonfullerene tandem organic solar cells with high performance of 14.11% *Adv. Mater.* **30** 1707508

[117] Ge H-T *et al* 2019 Low-temperature solution-processed hybrid interconnecting layer with bulk/interfacial synergistic effect in symmetric tandem organic solar cells *Org. Electron.* **75** 105423

[118] Meng L *et al* 2018 Organic and solution-processed tandem solar cells with 17.3% efficiency *Science* **361** 1094–8

[119] Chen W *et al* 2022 Monolithic perovskite/organic tandem solar cells with 23.6% efficiency enabled by reduced voltage losses and optimized interconnecting layer *Nat. Energy* **7** 229–37

[120] Zheng Z *et al* 2022 Tandem organic solar cell with 20.2% efficiency *Joule* **6** 171–84

[121] Liu G *et al* 2019 15% Efficiency tandem organic solar cell based on a novel highly efficient wide-bandgap nonfullerene acceptor with low energy loss *Adv. Energy Mater.* **9** 1803657

[122] Che X *et al* 2018 High fabrication yield organic tandem photovoltaics combining vacuum- and solution-processed subcells with 15% efficiency *Nat. Energy* **3** 422–7

[123] Zhang Y *et al* 2018 Thermally stable all-polymer solar cells with high tolerance on blend ratios *Adv. Energy Mater.* **8** 1800029

[124] Chen S *et al* 2018 A nonfullerene semitransparent tandem organic solar cell with 10.5% power conversion efficiency *Adv. Energy Mater.* **8** 1800529

IOP Publishing

Recent Advances in Solar Cells

N M Ravindra, Leqi Lin and Priyanka Singh

Chapter 13

Recent patents and disclosures

In this chapter, examples of recent patents, inventions and disclosures, from around the world and some national labs and industry, are summarized to grasp and appreciate the critical topics that are of interest from the perspective of research and development (R&D) and applications. Innovations in solar cell/module patents often focus on improving the efficiency, durability and cost-effectiveness of the technology. These include new materials for solar cell construction, novel manufacturing techniques, and advancements in energy conversion efficiency and innovations in storage and distribution of solar energy. Some of the research focuses on the multi-electricity generation solar cell/module systems that combine the solar cell system, energy storage system, and energy management system. Nowadays, beneficial policies from the governments around the world in supporting sustainable energy include a variety of incentives such as the US Government's Inflation Reduction Act (IRA). These incentives, combined with the increasing population and the efficiency of land use, have led to the thriving development of innovative solar energy systems. Examples of one hundred patents are summarized in this chapter to serve as references for scientists, engineers and industry practitioners of photovoltaics. These include patents on bifacial, cadmium sulfide, cadmium telluride, concentrator, gallium arsenide, homo and heterojunctions, indium phosphide, perovskites, silicon, space, tandem, and thin film solar cells as well as approaches to reducing the use of silver in solar cells and antireflection coatings for solar cell applications. Solar panel components such as junction box and system components such as inverters/micro-inverters and implementation schemes for addressing shade-associated losses are also addressed.

13.1 Introduction

The emphasis on the utilization of renewable energy sources, combined with the need to address the quality of the environment and hence human life, has taken the world by

doi:10.1088/978-0-7503-5994-8ch13

storm. There appear to be no boundaries between continents, countries and cultures in this endeavor. The number of discoveries, inventions and patents, in support of making the use of renewable energy sources efficient and affordable, is on the rise. China continues to be a major player in all aspects of renewable energy including the supply-chain of raw materials, R&D, technology transfer and manufacturing. It has been reported that the average annual growth rate of patent applications in China's photovoltaic industry sector has attained 23.1% [1]. In fact, according to the data released by China's National Intellectual Property Administration, China has filed 126 400 patent applications on solar cells, making China the first in the world in this sector [1]. In another report, according to the Journal *Nature*, 29% of the electricity in China is produced from renewable sources, up from 18% in 2010 [2]. According to the same report, China leads the world in scientific research output in renewable technologies [2]. This is illustrated in figure 13.1 [2].

Table 13.1 presents examples of the recent patents/patent applications in support of solar cell/module techniques, implementation methodologies, and industry R&D [3–5].

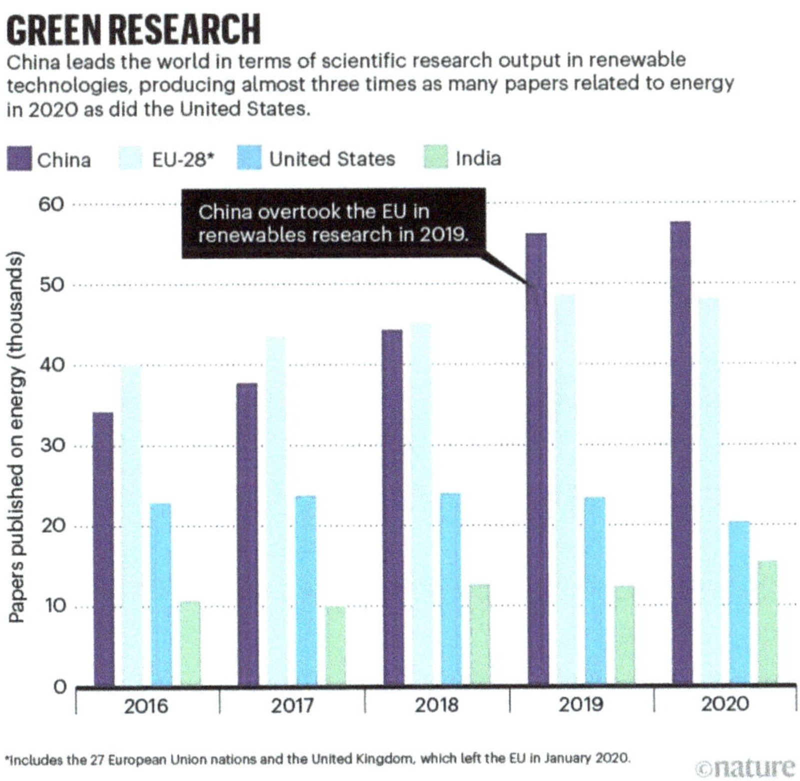

Figure 13.1. Illustration of a comparison of growth in scientific research output in renewable energy technologies in China, Europe, India and the United States. China dominates the world in its productivity in renewable energy technologies. Reproduced from [2] with permission from Springer Nature.

Table 13.1. Examples of recent patents/patent applications on solar cell/module techniques, implementation methodologies, and industry R&D.

#1	Description
Patent/patent application number	CN 115,207,137A
Title	Combined passivation back contact battery and preparation method thereof
Assignee	Jinyang Quanzhou New Energy Technology Co. Ltd
Published date	2022–10–18
Abstract	The invention belongs to the technical field of back contact batteries, and particularly relates to a combined passivation back contact battery and a preparation method thereof. Compared with the conventional heterojunction battery, the combined passivation back contact battery can obviously improve the fill factor *FF*, the photoelectric conversion efficiency, the yield and the like, and the preparation method is relatively simple, can be industrialized and is beneficial to improving the mass production efficiency.

#2	Description
Patent/patent application number	US 11,005,414 B2
Title	Solar module mounting
Assignee	Vivint Solar Inc
Date	2021–05–11
Abstract	Embodiments of the present disclosure are related to solar module mounting systems. A system may include an adhesion sheet configured to be secured to a roof of a structure via an adhesive. The system may further include at least one clamp configured for securing at least one solar module to the adhesion sheet. Other embodiments are related to methods of attaching one or more solar modules to a structure.

#3	Description
Patent/patent application number	CN 115,274,934A
Title	Preparation method of light flexible cadmium telluride film and light flexible cadmium telluride substrate
Assignee	Wu Weibing Kong, Wentao Li Chengxiang
Date	2022–11–01
Abstract	The invention relates to the field of solar power generation, in particular to a preparation method of a light flexible cadmium telluride thin film and a light flexible cadmium telluride substrate, which comprises the following steps: s1, selecting flexible ultrathin conductive glass with a cadmium sulfide layer deposited on the surface as an electroplating substrate, and using a plane titanium plate sucker to vacuum-adsorb the back of the conductive glass so as to fix the conductive glass; s2, placing the conductive surface of the conductive glass which is adsorbed and fixed downwards in a plating bath with a special structure, wherein the conductive glass is a cathode during plating, a titanium plate at the bottom of the plating bath is an anode, elastic needles which are arranged in an array mode are arranged on the titanium plate and are connected with cathode leads, and needle points are in contact with a conductive layer of the conductive glass; and S3, circularly introducing electroplating solution into the electrolytic cell, and starting electroplating to obtain the light flexible cadmium telluride thin film substrate. According to the preparation method of the flexible cadmium telluride thin film substrate, the prepared cell substrate has the characteristics of light weight, flexibility and uniform composition thickness, and meets the requirements of future photovoltaic building integration on flexible photovoltaic solar cells.

#4	Description
Patent/patent application number	CN 115,051,641A
Title	Solar cell module and manufacturing method
Assignee	Shanxi Installation Group Co Ltd
Date	2022–09–13
Abstract	The invention discloses a solar cell module and a manufacturing method thereof, relating to the technical field of solar cells. The invention also discloses a manufacturing method of the solar cell module, which comprises the following steps: the method comprises the following steps of firstly, preparing a power generation assembly, secondly,

preparing a light guide assembly, thirdly, preparing an outer frame, fourthly, splicing, and fifthly, connecting a circuit. According to the solar energy conversion device, the outer frame component is arranged, the solar energy conversion component and the sunlight guide component are arranged in the outer frame component, the solar energy generation module in the solar energy conversion component can convert sunlight into electric energy, and the sunlight guide component can guide sunlight from different directions to the solar energy generation module after refraction, so that the device can generate electricity through sunlight from multiple directions, the available illumination range of the device is further improved, and the power generation efficiency of the device is improved.

#5	Description
Patent/patent application nmber	CN 115,276,520A
Title	Intensive photovoltaic wind wall system
Assignee	Yu Peici Wang Jifei Zhu Pengbo Wang Shuai
Date	2022–11–01
Abstract	The invention discloses an intensive photovoltaic wind wall system, which is applied to the technical field of new energy and comprises the following components: the wind power generation device comprises a wind power generator, a rail, an automatic tracking device, a photovoltaic panel, a pressure sensing device, an adjusting device and a photoelectric detector, wherein the rail is an annular rail, the wind power generator is arranged at the center of the rail, the automatic tracking device is arranged on the rail, the photovoltaic panel is connected with the automatic tracking device through the adjusting device, and the pressure sensing device is arranged on the photovoltaic panel.

#6	Description
Patent/patent application nmber	CN 115,117,910A
Title	Multidirectional power generation system that multiple electricity generation mode combines
Assignee	Jiangsu Tianzheng Environmental Protection Technology Co. Ltd
Date	2022–09–27
Abstract	The invention discloses a multidirectional power generation system combining multiple power generation modes, which comprises an energy storage system, an energy management system and an output system, wherein the energy storage

(Continued)

Table 13.1. (*Continued*)

#6	Description
	system is respectively connected with a wind power generation system, a thermal power generation system, a photovoltaic power generation system and a water-hydrogen power generation system, the energy storage system is connected with the energy management system, and the energy management system is connected with the output system. The wind power generation system, the thermal power generation system, the photovoltaic power generation system and the water-hydrogen power generation system are adopted, the combined power generation of four groups of power generation systems combines the traditional energy, the limited improved power generation stability and the power generation output are realized, the energy management system and the output system are adopted, the electric energy distribution and the power generation capacity of each power generation system are adjusted, the multidirectional power generation is realized, various systems can be utilized to the maximum extent, clean energy equipment is not wasted, and the phenomenon that photovoltaic power generation and the like are abandoned is avoided.

#7	Description
Patent/patent application number	US 12,014,885
Title	Molecular doping enabled scalable blading of efficient hole transport layer-free perovskite solar cells
Assignee	Nutech Ventures, Inc.
Date	2024–06–18
Abstract	A method of forming a photoactive device includes steps of: forming a photoactive layer, the photoactive layer comprising a perovskite material and a dopant; wherein the photoactive device comprises a positive electrode and a negative electrode; wherein said photoactive layer is directly or indirectly in electronic communication with the positive electrode and directly or indirectly in electronic communication with the negative electrode; and wherein the photoactive device is free of a hole transport layer between the photoactive layer and the positive electrode.

#8	Description
Patent/patent application number	CN 115,172,478A
Title	Solar cell and photovoltaic module
Assignee	Hejiang Jinko Solar Co. Ltd Jinko Solar Co. Ltd
Date	2022–10–11
Abstract	The embodiment of the application relates to the field of photovoltaics, in particular to a solar cell and a photovoltaic module, which comprise a substrate, a tunneling layer, a doped conducting layer, a passivation film and an electrode, wherein the tunneling layer, the doped conducting layer, the passivation film and the electrode penetrate through the passivation film and are in contact with the doped conducting layer; the doped conducting layer comprises p-type doped layers and n-type doped layers which are alternately arranged; the p-type doping layer comprises a first doping layer, and one side, far away from the substrate, of the first doping layer is in contact with the passivation film; the n-type doped layer comprises a second doped layer, and one side, far away from the substrate, of the second doped layer is in contact with the passivation film; the electrodes include a first electrode formed in contact with the first doped layer through the passivation film, and a second electrode formed in contact with the second doped layer through the passivation film. The solar cell provided by the embodiment of the application can improve the passivation effect of the solar cell and improve the conversion efficiency of the cell.

#9	Description
Patent/patent application number	US 9,346,563 B1
Title	Solar powered space weapon
Assignee	Rick Martin
Date	2016–05–24
Abstract	A plurality of orbiting solar generators stay in constant touch. They can be congregated rapidly in space at any desired secret location. Once congregated they all discharge their energy to a death star. This death star could be a newly launched ICBM. A laser generator uses this newly acquired energy on the death star to project a non-nuclear laser ray to a target. The target could be a city, a ship or a satellite. In the event of an asteroid approaching earth, this system could destroy an asteroid. In peacetime the orbiting solar generators could supply electric power to an earth based power grid.

#10	Description
Patent/patent application number	CN 115,225,020A
Title	Thin film battery device with rotatable self-protection function
Assignee	Huaneng Clean Energy Research Institute
Date	2022–10–21
Abstract	The photovoltaic power generation system comprises a photovoltaic power generation system, a rotary supporting system, a foundation, an inverter box transformer substation, a power transmission line and an energy storage system. The photovoltaic power generation system adopts an advanced thin film battery for converting solar energy into electric energy for grid-connected power generation, and the rotary supporting system is used for fixedly supporting the photovoltaic power generation system and can drive the photovoltaic thin film battery to rotate. The basis plays and supports the whole rotation braced system of fixed mounting, and inverter box becomes the direct current that photovoltaic power generation system sent and becomes the alternating current and steps up, and transmission line produces the electricity with the electricity and carries to the electric wire netting end, and energy storage system stores the electricity that self film battery sent, provides the electric energy for rotation braced system, makes this set of system become 'zero carbon' environmental protection system. The device provided by the invention reduces the failure and replacement frequency of the photovoltaic power station under severe conditions, prolongs the service life of the photovoltaic module, and improves the economic benefit of the photovoltaic power station. The device is simple and convenient, and is quick and effective, reduces the operation and maintenance expenditure, removes manual operation from, reduces the labor cost, and ensures the personnel safety.

#11	Description
Patent/patent application number	CN 115,188,852A
Title	Photovoltaic module and photovoltaic system
Assignee	Shenzhen Saineng Digital Energy Technology Co. Ltd
Date	2022–10–14
Abstract	The invention is suitable for the technical field of photovoltaics, and provides a photovoltaic assembly and a photovoltaic system, wherein a cell array of the photovoltaic assembly comprises a plurality of cell string units which

are arranged along the longitudinal direction and are sequentially connected in series, a first cell string group of the cell string units comprises two first cell strings which are arranged at intervals along the longitudinal direction and are connected in series, a second cell string group comprises at least two second cell strings which are positioned between two adjacent first cell strings and are sequentially connected in series, and the first cell string and the second cell string both comprise a plurality of half cells which are arranged along the transverse direction and are connected in series. Therefore, even if the photovoltaic module forms the dust deposition belt at the lower part in the using process, the existence of the dust deposition belt only influences the power generation of the first battery string group at the lowest part in the longitudinal direction, and does not influence each battery string unit at the upper part, thereby improving the power generation efficiency of the photovoltaic module.

#12	Description
Patent/patent application number	CN 114,937,709A
Title	P-type PERC double-sided solar cell module
Assignee	Das Solar Co. Ltd
Date	2022–08–23
Abstract	The invention belongs to the technical field of solar cells, and particularly relates to a P-type PERC double-sided solar cell module which comprises double-sided cells, wherein a plurality of groups of upper plates are uniformly arrayed on one side of the tops of the double-sided cells, a plurality of groups of lower plates are uniformly arrayed on the other side of the tops of the double-sided cells, and the upper plates and the lower plates are matched; the bottom of the lower plate is symmetrically provided with two baffle plates along the vertical plane in the left-right direction of the lower plate, opposite side surfaces of the baffle plates are provided with butt joint grooves, the bottom of each baffle plate is provided with a supporting air bag, the bottom of each supporting air bag is provided with a supporting block, a connecting air passage is arranged inside each supporting block, one end of each connecting air passage is communicated with the corresponding supporting air bag, and the other end of each connecting air passage penetrates through the corresponding supporting block and is communicated with the outside through a hydraulic control one-way valve; the device concatenation is efficient, and the shock attenuation buffering is effectual, and grafting is fixed and the precision is high, and daylighting generating efficiency is high, and is good to the protective properties of two-sided battery piece, strong adaptability, and resistance to compression shock resistance is strong.

#13	Description
Patent/patent application number	CN 115,164,629A
Title	Solar-driven hydrothermal and electric-heating comprehensive output system
Assignee	University of Science and Technology of China USTC
Date	2022–10–11
Abstract	The invention relates to a solar-driven hydrothermal and electric heating comprehensive output system, and belongs to the field of water treatment technology and energy comprehensive utilization. The solar energy and heat energy combined solar energy and heat energy water heater comprises a photovoltaic and light heat assembly mechanism, a solar pond, a distiller, a balanced brine tank and a salt and heat water tank which are connected through pipelines to form a loop; the photovoltaic photo-thermal assembly mechanism is arranged in an inclined plane shape; a heat exchange coil is arranged in the solar pond; the distiller comprises a stepped base, a glass cover plate and a fresh water collecting box. According to the invention, the photovoltaic photo-thermal component and the solar pond are used for preheating the saline water, so that the production efficiency of the fresh water is improved. Meanwhile, the system has self-powered function, and can generate and store electric energy. The solar pond can be used for solar heat storage and can be used as a heat source for the system to run at night when no solar radiation exists. The heat generated by the photovoltaic photo-thermal component and the heat at the outlet of the distillation channel are taken away, so that the supply of a living heat source is realized. The invention not only solves the problem of low cost and high efficiency of the traditional salt water desalination technology, but also realizes the comprehensive output of solar-driven water and electricity heat.

#14	Description
Patent/patent application number	CN 115,085,513A
Title	Composite photovoltaic inverter and control method thereof
Assignee	Jiangsu Xumax Power Technology Co. Ltd Shenzhen Zhongxu New Energy Co. Ltd
Date	2022–09–20

Abstract	The invention discloses a composite photovoltaic inverter, which comprises a DC-DC boost conversion module, a DC-AC inversion module and a three-port circuit, wherein the three-port circuit comprises a first port, a second port and a third port, the first port is used for being connected with a first photovoltaic component string group in a photovoltaic array, the second port is an input port of the DC-DC boost conversion module and is used for being connected with a second photovoltaic component string group of the photovoltaic array, and the third port is an input port of the DC-AC inversion module; the composite photovoltaic inverter is also provided with a second control module; a bypass diode or a relay is connected in parallel to the DC-DC boost conversion module; the control modes of the DC-DC sub-control module of the second control module comprise a boosting compensation mode and a bypass mode.

#15	Description
Patent/patent application number	CN 217,641,367 U
Title	Connection mode of tin alloy main grid line electrode photovoltaic cell
Assignee	Alpha Solar Suzhou Co. Ltd
Date	2022–10–21
Abstract	The invention can maximize the overall efficiency of the photovoltaic array in a complex distributed photovoltaic power station scene, thereby not only reducing the cost, but also solving the problem of mismatch between component level and group cascade level. The application relates to a connection mode of a tin alloy main grid line electrode photovoltaic cell, which comprises the following steps: the battery piece unit is provided with at least two battery piece units which are sequentially stacked, each battery piece unit comprises a front surface and a back surface which are oppositely arranged, and at least part of the front surface of one of the two adjacent battery piece units is overlapped with at least part of the back surface of the other battery piece unit; the main grid line electrode is made of a tin alloy material and comprises a front tin alloy main grid line electrode arranged on the front side of the cell unit and a back tin alloy main grid line electrode arranged on the back side of the cell unit; the front tin alloy main grid line electrode of one of the two adjacent cell sheet units is overlapped with the back tin alloy main grid line electrode of the other cell sheet unit; and the front tin alloy main grid line electrode of one of the two adjacent cell sheet units is connected with the back tin alloy main grid line electrode of the other cell sheet unit in a melting way, so that the two cell sheet units are electrically connected in series.

#16	Description
Patent/patent application Number	US 9,234,843 B2
Title	On-line, continuous monitoring in solar cell and fuel cell manufacturing using spectral reflectance imaging
Assignee	Alliance for Sustainable Energy, LLC
Date	2016–01–12
Abstract	A monitoring system 100 comprising a material transport system 104 providing for the transportation of a substantially planar material 102, 107 through the monitoring zone 103 of the monitoring system 100. The system 100 also includes a line camera 106 positioned to obtain multiple line images across a width of the material 102, 107 as it is transported through the monitoring zone 103. The system 100 further includes an illumination source 108 providing for the illumination of the material 102, 107 transported through the monitoring zone 103 such that light reflected in a direction normal to the substantially planar surface of the material 102, 107 is detected by the line camera 106. A data processing system 110 is also provided in digital communication with the line camera 106.

#17	Description
Patent/patent application number	US 8,796,160 B2
Title	Optical cavity furnace for semiconductor wafer processing
Assignee	Alliance for Sustainable Energy, LLC
Date	2014–08–05
Abstract	An optical cavity furnace 10 having multiple optical energy sources 12 associated with an optical cavity 18 of the furnace. The multiple optical energy sources 12 may be lamps or other devices suitable for producing an appropriate level of optical energy. The optical cavity furnace 10 may also include one or more reflectors 14 and one or more walls 16 associated with the optical energy sources 12 such that the reflectors 14 and walls 16 define the optical cavity 18. The walls 16 may have any desired configuration or shape to enhance operation of the furnace as an optical cavity 18. The optical energy sources 12 may be positioned at any location with respect to the reflectors 14 and walls defining the optical cavity. The optical cavity furnace 10 may further include a semiconductor wafer transport system 22 for transporting one or more semiconductor wafers 20 through the optical cavity.

#18	Description
Patent/patent application number	CN 115.085,636A
Title	Composite cooperative power generation device, control method and control system
Assignee	Tianjin Binhai High Tech Zone Hegong Electric Appliance Technology Co. Ltd Hebei University of Technology
Date	2022–09–20
Abstract	The invention provides a composite cooperative power generation device, a control method and a control system, and relates to the field of new energy equipment. The present application achieves compactness of the overall structure and reduction in cost by a plurality of common structural members. The device mainly uses photovoltaic power generation in daytime, can realize the operation of 'photovoltaic + wind-powered electricity generation' mode, realizes the device maximum generated power. The device switches to run in 'wind' mode at night. When the vertical axis wind turbine exceeds the rated wind speed, the device can reduce the wind energy capture, and the speed is reduced by the electromagnetic brake until the brake is stopped.

#19	Description
Patent/patent application number	CN 115,148,831A
Title	Electron selective transmission material for photovoltaic device and application thereof
Assignee	Suzhou University
Date	2022–10–04
Abstract	The invention belongs to the technical field of photovoltaic devices, and particularly relates to an electron selective transmission material for a photovoltaic device and application thereof. The material has the characteristics of low work function and large forbidden band width, and can realize excellent electron selective transmission performance, and the material specifically comprises strontium oxide and strontium fluoride. The electron selective transmission material can effectively reduce the carrier recombination loss at the metal-crystalline silicon contact position, can obviously reduce the metal-crystalline silicon contact resistance, improves the extraction and

(Continued)

Table 13.1. (*Continued*)

#19	Description
	transportation performance of electrons, and improves the photoelectric conversion efficiency of a photovoltaic device. In addition, the electron selective transmission material has excellent thermal stability and environmental stability, and has wide application prospects in crystalline silicon solar cells, perovskite solar cells and other optoelectronic devices.

#20	Description
Patent/patent application number	WO 2022/066707 A1
Title	Methods and devices for integrated tandem solar module fabrication
Assignee	Caelux Corporation
Date	2022–03–31
Abstract	The present disclosure may provide methods and devices for tandem perovskite solar modules. The perovskite solar module may comprise a plurality of perovskite solar cells configured to match a voltage output of another solar module. The present disclosure may provide mixed composition perovskite layers. Solar cells fabricated using the mixed composition perovskite layers may demonstrate improved performance and stability.

#21	Description
Patent/patent application number	CN 114,977,314A
Title	Method for mutual compensation between photovoltaic electric energy and port power supply
Assignee	Jiangsu Yiyi Iot Technology Co. Ltd
Date	2022–08–30
Abstract	The embodiment of the invention relates to the technical field of port power supply, and particularly discloses a method for mutual compensation between photovoltaic electric energy and port power supply. The embodiment of the invention uses a space field, a shallow sea area and a building roof in a harbour area as solar bearing points to carry out the arrangement of solar photovoltaic components and the solar power generation; the port service area is used as a power generation unit to respectively provide power for a power grid; the port service area is used as a

hydrogen production unit, photovoltaic power generation is consumed, the hydrogen production station stores hydrogen, and the hydrogen is sent to a nearby hydrogen station. Therefore, distributed photovoltaic power generation is introduced into a port power supply system, a port wharf space field, a wharf edge seaside and a building roof are used as solar bearing points, a power station and a hydrogen generation station are built in an unoccupied field, photovoltaic electric energy and port power supply are mutually compensated, a carbon neutralization mode with characteristics such as a carbon neutralization forest, a carbon neutralization power station and a hydrogen generation station is developed, a green and clean port power supply mode is created, and elasticity and adjustability of port power supply loads are achieved.

#22	Description
Patent/patent application number	CN 115,276,533A
Title	Enhanced perovskite photovoltaic device
Assignee	Application filed by 国网甘肃省电力公司兰州供电公司, 甘肃远效科技信息咨询有限公司
Date	2022–11–01
Abstract	The invention belongs to the technical field of photovoltaic power generation equipment, and particularly relates to an enhanced type perovskite photovoltaic device. The photovoltaic cell comprises an installation base, a support frame, a support plate and a perovskite photovoltaic cell panel, wherein an installation sleeve is arranged at the upper part of the installation base, the support frame is sleeved in the installation sleeve, and the support frame can rotate in the installation sleeve; one side of the supporting frame is hinged with the back of the supporting plate, and the supporting plate is vertical to the horizontal plane of the mounting base; a supporting seat is arranged on one side of the side face of the supporting plate, which is far away from the supporting frame, the supporting seat is hinged with the back of the perovskite photovoltaic cell panel, and a plurality of perovskite photovoltaic cell panels are arranged; a converter is arranged on the same side of the supporting plate and the supporting frame and is connected with the perovskite photovoltaic cell panel through a lead; the converter is connected with a storage battery arranged on the mounting base through a lead. The invention relates to an enhanced perovskite photovoltaic device with high sunlight utilization rate and long service life.

#23	Description
Patent/patent application number	CN 217,685,922 U
Title	Solar photovoltaic photo-thermal comprehensive utilization device capable of adjusting heat gain
Assignee	University of Science and Technology of China USTC
Date	2022–10–28
Abstract	The utility model relates to an adjustable solar photovoltaic photo-thermal comprehensive utilization device who obtains heat belongs to solar photovoltaic photo-thermal technical field. The solar heat collector comprises a heat collecting mechanism, a rectangular frame, an upper cover plate adjusting mechanism and a lower cover plate adjusting mechanism; the heat collecting mechanism comprises a photovoltaic cell, a heat collecting plate, a heat exchange copper pipe and a collecting pipe; the improvement lies in that: the heat collecting plate is fixedly arranged in the rectangular frame; the high-transmittance flexible cover plate of the upper cover plate adjusting mechanism is wound on the upper scroll; or unfolding and sealing the top surface of the rectangular frame corresponding to the photovoltaic cell; the low-thermal-conductivity flexible cover plate of the lower cover plate adjusting mechanism is wound on the lower reel; or the bottom surface of the rectangular frame corresponding to the heat exchange copper pipe is unfolded and sealed. In winter, the high-transmittance flexible cover plate and the low-thermal-conductivity flexible cover plate are unfolded to respectively seal the upper part and the lower part of the heat collecting plate, so that heat preservation is realized; in summer, the high-transmittance flexible cover plate and the low-heat-conductivity flexible cover plate are collected through the reel, and the heat collecting plate is exposed in ambient air to realize cooling.

#24	Description
Patent/patent application number	CN 115,172,603A
Title	Full-printing perovskite-carbon silicon heterojunction laminated cell and laminated assembly thereof
Assignee	Hebei University
Date	2022–10–11

Abstract	The invention relates to the technical field of photovoltaic power generation, and provides a fully-printed perovskite-carbon silicon heterojunction laminated cell and a laminated assembly thereof. Through the technical scheme, the problems that the manufacturing cost of the laminated battery is high and the power loss of the assembly is large in the prior art are solved.

#25	Description
Patent/patent application number	CN 115,225,031A
Title	Solar photovoltaic-super capacitor coupling heat management device and method
Assignee	Shaanxi Coal New Energy Technology Co. Ltd Xian Jiaotong University
Date	2022–10–21
Abstract	The invention discloses a solar photovoltaic-supercapacitor coupling heat management device which is characterized by comprising a back plate channel and a shell structure containing a phase-change material; the back plate channel is used for transferring heat in the working process of the solar photovoltaic panel to the shell structure containing the phase-change material through the working medium flowing in the back plate channel so as to store the heat in the shell structure, and is used for transferring the heat based on the working medium so as to reduce the temperature of the solar photovoltaic panel; and the shell structure after heat storage is used for maintaining the super capacitor wrapped in the shell structure at a constant temperature.

#26	Description
Patent/patent application number	CN 115,180,080A
Title	Offshore solar platform
Assignee	Shanghai Jiaotong University
Date	2022–10–14
Abstract	The invention discloses an offshore solar platform, which relates to the field of solar energy and comprises a mooring component and six carrying units, wherein the carrying units are in isosceles triangle shapes, the carrying units are arranged in a regular hexagon shape, and the sides, close to each other, of the adjacent carrying units are hinged; a

(*Continued*)

Table 13.1. (*Continued*)

#26	Description
	plurality of photovoltaic panels are laid above each carrying unit and detachably connected with the carrying units; the mooring component is adapted to provide buoyancy to the piggyback unit. According to the offshore solar platform provided by the invention, the carrying units are arranged in a triangular shape and are arranged in a hexagonal shape, so that the carrying units can be spliced and expanded conveniently, the stability of a triangular mechanism is fully utilized, the supporting stability of a photovoltaic panel is improved, the mooring component comprises the floater, the mooring cable and the like, the floater is arranged below the platform, the mooring cable pulls the platform, the whole platform is positioned, the platform is prevented from being washed away by water flow, and the stability of the platform is improved.

#27	Description
Patent/patent application number	JP 714,875,3 B1
Title	Solar cell, photovoltaic module and method for manufacturing solar cell
Assignee	金井昇, 張彼克, シン ウ ヂァン
Date	2022–10–05
Abstract	The present application relates to the field of photovoltaics and provides a solar cell, a photovoltaic module and a method of manufacturing a solar cell. A solar cell has opposing front and back surfaces, the back surface including a textured area and a planar area adjacent to the textured area, a doped surface field within the base in the textured area, and a doping surface field within the doped surface field. A base in which the doping element is N-type or P-type; a tunnel dielectric layer located in a flat area on the back surface of the base; a doped conductive layer having a doping element, wherein the type of doping element in the doped conductive layer is the same as the type of doping element in the doping surface field; a back electrode contacting the doping surface field, at least to reduce the contact resistance of the solar cell.

#28	Description
Patent/patent application number	CN 114,864,724A
Title	Photovoltaic laminated tile assembly for preventing series disconnection of batteries
Assignee	Das Solar Co. Ltd
Date	2022–08–05
Abstract	The invention belongs to the technical field of photovoltaics, and particularly relates to a photovoltaic laminated tile assembly for preventing a battery from being broken; the battery pack comprises a back plate and a plurality of battery strings, wherein each battery string comprises a plurality of battery piece groups which are structurally connected in series and overlapped, each battery piece group comprises a first battery piece and a second battery piece which are overlapped end to end, the back surfaces of the first battery piece and the second battery piece are respectively provided with a positive electrode and a negative electrode which are symmetrically distributed, and the polarities of the electrodes on the back surfaces of the first battery piece and the second battery piece and on the same side are opposite; according to the invention, by arranging the conductive region and filling the conductive material, when the first welding strip or the second welding strip on the top of the conductive region is broken, the broken first welding strip or the broken second welding strip can still be electrically connected through the conductive material, so that the situation that the battery string is broken due to the breakage of the welding strips is avoided.

#29	Description
Patent/patent application number	JP 323,910,4 U
Title	A tandem type CVD diamond organic semiconductor thin film solar cell device installed in a gardening house
Assignee	五郎 五十嵐
Date	2022–09–15
Abstract	A CVD diamond semiconductor thin film type and an organic semiconductor thin film type are provided in tandem to provide a highly efficient and durable CVD diamond organic semiconductor thin film solar cell device that is configured to receive or shield sunlight. do. SOLUTION: Recoupling is suppressed in pn-type CVD diamond semiconductor thin film photoelectric conversion layers 3 and 4 of a top cell layer, a boron-doped p-type CVD diamond semiconductor thin film layer 3 and a nitrogen-doped n-type CVD diamond semiconductor thin film layer 4 junction. A bulk heterojunction structure in which an i-type intrinsic CVD diamond thin film layer 5 is provided,

(Continued)

Table 13.1. (*Continued*)

#29	Description
	and a p-type electron donor 6 and an n-type electron acceptor 8 are mixed in the bottom cell layer, or an organic semiconductor thin film photoelectric conversion layer 6 with a super-hierarchical nanostructure. 8 tandem type, or a p-type hole transport layer 6, a perovskite crystal layer, and an n-type electron transport layer 8-junction perovskite semiconductor thin film photoelectric conversion layer were provided in tandem type.

#30	Description
Patent/patent application number	CN 114,834,604 B
Title	Floating body array for photovoltaic on water
Assignee	Das Solar Co. Ltd
Date	2022–09–16
Abstract	The invention belongs to the technical field of overwater photovoltaics, and discloses a floating body array for overwater photovoltaics, which comprises a floating assembly and photovoltaic panels, wherein one end of the floating assembly is rotatably provided with a supporting assembly, one end of the supporting assembly, which is far away from the floating assembly, is rotatably connected onto the photovoltaic panels, the floating assembly comprises two floating panels, and a telescopic assembly for adjusting the distance between the two floating panels is fixedly arranged between the two floating panels; through the solar energy ware of following spot that sets up, realize the control to rotating part and supporting component turned angle to the realization is to the angular adjustment of photovoltaic board, and the turned angle through control supporting component realizes the angular adjustment of photovoltaic board under to the current season circumstances, realizes the seizure of photovoltaic board to the sun position at different moments every day through rotating part, with this realization photovoltaic board can both be abundant just to sunshine at the different moments in different seasons, improves the light conversion rate.

#31	Description
Patent/patent application number	CN 114,823,933A
Title	Solar cell structure and manufacturing method thereof
Assignee	Hengdian Group DMEGC Magnetics Co. Ltd
Date	2022–07–29
Abstract	The invention provides a solar cell structure and a manufacturing method thereof. The solar cell structure includes: the semiconductor device comprises a semiconductor substrate, wherein a first doped region and a second doped region are formed on the back surface of the semiconductor substrate; the first heavily doped region is formed in the semiconductor substrate, is arranged on one side of the first doped region far away from the back surface in a contact manner, is doped in a P type manner and is greater than the doping concentration of the first doped region; the second heavily doped region is formed in the semiconductor substrate, is arranged on one side of the second doped region far away from the back surface in a contact manner, and is doped in an N type manner and is greater than the doping concentration of the second doped region; the first passivation layer with fixed negative charges is formed on one side, far away from the first heavily doped region, of the first doped region; and a second passivation layer with fixed positive charges is formed on one side of the second doping region far away from the second heavily doped region. The solar cell structure has high conversion efficiency.

#32	Description
Patent/patent application number	CN 115,034,084A
Title	Parameter optimization method of multi-junction compound battery model for improving photoelectric conversion efficiency
Assignee	Chongqing Qinsong Technology Co. Ltd
Date	2022–09–09
Abstract	The scheme belongs to the technical field of solar photovoltaic power generation, and particularly relates to a parameter optimization method of a multi-junction compound cell model for improving photoelectric conversion efficiency. The method comprises the following steps: step A1: carrying out *I–V* characteristic test on the multi-junction compound battery through an experimental platform to obtain a plurality of groups of measured data samples of current and voltage of the battery; step A2: establishing a basic theoretical model equation of the multi-junction compound battery; step A3: determining an objective function of the optimization problem, step A4: and

(*Continued*)

Table 13.1. (*Continued*)

#32	Description
	(4) carrying out multiple iterations by a numerical iteration method until an optimal parameter value is obtained, and calculating the Root Mean Square Error (RMSE) of the optimal parameter value to judge the accuracy of the model. According to the method for extracting the optimal parameters of the multijunction compound battery model, the numerical method and the iteration method are combined, so that the accurate multijunction compound battery model can be constructed, and the output electrical characteristics of the battery under different conditions can be predicted and analyzed. This configuration results in an absorber system for two-stage solar concentration and spectral splitting for maximum solar energy.

#33	Description
Patent/patent application number	CN 114,978,023A
Title	Wind-resistant protection device of photovoltaic power station
Assignee	Dingbian Huanghe Solar Power Generation Co. Ltd
Date	2022–08–30
Abstract	The invention relates to the technical field of photovoltaic power stations, and discloses a wind-resistant protection device of a photovoltaic power station, which comprises a bottom plate, wherein a bottom rack is fixedly installed on the rear side of the upper end of the bottom plate, a top rack is fixedly installed on the upper side of the bottom rack, auxiliary reinforcing assemblies are arranged on the front sides of the top rack and the bottom rack, and a photovoltaic panel is fixedly installed on the front side of the top rack. This anti-wind protection device of photovoltaic power plant is provided with supplementary subassembly of strengthening, the auxiliary stay subassembly, the power transmission subassembly, prevent wind the board, backup pad and dwang, supplementary subassembly of strengthening and the auxiliary stay subassembly homoenergetic support the photovoltaic board, the board of preventing wind that the cross-section is 'V' style of calligraphy structure simultaneously, can cut apart the air current, the stability of preventing wind the board has been promoted, and the backup pad can carry out the auxiliary stay to preventing wind the board, effectively avoid preventing wind the board and take place to empty, prevent wind the board simultaneously and can protect the photovoltaic board, the rubble of avoiding in the sand blown by the wind beats on the photovoltaic board, the practicality of device has been promoted.

#34	Description
Patent/patent application number	JP 323,944,7 U
Title	Absorber system for harvesting solar energy
Assignee	Abdulwahab Ali A Almazroi, Anand Nayyar Monagi Hassan M Alkinani, Noor Zaman Jhanjhi
Date	2022–10–14
Abstract	An absorber system for two-stage solar concentration and spectral splitting for maximum solar energy harvesting. An absorber system for harvesting solar energy according to the present invention includes a frame (102), a plurality of conduits (104), and coupled to the conduits for concentrating solar rays and transmitting the rays. at least one optical amplifier (108) leading to said conduit; a plurality of axles (110) mechanically coupled to said inner core of said frame (102); a set of arms (114); a high concentration photovoltaic absorber (116) consisting of a periodic array of pyramidal nanostructures coated to absorb radiation. This configuration results in an absorber system for two-stage solar concentration and spectral splitting for maximum solar energy.

#35	Description
Patent/patent application number	CN 218,387,341 U
Title	Building photovoltaic integrated power generation device
Assignee	Wuhan Shijia New Energy Engineering Co. Ltd
Date	2023–01–24
Abstract	The utility model relates to a building photovoltaic integrated power generation device, which comprises a photovoltaic crystalline silicon battery component, a perovskite battery component, a direct current cable, a first inverter, a second inverter, an alternating current cable, a local grid-connected cabinet and a box-type transformer; the photovoltaic crystalline silicon battery component is arranged on the roof of a building, and the perovskite battery component is arranged on the wall of the building; the photovoltaic crystalline silicon battery component square matrix is connected into a first inverter through a direct current cable, the perovskite battery component is connected into a second inverter through a direct current cable, and the first inverter and the second inverter are

(*Continued*)

Table 13.1. (*Continued*)

#35	Description
	connected into a local grid-connected cabinet or a box-type transformer through alternating current cables. The utility model discloses make full use of crystal silicon battery absorbs the characteristics of highlight and perovskite battery absorption low light, installs respectively in the room item and the side of building, and it is complementary to form the advantage, on the basis that does not increase the building and take up an area of, the current roof of make full use of and outer wall area provide high efficiency, low-cost photovoltaic power generation device, provide green electricity, saving investment for the enterprise.

#36	Description
Patent/patent application number	WO 2022/105446 A1
Title	Single-axis angle tracking method and system for intelligent photovoltaic module
Assignee	深圳市中旭新能源有限公司
Date	2022–05–27
Abstract	Disclosed in the present invention are a single-axis angle tracking method and system for an intelligent photovoltaic module, relating to the field of sun-following automatic tracking of a support of photovoltaic power generation. The system comprises a photovoltaic module with a plurality of cell units and a power optimizer, a single-axis tracking support and a tracking control module, wherein the shadow on a back surface of the single cell unit is consistent with the shade of a front row and a rear row; and in order to avoid the mutual influence of the shading shadow among the cell units, the power optimizer makes the module operate at the maximum power, the parameters are shared with the tracking control module, and the single-axis tracking support is adjusted on the basis of an astronomical algorithm and controlled at the optimal tracking angle by using electric parameter information of the power optimizer, so that the optimal radiation quantity can be obtained at the incident angle of sunlight while the influence of shadow shielding on power can be reduced, and the optimal photovoltaic power generation power can be obtained. The single-axis tracking system can be effectively arranged in a large-scale photovoltaic power station, and the aim of significantly reducing the cost of the leveling degree can be achieved.

#37	Description
Patent/patent application number	CN 211,125,670 U
Title	Tower-type laminated tile solar photovoltaic module capable of being integrated on roof of automobile
Assignee	Seraphim Solar System Co. Ltd
Date	2020–07–28
Abstract	The utility model relates to a can integrate in tower pile of tiles solar PV modules at car roof, including the curved surface euphotic layer, can with the curved surface battery cluster of the lower bottom surface laminating of curved surface euphotic layer and can with the encapsulation backplate of the lower bottom surface laminating of curved surface battery cluster, curved surface euphotic layer and curved surface battery cluster are fixed through hot melt adhesive, and curved surface battery cluster and encapsulation backplate are also fixed through the hot melt adhesive, curved surface battery cluster includes a plurality of tower pile of tiles modules, all is equipped with on the tower pile of tiles module at the tower pile of modules at initiating terminal and the tower pile of tiles module at terminal and connects the lead wire, each tower pile of tiles module comprises a plurality of pile of tiles units, and each pile of tiles unit is formed by a plurality of battery board group fixed connection, and each adjacent battery board group part is range upon range of, and adjacent battery board group's length varies, and each battery board group includes at. The utility model discloses also effectually avoid breaking of battery piece because of stress emergence, safe practical in the effectual sunroof light transmissivity of assurance.

#38	Description
Patent/patent application number	CN 109,888,035 B
Title	Photovoltaic photo-thermal tile
Assignee	Flextech Co.
Date	2020–10–27
Abstract	The invention discloses a photovoltaic photo-thermal tile, and relates to the technical field of solar equipment. The battery board is connected with a junction box in series through a plurality of diodes; each battery pack sequentially comprises a front film layer, a first EVA layer, a battery layer, a second EVA layer, a back plate and a base material from top to bottom, wherein the base material comprises a base material coating and a hollow body, and a heat transfer medium is arranged in the hollow body. The invention is used for photovoltaic power generation equipment of building roofs.

#39	Description
Patent/patent application number	CN 106,571,771 B
Title	A kind of portable receipts stack-type solar-energy photo-voltaic cell system
Assignee	Flextech Co.
Date	2018–11–30
Abstract	A kind of portable receipts stack-type solar-energy photo-voltaic cell system, frame, multiple receipts stacked group parts, controller and battery are folded containing receiving, receiving stacked group part is the sliding stack structure of multi-disc, it is made of sliding equipment and photovoltaic cells, wherein sliding equipment is encapsulated on bottom plate by pulley assembly and slide assembly, photovoltaic cells, bottom plate is connect with sliding equipment, sliding equipment is the sliding rail lamination driving style of pull-down, and receiving stacked group part can suspend by any position, and in cambered surface towards sunlight; In photovoltaic cells, including the crystal silicon solar batteries piece that two rows are arranged in array, two panels crystal silicon solar batteries piece and metal-oxide-semiconductor in each column are arranged in parallel, and arranged in series are used between the column and the column, it is possible to prevent effectively from because of hot spot effect caused by partial occlusion. The present invention has the characteristics that structure is simple, system is reliable, is easy to use, and can be used as the blind system of building, also can be used in interim and field solar photovoltaic generation system.

#40	Description
Patent/patent application number	US 2023 028,768,2 A1
Title	Skylights with integrated photovoltaics and refractive light-steering
Assignee	Arizona State University ASU
Date	2023–09–14

Abstract	A skylight for a building includes a solar panel arranged within the skylight, the solar panel comprising one or more photovoltaic cells to collect direct radiation from rays of sunlight for conversion to electrical power, and an optical element to receive the direct radiation and refract it to the solar panel, and to receive the direct radiation and diffuse radiation scattered from the rays of sunlight and refract the direct radiation and the diffuse radiation through the skylight, bypassing the solar panel, to provide daylighting in the building.

#41	Description
Patent/patent application number	CN 110,931,584A
Title	Folded plate-shaped photovoltaic assembly, front glass used by same and photovoltaic system
Assignee	Guangdong Aiko Technology Co.
Date	2020–04–26
Abstract	The invention discloses a folded plate-shaped photovoltaic assembly, and front glass and a photovoltaic system used by the folded plate-shaped photovoltaic assembly. The photovoltaic module can improve the whole light receiving area of the photovoltaic module under the condition of limited area, and when sunlight irradiates the V-shaped unit of the front glass, the sunlight can be reflected and incident for many times between two inclined panels of the V-shaped unit, so that the energy absorption of the module to the sunlight is increased, and the generating capacity of the module is improved.

#42	Description
Patent/patent application number	WO 2021/008573
Title	Hot-spot-resistant single-plate photovoltaic module
Assignee	Jiangsu Coop&Inno Green Energy Technology Co. Ltd
Date	2023-05-31
Abstract	The present invention relates to the technical field of solar cell modules, and more particularly, to a hotspot-resistant single-plate photovoltaic module.

#43	Description
Patent/patent application number	CN 111,192,933A
Title	Solar photovoltaic module and building photovoltaic integrated module
Assignee	Shanghai Jinyangfang New Energy Technology Center LP
Date	2020–05–22
Abstract	The invention provides a solar photovoltaic module and a building photovoltaic integrated module, and belongs to the technical field of solar photovoltaic modules. The solar photovoltaic module comprises a solar photovoltaic module and a solar photovoltaic module, wherein the solar photovoltaic module is sequentially stacked from top to bottom: the solar cell comprises a glass layer, a first adhesive layer, a solar cell sheet, a second adhesive layer and a back plate; the solar cell comprises a plurality of sub-cells and welding strips; the end parts of two adjacent sub-battery pieces are mutually lapped; one end of the welding strip is arranged on the front surface of one sub-battery piece, and the other end of the welding strip is arranged on the back surface of the adjacent sub-battery piece; the thickness of the welding strip is less than or equal to 0.11 mm. The building photovoltaic integrated assembly increases the effective area of the battery, improves the power generation efficiency and has simple processing technology; rainwater can be effectively prevented from entering the assembly, so that the service life of the assembly is prolonged, and the assembly can be directly used as a roof and a car shed; when the solar photovoltaic modules are lapped, the installation is simple, the whole system is connected in a seamless mode, the combination is tight, no hole exists outside, no screws and other fasteners exist, and the waterproof performance is better.

#44	Description
Patent/patent application number	AU 2020 100,395, A4
Title	Foldable photovoltaic module
Assignee	FLEXTECH Co.
Date	2020–06–18
Abstract	Disclosed is a foldable photovoltaic module. The existing photovoltaic tile has a complicated structure, resulting in heavy weight, inconvenient transportation, insufficient flexibility, and high costs in installation and maintenance. In

	the disclosure, the foldable photovoltaic module includes a front membrane layer, a first adhesive membrane layer, a photovoltaic cell layer, a second adhesive membrane layer and a back membrane layer, wherein the front membrane layer, the first adhesive membrane layer, the photovoltaic cell layer, the second adhesive membrane layer and the back membrane layer are integrally formed. The disclosure is both a power generation device and a building element, which reduces the costs in structure, reinforcement and installation, is easy to maintain, and saves maintenance costs. The disclosure can be applied in a variety of forms, can be applied for arc-shaped surface, planar or elevation installations, and can be additionally installed an existing roof or used as a building element, i.e., for building integrated PV.

#45	Description
Patent/patent application number	CN 209,418,516 U
Title	A kind of regular hexagon MWT solar battery half and component
Assignee	Zhejiang Jingsheng Mechanical and Electrical Co. Ltd
Date	2019–09–20
Abstract	The utility model relates to technical field of solar batteries, it is specifically related to a kind of regular hexagon MWT solar battery half and component. A kind of regular hexagon MWT solar battery half, battery half are to draw sliver by laser bisection or the quartering by regular hexagon cell piece to be formed; Uniformly distributed several dot matrix holes on battery half light-receiving surface. A kind of MWT solar battery half component using above-mentioned battery half, including solar battery half and package assembling. The utility model improves the utilization rate to silicon wafer, saves the cost of production solar battery sheet, reduces environmental pollution and the wasting of resources, improve the output power of cell piece.

#46	Description
Patent/patent application number	US2014 013,084,2 A1
Title	Bussing for PV-module with unequal-efficiency bi-facial PV-cells
Assignee	Prism Solar Technologies Inc.

(*Continued*)

Table 13.1. (*Continued*)

#46	Description
Date	2014–05–15
Abstract	A PV module includes strings of serially electrically connected individual bifacial photovoltaic cells each of which is characterized by conversion efficiencies that are different for front and back sides of each cell. The module includes at least two of such strings which are electrically parallel to one another such that front sides of cells in one string and back sides of the cells in another string corresponding to the same side of the module. Each side of the module is thereby adapted to generate substantially the same amount of electrical power under otherwise equal circumstances. On a sunny day, the module generates as much electrical power before noon as after noon if the front side and the back side receive, aggregately, substantially the same amount of solar power incident thereon during the day.

#47	Description
Patent/patent application number	US 9,312,418 B2
Title	Frameless photovoltaic module
Assignee	Prism Solar Technologies Inc.
Date	2016–04–12
Abstract	A photovoltaic module employing an array of photovoltaic cells disposed between two optically transparent substrates such as to define a closed-loop peripheral area of the module that does not contain a photovoltaic cell. The module is sealed with a peripheral seal along the perimeter; and is devoid of a structural element affixed to an optically transparent substrate and adapted to mount the module to a supporting structure. The two substrates may be bonded together with the use of adhesive material and, optionally, the peripheral seal can include the adhesive material. The module optionally includes diffraction grating element(s) adjoining respectively corresponding PV-cell(s).

#48	Description
Patent/patent application number	CN 108,063,585A
Title	A kind of installation component for generating electricity on two sides photovoltaic module

Assignee	Suzhou Central Min Lai Solar Power Co. Ltd
Date	2018–05–22
Abstract	The present invention relates to a kind of installation components for generating electricity on two sides photovoltaic module, it is connected in including hook, bottom girder and tensile member, one end of hook on bottom girder, the other end is connected on roof, one end of tensile member is connected on bottom girder, and double-side assembly is mounted on by briquetting and tensile member on roof. Bottom girder provides mounting surface for double-side assembly, links up with overall structure being connected with roofing. Its advantage is as follows: The present invention by raise every row's double-side assembly compared with the height and inclination angle of roof mounting surface with realize the stairstepping of double-side assembly array arrange, light is made to inject the roof of module backside by the down suction between component front and rear row, utilize the diffusing reflection of roofing, module backside light is made to generate electricity, to improve the generated energy of system and generating efficiency. The installation component can make the uniform light power generation in the double-side assembly back side and pull open front and rear row spacing and cause the increase of the materials such as purlin, effectively promote the utilization rate of roofing, increase the installed capacity of system.

#49	Description
Patent/patent application number	CN 206,844,499 U
Title	Intelligent family **BIPV** roofs electricity generation system
Assignee	Zhejiang Jinbest Energy Secience and Technology Co., Ltd
Date	2018–05–22
Abstract	The utility model discloses a kind of intelligent family **BIPV** roofs electricity generation system, including the multiple **W** types tanks being arranged in along roof pitch direction on roof sleeper beam, multiple row **BIPV** photovoltaic modules are provided between adjacent **W** types tank, the end of **BIPV** photovoltaic modulies is erected on the support level of **W** types sink center; The junction of adjacent two row **BIPV** photovoltaic modules is provided with transverse drainage groove, and the end of transverse drainage groove is respectively erected on the outside-supporting edge of two adjacent **W** types tanks; The **BIPV** photovoltaic modules include glass panel, the first EVA film, multiple photovoltaic cells, the second EVA film, dimming glass bar and the glass back plate being sequentially

(*Continued*)

Table 13.1. (*Continued*)

#49	Description
	overlapped from top to bottom; The photovoltaic cell is in multirow or more column distributions, and the dimming glass bar is located at the position in the ranks or between row of photovoltaic cell. The utility model has that structural strength is high, good waterproof performance, and the advantages of intelligent lighting can be realized.

#50	Description
Patent/patent application number	CN 212,725,330 U
Title	Curved surface photovoltaic module
Assignee	Suzhou Wenjing New Energy Co. Ltd
Date	2021–03–16
Abstract	The utility model relates to a curved photovoltaic module, which comprises a front plate, a solar battery pack, a back plate and a packaging adhesive film; the front plate, the solar battery pack and the back plate are sequentially stacked and packaged together by a packaging adhesive film; the front plate, the solar battery pack and the back plate are all curved surfaces, and the undulation radian of the connecting surface is consistent; the curved surface photovoltaic module is an arched or wavy curved surface body. The curved surface photovoltaic module applies the silicon-based solar battery pack which is connected in series or in parallel by using the tiling technology, the splicing technology or the half-sheet technology to the curved surface photovoltaic module, thereby reducing the production cost and improving the power generation conversion rate of the module relative to the curved surface photovoltaic module using the amorphous silicon thin-film solar battery.

#51	Description
Patent/patent application number	CN 111,244,208 B
Title	Solar cell and application thereof
Assignee	Changzhou Shichuang Energy Co.
Date	2022–02–18

Abstract	The invention provides a solar cell, wherein the front side of the solar cell is provided with a pyramid suede structure; the projection of the pyramid vertex on the plane of the pyramid bottom surface is positioned on one side of the center of the pyramid bottom surface; and the vertexes of the pyramids on the cell sheet face the same direction. The invention also provides an application of the solar cell. Compared with the conventional battery piece adopting a regular rectangular pyramid suede structure, the reflectivity of the battery piece is lower; by adopting the assembly of the battery piece, the efficiency is higher; the assembly is suitable for vertical installation and horizontal installation; the components can be horizontally spliced into a square matrix, so that the floor area of the square matrix can be greatly reduced.

#52	Description
Patent/patent application number	CN 112,216,759A
Title	Three-terminal double-sided laminated solar cell and preparation process thereof
Assignee	CETC 18 Research Institute
Date	2021–01–12
Abstract	The invention discloses a three-terminal double-sided laminated solar cell and a preparation process thereof, belonging to the technical field of solar cells, wherein the three-terminal double-sided laminated solar cell at least comprises the following components: a metal electrode; the positive solar cell is positioned on the upper surface of the metal electrode; the reverse solar cell is positioned on the lower surface of the metal electrode; and the metal gate electrode is one of gold or silver or copper or germanium and gold, titanium and palladium and silver, nickel and copper and nickel, nickel and aluminum composite metal. By adopting the technical scheme, the solar cell can simultaneously solve the problem of acquiring the energy of reflected light of the sun and the earth, comprises a common anode or a common cathode of the positive and negative solar cells, and simultaneously converts the light energy from the positive and negative sides of the device into electric energy to supply power to a load.

#52	Description
Patent/patent application number	CN 216,290,815 U

(*Continued*)

Table 13.1. (*Continued*)

#52	Description
Title	Folding mechanism of light photovoltaic
Assignee	Aopu Shanghai New Energy Co.
Date	2022–04–12
Abstract	The utility model discloses a folding mechanism of a light photovoltaic, which comprises a folding assembly, a photovoltaic assembly, an upright post, a horizontal beam, a photovoltaic panel frame, a long push arm, a rotating pin, a short push arm, a pulley, a purlin, a side arm, a PET substrate and a battery panel. The utility model has the beneficial effects that: the power generation capacity of the photovoltaic module is expanded through the mechanical connection of the frame, and the application range of the light photovoltaic module is enlarged; the upright post is used as a photovoltaic relying mechanism, the photovoltaic frame is gradually unfolded depending on the upright post when being unfolded, and the horizontal supporting beam is fixed on the upright post and used for supporting the folded photovoltaic assembly when being folded, so that the structural stability of the assembly is ensured; the photovoltaic frame is only provided with the pulleys on the upright posts and the outermost side, the height of the assembly from the ground can be increased through the diameters of the pulleys, and the influence of soil and ground dust on the photovoltaic assembly is reduced; in addition, the number of the pulleys is reduced, so that the number of the middle photovoltaic panels can be increased, and the overall photovoltaic power generation capacity is improved; the pulley can ensure that the assembly can be moved to any required place for use.

#53	Description
Patent/patent application number	US 8,955,267 B2
Title	Hole-thru-laminate mounting supports for photovoltaic modules
Assignee	Maxeon Solar Pvt. Ltd
Date	2015–02–17
Abstract	A mounting support for a photovoltaic module is described. The mounting support includes a pedestal having a surface adaptable to receive a flat side of a photovoltaic module laminate. A hole is disposed in the pedestal, the hole adaptable to receive a bolt or a pin used to couple the pedestal to the flat side of the photovoltaic module laminate.

#54	Description
Patent/patent application number	CN 107,154,440A
Title	A kind of solar cell vacuum glazing
Assignee	Suzhou Austrian Energy Co. Ltd
Date	2017–09–12
Abstract	The invention discloses a kind of solar cell vacuum glazing, including solar cell panel assembly, glass plate, sealing ring, framework, described solar cell panel assembly is be arranged in parallel by framework with glass plate, described solar cell panel assembly is tightly connected by sealing ring with glass plate, described solar cell panel assembly, glass plate, sealing ring formation confined space. Windowpane of the present invention has low cost, high generation efficiency, long-life, heat-insulating sound-insulating, strong energy-saving effect.

#55	Description
Patent/patent application number	CN 214,753,804 U
Title	Novel electricity generation building materials of structure
Assignee	Heliou New Energy Technology Shanghai Co. Ltd Helio New Energy Co Ltd
Date	2021–11–16
Abstract	The utility model particularly relates to a power generation building material with a novel structure, which comprises a transparent cover plate, a first sealing layer, a power generation circuit layer, a second sealing layer and a metal back plate from top to bottom in sequence; the metal back plate is rectangular, two sides in the long edge direction are wavy curved surfaces, the middle of the metal back plate is a plane, and the transparent cover plate, the first sealing layer, the power generation circuit layer and the second sealing layer are all fixed in the middle of the metal back plate; when the adjacent power generation building materials are in lap joint, the wave-shaped curved surfaces of the adjacent metal back plates are overlapped and clamped, and when the power generation building materials are in lap joint with the color steel tiles, the wave-shaped curved surfaces of the power generation building materials cover the

(*Continued*)

Table 13.1. (*Continued*)

#55	Description
	wave-shaped curved surfaces of the color steel tiles. The utility model discloses a power generation building materials, the installation is convenient, can directly lay in the roofing, and is high with the degree of combination of former roofing, need not demolish original various steel tile roofing, and adopts MM overlap joint mode, does not need sticky can gain good waterproof performance.

#56	Description
Patent/patent application number	CN 213,717,907 U
Title	Single-shaft angle tracking system of intelligent photovoltaic module
Assignee	Guangzhou Zhongxu New Energy Co. Ltd
Date	2021–07–16
Abstract	The utility model discloses a single-shaft angle tracking system of an intelligent photovoltaic assembly, which relates to the field of the automatic sun tracking of a support of photovoltaic power generation, and comprises a photovoltaic assembly with a plurality of battery units and a power optimizer, a single-shaft tracking support and a tracking control module, the back shadow and the front and back rows of the single battery unit are consistent in shielding, the mutual influence of the shielding shadows among the battery units is avoided, the power optimizer enables the assembly to operate at the maximum power, the parameters of the assembly are shared in the tracking control module, the single-shaft tracking support is adjusted and controlled by the optimal tracking angle on the basis of an astronomical algorithm by utilizing the electrical parameter information of the power optimizer, the optimal irradiation amount of the incident angle of sunlight can be obtained while the influence of the shadow shielding on the power is reduced, the optimal photovoltaic power generation power can be obtained, and the single-shaft tracking system can be effectively equipped in a large photovoltaic power station, the purpose of greatly reducing the cost of the leveling degree is achieved.

#57	Description
Patent/patent application number	CN 206,349,963 U

Title	A kind of solar energy installation system
Assignee	Shanghai Solar Power Technology Co. Ltd
Date	2017–07–21
Abstract	The utility model discloses a kind of solar energy installation system, including multiple columns, crossbeam, curb girder, brace, supporting plate, reflective mirror and solar cell module. Crossbeam, curb girder and column are connected with each other by described column as the main support point of whole installation system using the method for welding, and brace is connected in the rectangle of crossbeam, curb girder and column composition, plays a part of reinforcing. Multiple supporting plates connect into a cuboid, and the one side of cuboid and ground are in certain inclination angle so that on sunshine vertical irradiation to this face. Two pieces of reflective mirrors are respectively placed in the front and back of cuboid, angled with two sides on adjacent cuboid respectively so that on the reflected vertical irradiation of sunshine to corresponding plane. Solar cell module is installed on three faces of cuboid. The utility model adds the component installation in unit space, takes full advantage of sunshine, improves the utilization rate of solar energy.

#58	Description
Patent/patent application number	US 2012 008,539,5 A1
Title	Solar module attachment device and mounting method
Assignee	Bengbu Design and Research Institute for Glass Industry
Date	2014–07–01
Abstract	An attachment device for a module for collecting energy originating from solar radiation to a structure, such as a roof, a facade, or a mounting structure of a ground-mounted structure, wherein the module includes on its rear face at least one reinforcing profiled section. The attachment device includes at least one support secured to the structure. The support includes a snap-fastening mechanism with respect to the reinforcing profiled section of the module, which snap-fastening mechanism can be activated by applying a one-way thrust force pushing the module in the direction of the structure.

#59	Description
Patent/patent application number	KR 102,379,412 B1
Title	Bifacial solar system
Assignee	영남대학교 산학협력단
Date	2022–03–29
Abstract	The present invention relates to a double-sided light-receiving photovoltaic module frame and a photovoltaic system to which the frame is applied. Solves the problem of low power generation efficiency due to the shadow caused by the wing of the frame in the sunlight incident on the rear side.

#60	Description
Patent/patent application number	CN 111,868,936A
Title	Solar cell module and solar power generation system
Assignee	Sharp Corp.
Date	2020–10–30
Abstract	In a solar module housed in a housing (3), bypass diodes (10a) are connected to the start and end of a solar cell string (4A) in which solar cells are arranged in a row, and bypass diodes (10BC) are connected to a solar cell string (4BC) in which solar cells are arranged in two rows. Further, the solar cell string (4A) is disposed on either the ridge side or the side on the ridge side on which shadows are easily cast.

#61	Description
Patent/patent application number	CN 219,642,848 U
Title	Photovoltaic high-reflection black grid glass and photovoltaic cell panel
Assignee	Jiangsu Haiborui Photovoltaic Technology Co. Ltd
Date	2023–09–05

Abstract	The utility model discloses a photovoltaic high-reflection black grid glass and a photovoltaic cell panel, wherein the black grid glass plate and a high-reflection coating component are arranged on the surface of the glass plate, the high-reflection coating component are crisscrossed on the glass plate to form a plurality of rectangular grids, and a cell is positioned in the black grid; the high-reflection coating component comprises a white substrate positioned in the middle and organic black coatings positioned on the upper side and the lower side of the white substrate, and sunlight passes through the organic black coatings and then is reflected to the battery piece through the white substrate and the glass plate; the photovoltaic cell component has high utilization rate of the front and back light rays, converts infrared light into electric energy, has high power, reduces the heating of the component, and prolongs the service life of the component.

#62	Description
Patent/patent application number	CN 108,321,233 B
Title	Dual-glass cadmium telluride solar cell module and preparation method thereof
Assignee	Econess Energy Co. Ltd
Date	2023–11–10
Abstract	The application relates to a double-glass cadmium telluride solar cell module, which mainly comprises toughened glass, a polyurethane adhesive layer, a PVC transparent sealing adhesive film, a toughened glass back plate, a junction box and outgoing lines, wherein the toughened glass, the polyurethane adhesive layer and the polyurethane adhesive layer are sequentially arranged from top to bottom, the polyurethane adhesive layer is coated on one side of the glass, the cadmium telluride photoelectric material film layer is coated on one side of the polyurethane adhesive layer, the PVC transparent sealing adhesive film, the toughened glass back plate and the back plate are fixed and connected by outgoing lines, an aluminum alloy frame is arranged after the back plate is packaged and fixed, an insulating groove is arranged at the right lower part of the aluminum alloy frame so as to be beneficial to placing a transmission wire, wherein the glass cadmium telluride solar cell is formed by preparing polyurethane and acetone adhesive slurry on one side of the glass according to a formula by adopting an anilox roller coating process, preparing the polyurethane and the adhesive slurry of cadmium telluride and the polyurethane and the acetone to be used as a bottom layer, and then printing grid wires on the cadmium telluride film layer. The solar cell module fills the domestic blank, and has the characteristics of low production cost, less terminal investment, large popularization and application space, high performance, environmental protection and long service life.

#63	Description
Patent/patent application number	CN 106,601,848A
Title	Crystalline silicon photovoltaic module
Assignee	梁结平
Date	2017–04–26
Abstract	The invention relates to a crystalline silicon photovoltaic module, which comprises photovoltaic glass, a front-layer film, a plurality of crystalline silicon cells arranged in an array at intervals, a rear-layer film and a photovoltaic backplane, wherein the photovoltaic glass, the front-layer film, the crystalline silicon cells, the rear-layer film and the photovoltaic backplane are sequentially laminated and glued; the photovoltaic glass is high-transmission glass, a pair of reflection parts and a pair of refraction parts are arranged in the photovoltaic glass between two adjacent crystalline silicon cells, the pair of reflection parts is located between the pair of refraction parts, and incident light is reflected through the reflection parts, then irradiates the refraction parts, is then refracted through the refraction parts and then irradiates the crystalline silicon cells. The crystalline silicon photovoltaic module greatly improves the photon utilization rate and the crystalline silicon photovoltaic module output power, and the photoelectric conversion efficiency of the crystalline silicon photovoltaic module is high.

#64	Description
Patent/patent application number	CN 203,553,191 U
Title	Novel and efficient solar panel
Assignee	Shenzhen Suoyang New Energy Co. Ltd
Date	2014–04–16
Abstract	The utility model relates to an energy conversion device, and provides a novel and efficient solar panel. The novel and efficient solar panel comprises a transparent glass plate used as a front substrate, a plurality of solar photovoltaic cells, a TPT (Tedlar/PET/Tedlar) film used as a back substrate, an installation border and at least one junction box, wherein the plurality of solar photovoltaic cells are connected with the transparent glass plate by a first EVA (ethylene/vinyl acetate copolymer) thin film; the TPT film is connected with the plurality of solar photovoltaic cells by a second EVA thin film; the installation border is connected with the TPT film; and the junction box is arranged

	on the installation border, and the junction box is connected with the plurality of solar photovoltaic cells by positive and negative electrode lines. The solar panel disclosed by the utility model is high in conversion efficiency and wide in power output range.

#65	Description
Patent/patent application number	CN 102,104,346 B
Title	A kind of light-concentrating photovoltaic-temperature difference power-generating integrated device
Assignee	SHANGHAI CHAORI SOLAR ENGINEERING Co. Ltd Suzhou Gcl System Integration Technology Industrial Application Research Institute Co. Ltd Zhangjiagang Xiexin Integrated Technology Co. Ltd GCL System Integration Technology Co. Ltd GCL System Integration Technology Suzhou Co. Ltd
Date	2015–12–02
Abstract	The invention belongs to a kind of light-concentrating photovoltaic-temperature difference power-generating integrated device. Comprise concentrator cell assembly, liquid cooling apparatus, heat collector, bismuth telluride-based thermoelectric electric organ forms, it is characterized in that: concentrator cell assembly is formed by monocrystaline silicon solar cell with at the column collector lens on monocrystaline silicon solar cell top, binded on one piece of ceramic substrate by heat-conducting silica gel sheet in the bottom of monocrystaline silicon solar cell, ceramic substrate is connected with liquid cooling apparatus, liquid cooling apparatus is by circulating cooling pipeline, water pump and radiator composition, the side of circulating cooling pipeline connects the ceramic substrate bottom monocrystaline silicon solar cell, opposite side connects radiator, circulating cooling pipeline connects water pump, the top ceramic sheet of thermoelectric generator sticks a slice heat-conducting silica gel sheet be bonded on the radiator of liquid cooling apparatus. Advantage of the present invention to reclaim the waste heat in monocrystaline silicon solar cell generation in concentrating photovoltaic power generation, and the efficiency of whole concentrating photovoltaic power generation is increased substantially.

#66	Description
Patent/patent application number	CN 114,798,690 B
Title	Method for separating and recycling waste crystalline silicon photovoltaic panels
Assignee	Institute of Process Engineering of CAS Ganjiang Innovation Academy of CAS
Date	2023–08–04
Abstract	The invention relates to a method for separating and recycling waste crystalline silicon photovoltaic plates, which comprises the steps of disassembling the waste crystalline silicon photovoltaic plates to obtain an aluminum frame, a junction box and a photovoltaic module laminated layer, dividing the photovoltaic module laminated layer, immersing the divided laminated layer into specific organic solution for interlayer separation, and then carrying out solid-liquid separation, washing and drying on the system to separate and recycle a backboard, a battery piece and glass; the organic solvent comprises a main solvent with the boiling point more than 150 °C, and the main solvent contains double bonds and/or triple bonds, compared with solvents such as chlorobenzene with low boiling point and the like used in the prior art, the organic solvent is more environment-friendly due to no toxicity and no pollution, and can be recycled after standing after being used, thereby being beneficial to reducing the cost; when interlayer separation is carried out in the organic solvent, the expansion rate of EVA is small, the quality of the recovered battery piece is high, and no crack exists.

#67	Description
Patent/patent application number	US 2022 024,734,3 A1
Title	Thermoelectric active storage embedded hybrid solar thermal and photovoltaic wall module
Assignee	Yonghua Wang
Date	2022–08–04

Abstract	Solar collection and storage module systems as building blocks are provided to build walls or shingles of buildings to transform any buildings into stabilized power generation stations and tie to power grid to form power grid-interactive efficient buildings. The solar collection and storage module system comprises a hybrid photovoltaic and thermal panel, thermoelectric modules, thermal storage package, control system, and battery storage. The incident sunlight is partially converted into electricity directly by the photovoltaic part of the system directly, and rest part is transformed into heat which is extracted, boosted to high temperature, and stored into the thermal storage package by the thermoelectric modules operating in cooler mode at this movement. At night or in cloudy days, the stored heat flow through the thermoelectric modules, which are switched to generator mode by the control system, generating electricity. In the module system, the cogenerated heat is stored in thermal energy format and outputted in electrical energy format; the total conversion efficiency of the module system is significantly improved. When the module systems are used as wall modules or shingles to build buildings, the encapsulation properties of the buildings are substantially improved.

#68	Description
Patent/patent application number	CN 217,306,526 U
Title	Double-sided inflation type honeycomb runner PVT assembly
Assignee	Dalian Qunzhi Technology Co. Ltd
Date	2022–08–26
Abstract	The utility model provides a two-sided inflation formula honeycomb type runner PVT subassembly comprises aluminium system inflation formula heat transfer board, EVA glued membrane, black photovoltaic backplate, EVA glued membrane, photovoltaic cell piece subassembly, EVA glued membrane, toughened glass board from bottom to top, and aluminium system inflation formula heat transfer board, black photovoltaic backplate, photovoltaic cell piece subassembly, toughened glass board lean on the EVA glued membrane to press the bonding under the high temperature to aluminium alloy frame encapsulation shaping. The utility model discloses the subassembly can produce the heat in the electricity production, and the temperature of photovoltaic cell piece has been reduced again to the heat-producing process, and then has improved the electrical efficiency of photovoltaic cell piece. The turbulence degree of the internal circulating working medium is increased by the double-sided inflation type

(*Continued*)

Table 13.1. (*Continued*)

#68	Description
	honeycomb flow channel, the heat absorption capacity of the aluminum inflation type heat exchange plate is improved, the temperature of the photovoltaic cell is reduced to a great extent, and the electrical efficiency of the photovoltaic cell is further improved. The utility model discloses the subassembly has carried out degree of depth development and has utilized solar energy, has improved the thermoelectric efficiency that solar energy utilized.

#69	Description
Patent/patent application number	US 9,701,696 B2
Title	Methods for producing single crystal mixed halide perovskites
Assignee	Alliance for Sustainable Energy, LLC, Golden, CO (US); Shanghai Jiaotong University, Shanghai (CN)
Date	2017–07–11
Abstract	An aspect of the present invention is a method that includes contacting a metal halide and a first alkylammonium halide in a solvent to form a solution and maintaining the solution at a first temperature, resulting in the formation of at least one alkylammonium halide perovskite crystal, where the metal halide includes a first halogen and a metal, the first alkylammonium halide includes the first halogen, the at least one alkylammonium halide perovskite crystal includes the metal and the first halogen, and the first temperature is above about 21 °C.

#70	Description
Patent/patent application number	US 10,910,569 B2
Title	Organo-metal halide perovskites films and methods of making the same
Assignee	Alliance for Sustainable Energy, LLC, Golden, CO (US); Brown University—Technology Ventures Office
Date	2021–02–02
Abstract	An aspect of the present disclosure is a method that includes applying a solution that includes a first solvent, a halogen-containing precursor, and a metal halide to a substrate to form a coating of the solution on the substrate, contacting the coating with a second solvent to form a first plurality of organo-metal halide perovskite crystals on the substrate, and thermally treating the first plurality of organo-metal halide perovskite crystals, such that at least a portion of the

	first plurality of organo-metal halide perovskite crystals is converted to a second plurality of organo-metal halide perovskite crystals on the substrate. The halogen-containing precursor and the metal halide are present in the solution at a molar ratio of the halogen-containing precursor to the metal halide between about 1.01:1.0 and about 2.0:1.0, and a property of the second plurality of organo-metal halide perovskite crystals is improved relative to a property of the first plurality of organo-metal halide perovskite crystals.

#71	Description
Patent/patent application number	US 10,910,569 B2
Title	Oriented perovskite crystals and methods of making the same
Assignee	Alliance for Sustainable Energy, LLC, Golden, CO (US)
Date	2020–04–07
Abstract	An aspect of the present disclosure is a method that includes combining a first organic salt (A^1X^1), a first metal salt ($M^1(X^2)_2$), a second organic salt (A^2X^3), a second metal salt (M^2Cl_2), and a solvent to form a primary solution, where A^1X^1 and $M^1(X^2)_2$ are present in the primary solution at a first ratio between about 0.5 to 1.0 and about 1.5 to 1.0, and A^2X^3 to M^2Cl_2 are present in the primary solution at a second ratio between about 2.0 to 1.0 and about 4.0 to 1.0. In some embodiments of the present disclosure, at least one of A^1 or A^2 may include at least one of an alkyl ammonium, an alkyl diamine, cesium, and/or rubidium.

#72	Description
Patent/patent application number	US 11,876,484 B2
Title	Protecting solar panels from damage due to overheating
Assignee	Renewable Energy Products Manufacturing
Date	2024–01–16
Abstract	Systems and methods are provided for protecting solar panels from damage due to overheating. A system comprises a solar panel and a control system. The solar panel comprises a plurality of solar cells, and a plurality of thermochromic temperature sensors thermally coupled to different areas of the solar panel. The thermochromic

(Continued)

Table 13.1. (*Continued*)

#72	Description
	temperature sensors are configured to change color in response to heat generated by the solar cells in the different areas of the solar panel. The control system is configured to detect colors of the thermochromic temperature sensors, determine a temperature of each area of the solar panel based on the detected colors of the thermochromic temperature sensors, and cause the solar panel to shut down in response to determining that the temperature of at least one area of the solar panel exceeds a predetermined temperature threshold.

#73	Description
Patent/patent application number	US 7,897,867 B1
Title	Solar cell and method of manufacture
Assignee	SunPower Corporation; Maxeon Solar Technologies
Date	2011–03–01
Abstract	Solar cell that is readily manufactured using processing techniques which are less expensive than microelectronic circuit processing. In preferred embodiments, printing techniques are utilized in selectively forming masks for use in etching of silicon oxide and diffusing dopants and informing metal contacts to diffused regions. In a preferred embodiment, p-doped regions and n-doped regions are alternately formed in a surface of the wafer through use of masking and etching techniques. Metal contacts are made to the p-regions and n-regions by first forming a seed layer stack that comprises a first layer Such as aluminum that contacts silicon and functions as an infrared reflector, second layer Such titanium tungsten that acts as diffusion barrier, and a third layer functions as a plating base. A thick conductive layer Such as copper is then plated over the seed layer, and the seed layer between plated lines is removed. A front surface of the wafer is preferably textured by etching or mechanical abrasion with an IR reflection layer provided over the textured surface. A field layer can be provided in the textured surface with the combined effect being a very low surface recombination velocity.

#74	Description
Patent/patent application number	US 11,870,002 B2
Title	Methods and systems for use with photovoltaic devices
Assignee	First Solar Inc.
Date	2024–01–09
Abstract	According to embodiments provided herein, the performance of photovoltaic device can be improved by rapidly heating an absorber layer of a device in open-circuit to a high temperature for a short period of time followed by rapid quenching. The rapid heating may be accomplished by one or more pulses of high intensity electromagnetic energy. The energy may be visible light. The energy may be absorbed primarily in the absorber layer, such that the absorber layer is preferentially heated, promoting chemical reactions of dopant complexes. The dopant chemical reactions disrupt compensating defect complexes that have formed in the device, and regenerate active carriers.

#75	Description
Patent/patent application number	US 11,929,447 B2
Title	Annealing materials and methods for annealing photovoltaic devices with annealing materials
Assignee	First Solar Inc.
Date	2024–03–12
Abstract	A method for annealing an absorber layer is disclosed, the method including contacting a surface of the absorber layer with an annealing material provided as a gel. The annealing material comprises cadmium chloride and a thickening agent. A viscosity of the gel of the annealing material is greater than or equal to 5 millipascal seconds.

#76	Description
Patent/patent application number	US 11,866,817 B2
Title	Thin-film deposition methods with thermal management of evaporation sources
Assignee	First Solar Inc.
Date	2024–01–09
Abstract	An evaporation system comprises an evaporation chamber having an interior enclosed by one or more chamber walls; an evaporation source comprising (i) a source body for containing a feedstock material, and (ii) an evaporation port

(*Continued*)

Table 13.1. (*Continued*)

#76	Description
	fluidly coupling the source body with an interior of the evaporation chamber; an insulation material; and a computer-based controller for configuring the insulation material in (i) a first configuration in which the insulation material is disposed snugly around the source body and (ii) a second configuration in which at least a portion of the insulation material is spaced away from the source body and at least a second portion of the insulation material is disposed snugly around the source body; wherein the insulation material does not cover an opening of the evaporation port in the first configuration and the second configuration.

#77	Description
Patent/patent application number	US 8,592,249 B1
Title	Photovoltaic solar cell
Assignee	National Technology and Engineering Solutions of Sandia LLC
Date	2013–11–26
Abstract	A photovoltaic solar cell for generating electricity from sunlight is disclosed. The photovoltaic solar cell comprises a plurality of spaced-apart point contact junctions formed in a semiconductor body to receive the sunlight and generate the electricity therefrom, the plurality of spaced-apart point contact junctions having a first plurality of regions having a first doping type and a second plurality of regions having a second doping type. In addition, the photovoltaic solar cell comprises a first electrical contact electrically connected to each of the first plurality of regions and a second electrical contact electrically connected to each of the second plurality of regions, as well as a passivation layer covering major surfaces and sidewalls of the photovoltaic solar cell.

#78	Description
Patent/patent application number	US 11,145,774 B2
Title	Configurable solar cells
Assignee	C3ip LLC
Date	2021–10–12

Abstract	A photovoltaic cell may include a substrate configured as a single light absorption region. The cell may include at least one first semiconductor region and at least one second semiconductor region arranged on or in the substrate. The cell may include a plurality of first conductive contacts arranged on the substrate and physically separated from one another and a plurality of second conductive contacts arranged on the substrate and physically separated from one another. Each first conductive contact may be configured to facilitate electrical connection with the at least one first semiconductor region. Each second semiconductor conductive contact may be configured to facilitate electrical connection with the at least one second semiconductor region. Each of the first conductive contacts may form at least one separate cell partition with at least one of the second conductive contacts, thereby forming a plurality of cell partitions on or in the substrate.

#79	Description
Patent/patent application number	US 11,888,322 B2
Title	Photovoltaic system and maximum power point tracking control method for photovoltaic system
Assignee	Huawei Digital Technologies Co. Ltd Huawei Digital Power Technologies Co. Ltd
Date	2024–01–30
Abstract	This application provides a photovoltaic system and a maximum power point tracking control method for a photovoltaic system. The photovoltaic system includes an MPPT controller and a power converter, and the MPPT controller is connected to the power converter. The MPPT controller is configured to: be connected to a photovoltaic array, and track a global maximum power point MPP of the photovoltaic array. The MPPT controller may be further configured to obtain, when there is a periodic shade for the photovoltaic array, a multi-peak search start moment of global MPPT of the photovoltaic array based on a status of tracking the global MPP of the photovoltaic array in a target time period, so that when the multi-peak search start moment in each MPPT period arrives, the global MPPT of the photovoltaic array is started, to output a working point of the global MPP of the photovoltaic array to the power converter. According to this application, efficiency of obtaining the working point of the global MPP of the photovoltaic array can be improved, and precision of controlling the global MPPT of the photovoltaic array can be improved.

#80	Description
Patent/patent application number	KR 102,172,004 B1
Title	Micro inverter for photovoltaic power generation and photovoltaic power generation system using the same
Assignee	롯데에너지 주식회사
Date	2020–10–30
Abstract	The present invention relates to a micro inverter for a photovoltaic power generation. The micro inverter for a photovoltaic power generation comprises: a lower case plate having a plate shape; a case cover to cover the lower case plate; and a substrate installed at the lower case plate. The substrate includes a first conductor connected with a first solar cell module in parallel; a second conductor connected with a second solar cell module in parallel; a first switch connected with the first solar cell module and the first conductor in parallel; a second switch connected with the second solar cell module and the second conductor in parallel; a shuffling inductor connected between the first and second conductor and the first and second switches; a boost inductor connected with the first solar cell module, the first conductor, and the first switch; a third switch connected with the boost inductor, the second solar cell module, the second conductor, and the second switch; and an MPPT controller to control an operation tracking a maximum power point based on each voltage of the first solar cell module and the second solar cell module. The first switch, the second switch, and the third switch are operated by the MPPT controller.

#81	Description
Patent/patent application number	KR 102,126,790 B1
Title	Shingled solar cell module
Assignee	선파워 코포레이션
Date	2020–06–25
Abstract	High efficiency configurations for solar cell modules include solar cells that are electrically coupled to each other in a shingled manner to form super cells, which effectively utilize the area of the solar module, reduce series resistance, and reduce module efficiency. It can be arranged to increase. The front metallization patterns on the solar cells can be configured to enable single step stencil printing, which is possible by the overlapping configuration of solar cells in the super cells. The photovoltaic system can include two or more of these high voltage solar cell modules that are

	electrically connected in parallel to each other and the inverter. Solar cell cutting mechanisms and solar cell cutting methods apply a vacuum between the bottom surfaces of the solar cell wafer and the curved support surface to bend the solar cell wafer against a curved support surface, thereby a plurality of solar cells The solar cell wafer is cut along one or more prefabricated scribe lines to provide. The advantage of these cutting tools and cutting methods is that they do not require physical contact with the top surfaces of the solar cell wafer. Solar cells are manufactured with reduced charge recombination losses at the edges of the solar cell, for example, without truncated edges that promote charge recombination. The solar cells can have narrow rectangular gigascopic structures and can advantageously be employed in shingled (overlapping) arrangements to form super cells.

#82	Description
Patent/patent application number	KR 202,400,404,43A
Title	Photovoltaic power generation system using micro inverter
Assignee	와이에이치에너지(주)
Date	2024–03–28
Abstract	The present invention relates to a solar power generation system using a micro inverter. A solar power generation system using a micro inverter according to an embodiment of the present invention includes a micro inverter for solar power generation, a solar panel on which a plurality of solar cell modules are installed, and a micro inverter for solar power generation installed on the rear of the solar panel. In the solar power generation system using a micro inverter including a support part for supporting, the micro inverter for solar power generation includes: a lower case plate formed in a plate shape; a case cover configured to cover the case lower plate; It includes a substrate installed on the lower plate of the case, and the support part can support the micro-inverter for solar power generation to be disposed inside the solar panel.

#83	Description
Patent/patent application number	US 8,642,880 B2
Title	Interchangeable and fully adjustable solar thermal-photovoltaic concentrator systems

(Continued)

Table 13.1. (*Continued*)

#83	Description
Assignee	Chia-Chin Cheng John W. Holmes
Date	2014–02–04
Abstract	An interchangeable and fully adjustable solar thermal-photovoltaic concentrator system is provided, comprising: one or more heat collecting elements; one or more primary reflectors having one or more openings; one or more sunray directing optical mechanisms at the sun-collecting side of primary reflectors and between the heat collecting elements and openings in the primary reflectors; one or more PV cell modules disposed at the non-sun-collecting side of the primary reflectors; and one or more sunray distributing optical mechanisms disposed at the non-sun-collecting side of the primary reflectors. Wherein after sunrays irradiate to the primary reflectors, a proportion of sunrays ranging from 0% to 100% are reflected to one or more heat collecting elements and the remaining sunrays are directed by the sunray directing optical mechanisms, through openings in the primary reflectors, and distributed by the sunray distributing optical mechanisms to one or more PV cell modules.

#84	Description
Patent/patent application number	US 8,697,481 B2
Title	High efficiency multijunction solar cells
Assignee	Cactus Materials Inc.
Date	2014–04–15
Abstract	Multijunction solar cells having at least four subcells are disclosed, in which at least one of the subcells comprises a base layer formed of an alloy of one or more elements from group III on the periodic table, nitrogen, arsenic, and at least one element selected from the group consisting of Sb and Bi, and each of the subcells is substantially lattice matched. Methods of manufacturing solar cells and photovoltaic systems comprising at least one of the multijunction solar cells are also disclosed.

#85	Description
Patent/patent application number	US 11,121,282 B2
Title	Method for producing a CdTe thin-film solar cell
Assignee	China Triumph International Engineering Co. Ltd CTF Solar GmbH
Date	2021–09–14
Abstract	The present invention describes a method for producing CdTe thin-film solar cells, in which special parameters of different processing steps and a special sequence of processing steps result in improved characteristics of the produced CdTe solar cells.

#86	Description
Patent/patent application number	US 9,590,131 B2
Title	Systems and methods for advanced ultra-high-performance InP solar cells
Assignee	Alliance for Sustainable Energy LLC.
Date	2017–03–07
Abstract	Systems and Methods for Advanced Ultra-High-Performance InP Solar Cells are provided. In one embodiment, an InP photovoltaic device comprises: a p-n junction absorber layer comprising at least one InP layer; a front surface confinement layer; and a back surface confinement layer; wherein either the front surface confinement layer or the back surface confinement layer forms part of a High-Low (HL) doping architecture; and wherein either the front surface confinement layer or the back surface confinement layer forms part of a heterointerface system architecture.

#87	Description
Patent/patent application number	US 8,614,395 B1
Title	Solar cell with back side contacts
Assignee	Alliance for Sustainable Energy, LLC., National Technology and Engineering Solutions of Sandia LLC

(*Continued*)

Table 13.1. (*Continued*)

#87	Description
Date	2013–12–24
Abstract	A III–V solar cell is described herein that includes all back side contacts. Additionally, the positive and negative electrical contacts contact compound semiconductor layers of the solar cell other than the absorbing layer of the solar cell. That is, the positive and negative electrical contacts contact passivating layers of the solar cell.

#88	Description
Patent/patent application number	US 9,960,307 B1
Title	Method for producing thin-film solar cells
Assignee	China Triumph International Engineering Co., Ltd (Shanghai), CTF SOLAR GMBH (Dresden)
Date	2018–05–01
Abstract	A method to produce thin film solar cells in superstrate or substrate configuration is an efficient way to minimize the loss due to absorption in CdS layer and to eliminate the $CdCl_2$ activation treatment step. This is achieved by applying a sacrificial metal-halide layer between the CdS-layer and the CdTe-layer of the solar cells.

#89	Description
Patent/patent application number	CN 114,267,753A
Title	TOPCon solar cell, preparation method thereof and photovoltaic module
Assignee	Haining Astronergy Technology Co. Ltd
Date	2022–04–01
Abstract	The application discloses a preparation method of a TOPCon solar cell, which comprises the steps of forming a silicon oxide layer on the back surface of a silicon wafer; forming a first doped amorphous silicon layer on the surface of the silicon oxide layer deviating from the silicon wafer in an in-situ doping mode; forming an intrinsic amorphous silicon layer on the surface of the first doped amorphous silicon layer, which is far away from the silicon oxide layer; doping the intrinsic amorphous silicon layer to form a second doped amorphous silicon layer; and simultaneously crystallizing the first doped amorphous silicon layer and the second doped amorphous silicon layer to

	correspondingly form a first doped polycrystalline silicon layer and a second doped polycrystalline silicon layer to obtain the TOPCon solar cell. The method divides the doped polycrystalline silicon layer into a first doped polycrystalline silicon layer and a second doped polycrystalline silicon layer, the first doped polycrystalline silicon layer is formed in an in-situ doping mode, the second doped polycrystalline silicon layer is formed in an ex-situ doping mode, the ex-situ doping mode can shorten the preparation time, and the using amount of gas required in the preparation process can be reduced.

#90	Description
Patent/patent application number	US 2024 028,287,2 A1
Title	Method for producing at least one photovoltaic cell for converting electromagnetic radiation into electrical energy
Assignee	Fraunhofer-Gesellschaft zur Förderung der angewandten Forschung e.V.
Date	2022–06–08
Abstract	A method for producing at least one photovoltaic cell for converting electromagnetic radiation into electrical energy, having the method steps of: (A) providing a superstrate in the form of a semiconductor substrate; (B) applying photovoltaic cell semiconductor layers for forming at least one photovoltaic cell to a rear face of the superstrate indirectly or directly, and the photovoltaic cell semiconductor layers have at least one absorber layer formed from a direct semiconductor. The superstrate is in the form of a current conducting layer having a thickness greater than 10 μm and, in method step B, the photovoltaic cell semiconductor layers are formed with an electrically conductive connection to the current conducting layer and wherein the band gap of the current conducting layer is larger by at least 50 meV than the band gap of the absorber layer.

#91	Description
Patent/patent application number	US 2023 041,154,6 A1
Title	Solar-cell module
Assignee	Fraunhofer-Gesellschaft zur Förderung der angewandten Forschung e.V.
Date	2023–12–21

(*Continued*)

Table 13.1. (*Continued*)

#91	Description
Abstract	A solar cell module, having module segments, each with photovoltaic solar cells interconnected in series. The cells of the module segments are arranged on or in a curved planar carrier element. Each cell has a solar cell normal vector, and the module has a module normal vector, corresponding to the vectorial mean value of the cell normal vectors. Each cell has a tilt angle, corresponding to the angle between the cell normal vector and the module normal vector. Each module segment is assigned a tilt angle range having limits defined by minimum and maximum tilt angles of the cells of the module segment. The tilt angle ranges of at least two module segments are disjoint. The module segments are interconnected in parallel, each module segment has the same number of cells, and each cell of a module segment is arranged directly adjacent to a further cell of that module segment.

#92	Description
Patent/patent application number	CN 212,967,726 U
Title	A novel solar panel using phase change inhibiting materials for heat dissipation
Assignee	PCI Green Tech PTY Ltd
Date	2020–06–04
Abstract	The utility model discloses a new type of solar cell panel for dissipating heat by utilizing phase-change inhibiting materials, which comprises a solar cell panel. In addition, a phase-change suppression heat-dissipating plate is provided. Since the heat-dissipating plate has extremely high heat flux density and thermal conductivity, it can effectively transfer the heat inside the solar cell module to the outside of the heat source, greatly improving the photoelectric conversion efficiency, while prolonging the The life of the battery components and does not consume additional energy.

#93	Description
Patent/patent application number	US 8,153,282 B2
Title	Solar cell with antireflective coating with graded layer including mixture of titanium oxide and silicon oxide
Assignee	Guardian Industries Corp.

Date	2012–04–10
Abstract	There is provided a coated article (e.g., solar cell) that includes an improved anti-reflection (AR) coating. This AR coating functions to reduce reflection of light from a glass substrate, thereby allowing more light within the solar spectrum to pass through the incident glass substrate. In certain example embodiments, the AR coating includes a graded base layer having a varying refractive index value, and an overcoat layer which may be provided for destructive interference purposes.

#94	Description
Patent/patent application number	CN 109,585,657A
Title	A perovskite solar cell module
Assignee	Guilin University of Technology
Date	2019–11–10
Abstract	The invention discloses a perovskite solar cell assembly. Its structure is a sandwich structure of 'conductive glass/electron transporter//multifunctional nickel mesh//electron transporter/conductive glass', which specifically includes upper conductive glass, lower conductive glass, multifunctional nickel mesh, upper electron transporter, and lower layer. Electron transporters, electrodes and glass binders. This structure allows the perovskite absorber layer to fully contact the electron transporter, so that electrons can be quickly exported from the conductive glass on both sides, which greatly improves the transmission efficiency of electrons; the network structure can increase the perovskite absorber layer/hole transport The interfacial contact of the layers promotes the hole transport efficiency. In addition, the entire perovskite absorber layer is completely closed by tightly encapsulating the battery assembly, which improves the environmental stability of the entire battery assembly. The invention has the advantages of simple structure, reasonable design, high practical value, low cost, convenient assembly of battery components, and the like, making it particularly suitable for use as a perovskite solar battery component.

#95	Description
Patent/patent application number	KR 2018 006,468,6A

(Continued)

Table 13.1. (*Continued*)

#95	Description
Title	Composition of cleaning solution for solar panel glass
Assignee	소순영
Date	2018–06–15
Abstract	The present invention relates to a glass cleaning solution composition for a solar panel, which is to remove contamination such as dust or stain bonded to the surface of glass for a solar panel. According to the present invention, the glass cleaning solution composition for a solar panel comprises: 0.01–5 wt% of nonionic silicon-based surfactant; 1–20 wt% of one or more anion-based surfactants; and an aqueous solution including 0.5–20 wt% of one or more solvents selected in a group comprising diethylene glycol-n-butyl ether, 3-methoxy-methyl-1-butanol, ethylene glycol monobutyl ether, ethyl alcohol, butanol, isopropyl alcohol, and polypropylene glycolmethylether, wherein the aqueous solution also includes 0.01–5 wt% of one or more silicate compounds selected from a group comprising sodium silicate and potassium silicate and pH of the aqueous solution is pH6 to pH10. According to the present invention, a glass cleaning solution for a solar panel invented by the present invention enables residues to rarely remain after cleaning the glass by having an excellent cleaning force and, also, can improve the durability of the effect after washing.

#96	Description
Patent/patent application number	US 9,059,350 B2
Title	Junction box for a photovoltaic solar panel
Assignee	Multi Holding AG, Applied Materials Inc.
Date	2015–06–16
Abstract	A junction box for a solar panel having a housing, a lid, a first connector and a second connector. The housing comprises sidewalls and a top wall defining an interior space. The first coupling having a first contact element and the second coupling having a second contact element. The contact elements penetrate at least one of the sidewalls, so that the contact elements provide an electrical contact from external contact elements to internal contact elements, such as solder tails. Internal contact elements are arranged a, least partially in the interior space. The top wall having an opening extending only partially in the top wall. The opening is located such in the top wall that access to the solder tails in a substantially perpendicular direction to the surface of a solar panel for connecting the solder tail to the solar panel is provided.

#97	Description
Patent/patent application number	EP 013,729,1 A2
Title	Amorphous silicon solar cells
Assignee	Komatsu Ltd
Date	1985–04–17
Abstract	An amorphous silicon solar cell comprising a glass substrate, a transparent conductive film formed on the glass substrate on one side thereof and having micro columns or fine crystals irregularly formed on on the other side, a plurality of amorphous silicon layers superposed on the other side of the conductive film, and a metal electrode formed on the superposed silicon layer. At the interface between the transparent conductive film and the amorphous silicon layer is formed an intermediate layer in which both materials of the conductive film and the silicon layer are mixed. The intermediate layer has a refractive index between the conductive film and the silicon layer. The glass substrate may be substituted with a metal substrate, in which case the plurality of silicon layers are formed directly on the metal substrate, on which the transparent conductive film having an irregular surface on the side opposite to the side where the silicon layers are formed and a metal electrode are formed in this order.

#98	Description
Patent/patent application number	US 6,452,086 B1
Title	Solar cell comprising a bypass diode
Assignee	Airbus Se
Date	2002–09–17
Abstract	The invention relates to a production method for a solar cell and to the solar cell itself which comprises an integrated bypass diode on the side facing away from the incidence of light and which can be produced in a simple manner by diffusion. A one-piece electric conductor serves to connect two successive solar cells in series and simultaneously effects the contacting of the corresponding bypass diode.

#99	Description
Patent/patent application number	US 8,257,998 B2
Title	Solar cells with textured surfaces
Assignee	Massachusetts Institute of Technology
Date	2012–09–04
Abstract	Semiconductor photovoltaic cells have surfaces that are textured for processing and photovoltaic reasons. The absorbing regions may have parallel grooves that reduce loss of solar energy that would otherwise be lost by reflection. One form of texturing has parallel grooves and ridges. The cell also includes regions of metallization for collecting the generated electrical carriers and conducting them away, which may be channels. The topography is considered during production, using a process that takes advantage of the topography to govern what locations upon will receive a specific processing, and which locations will not receive such a processing. Liquids are treated directly into zones of the cell. They migrate throughout a zone and act upon the locations contacted. They do not migrate to other zones, due to impediments to fluid flow that are features of the surface texture, such as edges, walls and ridges. Blocking liquid may also be deposited and migrate within a zone, to block or mask a subsequent activity, such as etching.

#100	Description
Patent/patent application number	US 10,074,761 B2
Title	Solar cell assembly and method for connecting a string of solar cells
Assignee	Azur Space Solar Power GmbH
Date	2018–06–15
Abstract	The invention relates to a solar cell assembly comprising at least one first solar cell and at least one discrete protective diode (101) that is connected to the solar cell. The aim of the invention is to comprehensively protect a solar cell, a solar cell composite or a string of cells by means of one or more protective diodes, without resorting to the use of the material of the solar cells. To achieve this, in addition to a front and a rear contact (13, 15), the protective diode comprises an additional contact (14) that is placed at a distance from the front contact and is electrically connected to said contact via a p/n junction. A connector leads from the additional contact to a second solar cell, the latter in turn being connected to the first solar cell in a string.

13.2 Conclusions

Examples of 100 patents, patent applications and inventions have been summarized in the above chapter. These include materials, components, cells and modules. Perovskite solar cells, with the promise of excellent match with the solar spectrum and thus high efficiency, will continue to be of interest to the photovoltaics and materials science community. Once the stability of perovskites is well understood and established, the technology transfer from prototyping to manufacturing of cells should occur seamlessly. 2D materials such as the transition metal dichalcogenides and graphene are expected to play significant roles in addressing a variety of functionalities in solar cells.

References

[1] China's patent applications for solar cells rank first in the world https://news.cgtn.com/news/2024-01-05/China-s-patent-applications-for-solar-cells-rank-first-in-the-world-1q780u7KCWs/p.html (accessed 29 August 2024)

[2] Tay A 2022 By the numbers: China's net-zero ambitions, spotlight, part of nature spotlight on China's net-zero ambitions *Nature* 5 April 2022 https://doi.org/10.1038/d41586-022-00802-3

[3] https://patents.google.com/

[4] https://uspto.gov/patents/search

[5] https://nrel.gov/pv/perovskite-patent-portfolio.html

IOP Publishing

N M Ravindra, Leqi Lin and Priyanka Singh

Chapter 14

Conclusions

In the previous chapters, the book described recent developments in emerging photovoltaic (PV) technologies such as cadmium telluride (CdTe), copper indium gallium selenide (CIGS), copper zinc tin sulphide (CZTS), perovskite solar cells (PSCs), polymer solar cells, tandem solar cells (TSCs), concentrator solar cell systems including recent patents and disclosures. In addition, various computational methods such as instantaneous and long-term performance measurements, influence of weather conditions, approaches to hot-spot reduction and shade loss minimization were discussed for improving efficiency, reducing costs, and enhancing the versatility of solar cells. Furthermore, some critical observations and the analysis of the results were provided for a significant insight into the physics and functioning of solar cells. This chapter will discuss the evolution of solar cells through various generations. Some notable projects and highlights in solar cell technologies have been addressed globally. The contribution of research and development (R&D) labs and international organizations to the advancement of solar cells is explored. An overview of the industries that produce terrestrial solar cells based on silicon, emerging technologies in thin-films and space solar cells, based on III–V semiconductors, is presented. Specific areas where improvements to solar cell technology may have significant effects on their applications and efficacy are explored.

14.1 Solar cell generations

Since the first demonstration of the photovoltaic effect by Edmond Becquerel in 1839, progress in solar cells has evolved over the span of two centuries. The developments involve numerous scientific breakthroughs and technological advancements. In 1954, Bell Laboratories produced the first practical silicon solar cell with an efficiency of ~6% [1]. This milestone is considered to be the birth of modern photovoltaics. Later on, Vanguard I satellite utilized solar cells to power its radio transmitter, marking the first use of solar power in space applications [2]. During the 1980s, solar technology began its wider commercial use, especially in remote and off-grid applications with

doi:10.1088/978-0-7503-5994-8ch14

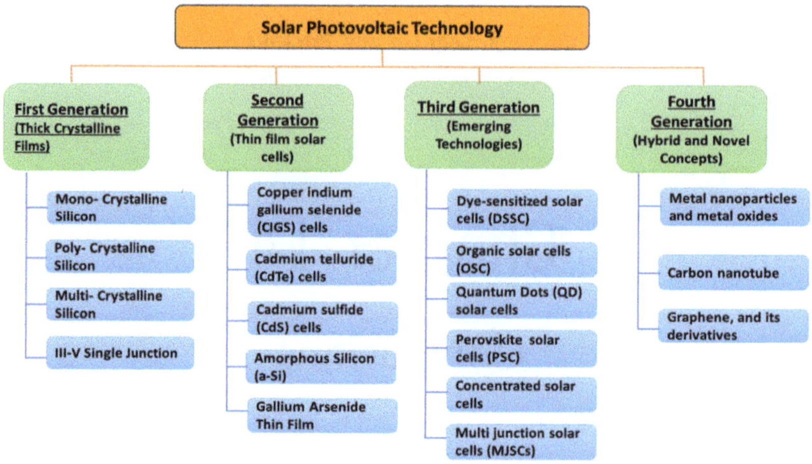

Figure 14.1. Various types and generation of solar cells and recent advancements in photovoltaic technologies [7, 21]. Reproduced from [7]. CC BY 4.0.

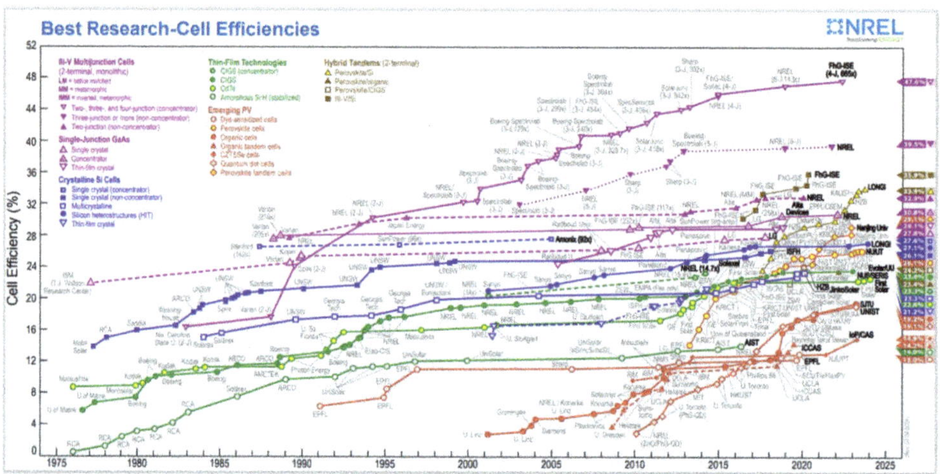

Figure 14.2. NREL best research-cell efficiencies chart. Reproduced with permission from [10] credit: NREL.

improving efficiencies and gradually decreasing costs. Since then, solar cells have developed through several generations, each with advances in PV technology and efficiency [3–5]. This is illustrated in figures 14.1 and 14.2. PV technology can be categorized into four major generations [3–7].

14.1.1 First generation

The first PV generation is based on crystalline silicon (c-Si), multicrystalline silicon (mc-Si), polycrystalline silicon (pc-Si), and gallium arsenide (GaAs) [7–9]. Si-based PV cells were first to enter the PV sector and currently account for more than 80% of

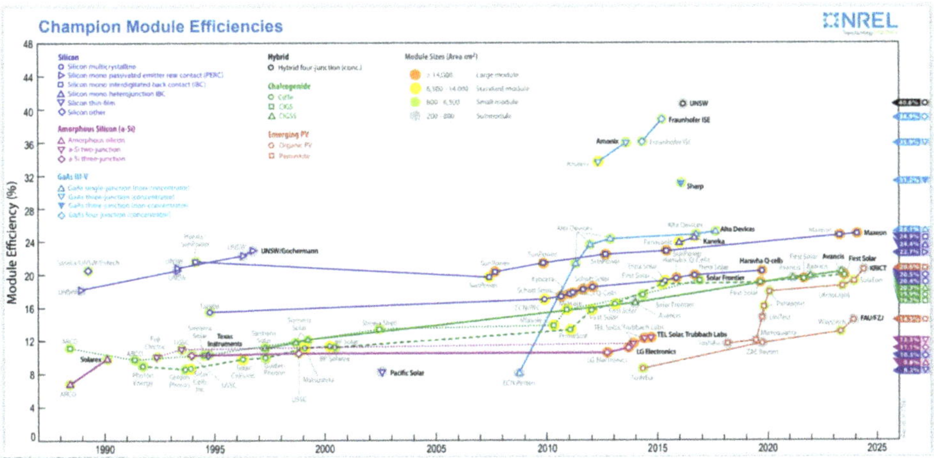

Figure 14.3. Champion photovoltaic module efficiency chart. Reproduced with permission from [15] credit: NREL.

the global installed capacity and 90% of the market share. Silicon solar cells have remained the most commonly utilized due to their relatively high efficiency and low cost. At the laboratory level, efficiency of Si-based cells is usually around 21%–27% [10]. Passivated emitter rear cell (PERC) [11, 12] continues to dominate the market due to its high efficiency and relatively low manufacturing costs. PERC silicon solar cells are seeing incremental efficiency improvements and widespread adoption. High efficiencies and good temperature performance are achieved by combining thin amorphous silicon (a-Si) layers with c-Si wafers in heterojunction with intrinsic thin layer (HIT) solar cells [13, 14]. As shown in figure 14.3 commercial modules have a maximum power conversion efficiency of 8%–24% [15]. Despite a well-established technology, high efficiency, and long-term reliability, it is expensive due to the requirement of high purity silicon.

14.1.2 Second generation

Second generation PV cells are based on thin films such as a-Si, CdTe, CZTS, and CIGS, and are intended to provide a lower-cost alternative to c-Si cells [7, 16]. The light-absorbing thickness of the second-generation PV cells is ~10 μm, compared to ~200–300 μm in the first generation [3–6, 16]. Thus, this has made it possible to develop the second-generation solar cells with flexible substrates, less material, and lower manufacturing costs. Although the second-generation solar cells have superior mechanical qualities that are suited for flexible substrate applications, efficiency is generally lower, usually ~14%–23% [10]. Thin film technology development required new growth techniques; it has since advanced and this has allowed for new opportunities in the industry [17]. The module efficiencies, based on these cells, range from ~9.8% to 20 % [15].

14.1.3 Third generation

PV cells of the third generation represent tandem, perovskite, organic, dye-sensitized, and emerging concepts. This generation encompasses a wide range of methods, from low-cost, low-efficiency cells (dye-sensitized, organic solar cells) to high-cost, high-efficiency systems (III–V multi-junction cells) for space applications [7]. Despite the fact that some third-generation PV cells have been studied for more than 25 years, third-generation PV cells are frequently referred to as 'emerging concepts' due to their low market penetration [18, 19]. PV cells in this category have recorded efficiencies ranging from 13%–26% [10], whereas multi-junction solar cells (MJSCs) have the potential to achieve efficiencies as high as >40% [10], making them suitable for specialized applications such as concentrated solar power (CSP) and space missions. PSCs have demonstrated a strong potential for improvement, achieving efficiencies of more than 25% [10]. While third-generation photovoltaic cells are lightweight and flexible, and they have the potential to be more efficient than earlier generations at lower production costs, scaling up manufacturing is hindered by stability problems (particularly with perovskites) and toxicity issues with certain materials. Although the third-generation solar PV cell-based PV modules are still relatively new and could soon make their way onto the market, NREL (National Renewable Energy Laboratory) has reported the module efficiencies for PSCs to be ~20.6% [15] and for organic solar cells are ~14.5% [15].

14.1.4 Fourth generation

Fourth-generation PV cells are also referred to as hybrid inorganic cells as they combine the low cost and flexibility of polymer thin films with the stability of organic nanostructures such as carbon nanotubes, graphene, metal nanoparticles and oxides, and their derivatives [3, 5, 7]. Fourth-generation photovoltaics are sometimes known as 'nanophotovoltaics,' as they incorporate innovative ideas that make use of cutting-edge nanotechnology. These concepts have significant potential for the future of photovoltaics [7, 20].

Each generation marks a significant advancement in solar cell technology in terms of efficiency, cost, and adaptability, with the goal of making solar energy a more practical and widespread source of renewable energy. Figure 14.1 illustrates examples of solar cell types for each generation.

14.2 Global solar cell activity and highlights

Solar cell activities and highlights around the world showcase the rapid growth and technological advancements in the solar energy sector. Some notable activities and highlights from various regions are described in the following section.

The United States continues to develop some of the world's largest solar farms, such as the Topaz Solar Farm in California [22] and the Copper Mountain Solar Facility in Nevada [23]. Tesla is advancing solar technology with integrated solar roof tiles. Germany has been a global leader in solar energy adoption, driven by strong government incentives and policies. Spain hosts several large solar power plants, including the Gemasolar Thermosolar Plant, which uses CSP technology [24]. China is the largest producer and installer of solar panels globally. Companies like JinkoSolar [25] and LONGi [26] are major players in the solar industry. Strong government policies and subsidies have fuelled the rapid expansion of solar power in China. India has set ambitious renewable energy projects like the Bhadla Solar Park [27], one of the largest in the world, highlighting India's commitment to solar energy. Dubai aims to generate 75% of its energy from clean sources by 2050, with solar playing a central role. Morocco's Noor Ouarzazate Solar Complex [28] is one of the largest CSP plants in the world. Australia has one of the highest rooftop solar installations in the world. A project like the Sun Cable [29], is aiming to export solar energy from Australia to Singapore. Institutes at ANU (Australian National University) and UNSW (University of New South Wales) are at the forefront of solar research and development. China, Japan, and the Netherlands are leading in the development of floating solar farms [30], which utilize bodies of water to deploy solar panels. Bifacial panels, which capture sunlight on both sides, are gaining traction due to their higher energy yield and efficiency. Federal and state incentives, including the US Government sponsored Inflation Reduction Act (IRA), tax credits and renewable portfolio standards, support the growth of solar energy.

14.2.1 Research and Development laboratories in solar cell technology

R&D laboratories that are focused on the development of solar cell technology play a pivotal role in advancing the efficiency, reliability, and cost-effectiveness of solar energy. These labs are often associated with universities, government agencies, research institutes, and private companies that are dedicated to pushing the boundaries of solar cell innovation.

14.2.1.1 National Renewable Energy Laboratory (NREL)

NREL [(elevated to a National Laboratory on September 16th 1991), the former Solar Energy Research Institute (SERI—opened on July 5th 1977)], is a prominent U. S. Department of Energy (US DOE) laboratory conducting research and development in renewable energy, including solar PV. Advanced materials for solar cells, such as thin-films of CdTe and CdS, silicon heterojunction, emerging PSCs, CSP and systems are some of the research areas of interest to NREL. NREL is pioneering tandem solar cell technologies, e.g. perovskite and silicon to capture a broader spectrum of sunlight for efficient solar cells. The laboratory works on advanced manufacturing techniques to reduce the cost and improving their scalability. NREL is also involved in integrating solar cells into larger energy systems, including solar-plus-storage solutions and grid integration studies to maximize the efficiency and reliability of solar power. Along with these initiatives, since its inception, NREL has been publishing a chart of the highest confirmed conversion efficiencies for research cells for a range of PV technologies (figure 14.2) [10]. The chart depicts the efficiencies for c-Si, thin-film technologies, emerging photovoltaics, single and gallium arsenide MJSCs. Recently, the NREL team added a new category to the chart called 'hybrid tandems' in response to the increasing interest in tandem PV. While perovskite/Si and perovskite/CIGS were already categorized in 'Emerging PV,' III–V/Si and perovskite/organic are newer subcategories of hybrid tandems [10]. Although all the tandem subcategories have been moved into the new hybrid tandems category, perovskite/perovskite tandem is listed under 'Emerging PV'. Besides, since 1988 to the present, NREL (earlier SERI) has also maintained a chart showing the highest confirmed conversion efficiencies for champion modules for a variety of PV technologies [15]. Module efficiencies are shown for the semiconductors: emerging photovoltaics, silicon, a-Si, GaAs III–V, hybrid, and chalcogenide in these charts. In each chart (figures 14.2 and 14.3), the most recent world record for a specific technology is represented by a flag on the right side with the efficiency and technology symbol [10, 15].

14.2.1.2 Fraunhofer Institute for Solar Energy (Fraunhofer, ISE)

Fraunhofer ISE, based in Germany and founded in 1981, is one of the largest solar energy research institutes in Europe. Research areas at ISE are based on silicon PV (c-Si and silicon thin-films), III–V compound semiconductors, PSCs, module technologies, PV system analysis, and energy economics. ISE is known for its contributions to high-efficiency PERC & TopCon (Tunnel Oxide Passivated Contact) silicon solar cells as well as for significant advancements in bifacial solar cell technology [31]. ISE focuses on both laboratory-scale innovations and scalable industrial processes and is actively developing building-integrated photovoltaics (BIPV) technologies, which integrate solar cells into building materials such as windows and facades.

14.2.1.3 Institut National de l'Energie Solaire (INES)

INES was started in Grenoble, France in February 2005. INES focuses on R&D, innovation and education in advanced photovoltaic technology. In particular, it also addresses the integration of photovoltaics into electrical systems as well as management of energy sources and systems.

These R&D labs not only drive technological advancements in solar cell efficiency and reliability, but also contribute to global efforts in reducing the cost of solar energy and integrating renewables into the energy mix.

In addition to these R&D labs, international organizations such as International Energy Agency (IEA) provide comprehensive reports and analysis on the global status and future outlook of solar photovoltaics. Ernst & Young's 'Renewable Energy Country Attractiveness Index' (RECAI) evaluates the attractiveness of renewable energy investment and deployment opportunities in various countries.

14.2.2 International Energy Agency

IEA is an international organization that provides data and research on energy, including solar energy [32]. The IEA Photovoltaic Power Systems Programme (PVPS) is a collaborative research and information dissemination initiative that focuses on the development and deployment of PV solar energy technologies globally. It includes analysis of market trends, policy developments, and technology advancements. IEA publishes comprehensive reports on the global solar market and provides valuable insights for policymakers, industry stakeholders, and researchers. According to IEA's reports, the global installed capacity of solar PV has been increasing rapidly, driven by declining costs, technological advancements, and supportive government policies. IEA projects that additions to renewable power capacity will increase over the next five years, with solar PV and wind power accounting for a record 96% of this increase, due to their lower generation costs than those of both fossil and non-fossil alternatives in most countries, and policies continuing to support them. As shown in figure 14.4, in comparison to 2022, solar PV and wind additions are expected to more than double by 2028, consistently surpassing records to reach almost 710 GW. Figure 14.5 [33] illustrates global renewable electricity capacity growth from 2017 to 2030, highlighting regional contributions. China leads significantly in renewable expansion, followed by the US and the EU, which also show strong policy-driven growth. India emerges as a key player, emphasizing solar and wind energy. Latin America, ASEAN (Association of Southeast Asian Nations), and MENA (Middle East and North Africa) regions demonstrate moderate growth, while Sub-Saharan Africa and other countries lag behind. The chart emphasizes the global transition to renewables, with China as the dominant force and varying contributions from other regions [33].

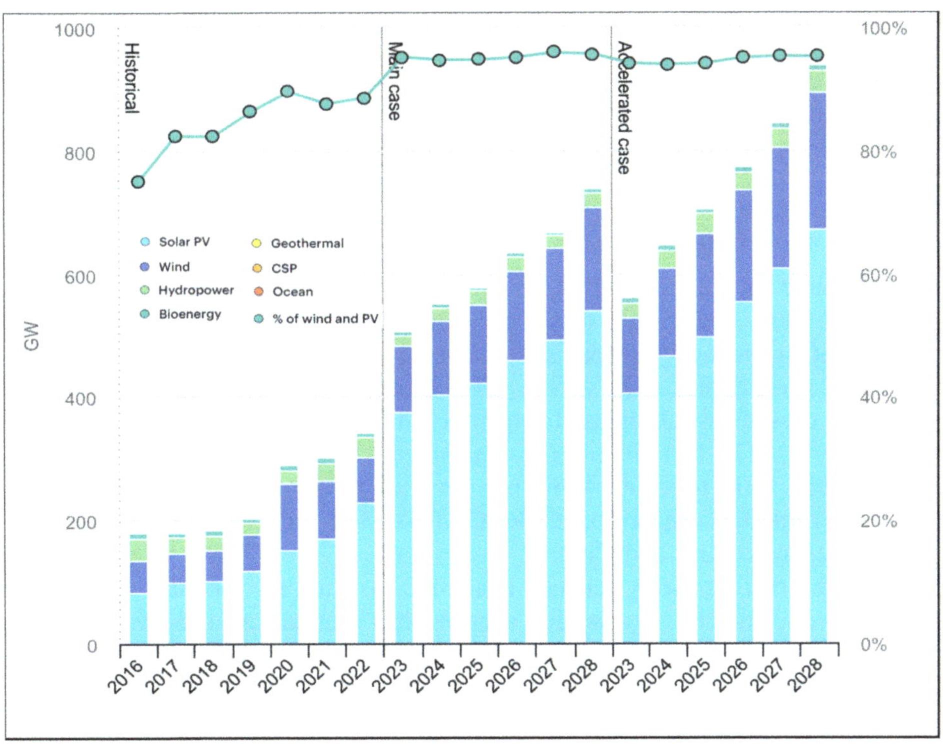

Figure 14.4. Renewable electricity capacity additions by technology and segment, 2016–28. Reproduced from [32] CC BY 4.0.

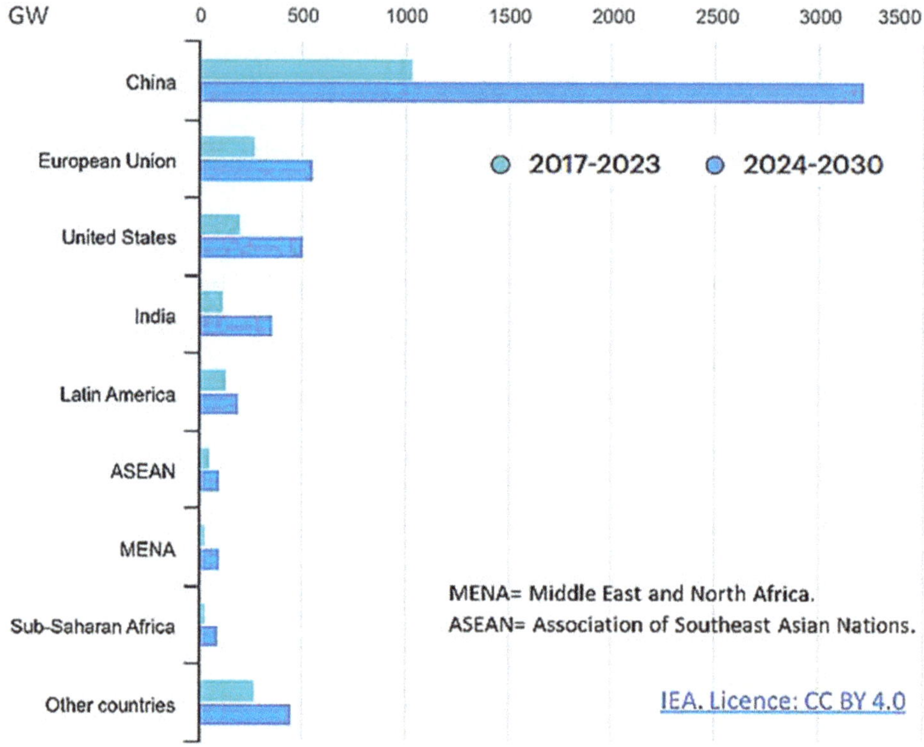

Figure 14.5. Projected renewable electricity capacity growth by country/region (2017–2030). Reproduced with permission from [33], copyright 2023 EYGM Limited. All Rights Reserved.

14.2.3 Renewable Energy Country Attractiveness Index (RECAI)

Ernst & Young introduced the 'Renewable Energy Country Attractiveness Index' (RECAI) in 2003, which publishes the bi-annual ranking of the top 40 nations in terms of REC investment prospects [34]. The countries featured in the index belong to different geographies (e.g., USA, UK, India, UAE, Kenya), report versatile growth, are diverse in ecological footprints, and differ with regard to capital and labour intensity. The top 10 countries that invest in renewable energy sources are listed in table 14.1, together with their overall information, technology-specific scores, and rankings. According to RECAI 62 [34], the top three economies are United States, Germany and China.

Moreover, collaborations between academic institutions, industry partners, and government agencies are crucial for accelerating innovation and overcoming the technical challenges in the renewable energy sector. These collaborating institutions play crucial roles in advancing solar cell technology, promoting the adoption of solar energy, and shaping the future of renewable energy on a global scale. Their combined efforts are driving significant progress in the efficiency, cost reduction, and sustainability in the solar industry.

Table 14.1. List of top ten countries that invest in renewable energy sources together with their overall information, technology-specific scores, and rankings [34].

RECAI 62 score			Technology-specific scores							
Rank	Market	Score	Onshore wind	Offshore wind	Solar PV	Solar CSP	Biomass	Geothermal	Hydro	Marine
1	US	73.9	58.4	59.5	58.3	46.8	40.8	47.4	39.6	20.7
2	Germany	71.4	53.4	52.3	56.1	32.4	50.7	38.1	41.4	20.9
3	China Mainland	71.4	52.5	55.7	61.5	55.0	49.3	24.8	51.0	17.9
4	France	70.6	56.2	52.3	54.8	23.9	46.9	39.8	42.1	38.5
5	Australia	70.2	53.5	42.4	57.2	47.2	41.8	15.8	27.1	25.7
6	India	69.2	53.7	28.7	62.7	34.3	43.6	24.8	48.9	20.0
7	UK	68.3	57.6	57.6	48.0	15.1	54.8	36.8	39.3	34.8
8	Spain	67.1	54.0	35.6	54.0	29.2	40.0	15.4	23.2	23.0
9	Denmark	66.3	54.1	51.5	47.0	17.4	45.4	16.4	22.3	22.1
10	Netherlands	66.1	53.5	48.3	48.8	16.0	51.7	24.8	27.7	16.8

14.2.4 Solar cell industries

The solar cell market is diverse and includes a wide range of players, from manufacturers of solar cells and modules to developers of solar power projects and providers of related technologies and services. Some of the top industries are discussed in this section.

14.2.4.1 Terrestrial solar cell industries

Several leading industries are at the forefront of manufacturing and developing terrestrial solar cells. These companies are known for their innovation, production capacity, and contributions to the solar energy market.

LONGi, China is one of the largest manufacturers of c-Si wafers and solar panels. The company is known for its high-efficiency solar products and significant investment in R&D. LONGi [26] has pioneered several advancements in PERC, TOPCon technology and bifacial solar panels. JinkoSolar, China is another leading global solar module manufacturer [25]. The company has a vast production capacity and a strong international presence. JinkoSolar focuses on high-efficiency c-Si solar modules and has set multiple records for module efficiency. Trina Solar Limited, China is a major player in the solar industry, known for its high-quality PV modules and comprehensive solar solutions. The company has developed several advanced technologies, including Vertex series panels with high power output and multi-busbar design. Canadian Solar, Canada [25] is one of the largest solar technology and renewable energy company. Canadian Solar has been a leader in high-efficiency PERC and bifacial solar panels and has expanded its focus to include energy storage solutions. The PV sector of India is growing rapidly to achieve 50% renewable energy by 2030, driven by several companies (listed in table 14.2) investing in solar power, innovation, and large-scale projects. With advancements and government support, India is moving toward a sustainable solar future [35]. First Solar [25, 36], USA specializes in manufacturing thin-film CdTe solar panels, silicon-based

Table 14.2. Examples of leading industries involved in the manufacture of terrestrial solar cells based on silicon, thin film, emerging technologies and III–V semiconductor based space solar cells.

Terrestrial solar cell/module/panel		Space solar cells/module/panel
Silicon/thin film	Emerging technologies/tandem solar cell	III–V semiconductor
LONGi, China mc-Si modules PERC, TOPCon, bifacial panels [26]	LONGi, China Perovskite/silicon TSCs [26]	Spectrolab, USA Triple-junction and quadruple-junction solar cells [50]
JinkoSolar, China High-efficiency mc-Si and pc-Si solar modules [25]	Oxford PV, UK Perovskite solar cells, high-efficiency perovskite/silicon TSCs [41]	SolAero, New Mexico, USA High-efficiency multi-junction solar cells, solar panels, and arrays [51]

Trina Solar Limited, China Vertex series panels, multi-busbar design modules [25]	Saule Technologies, Poland Inkjet-printed perovskite solar cells, flexible and transparent solar panels [42]	Azur Space Solar Power GmbH, Germany GaAs-based multi-junction solar cells [52]
Tongwei Solar, China Monocrystalline and multicrystalline silicon & solar cells [39]		
DAS Solar, China PV cells & modules [39]		
Canadian Solar, Canada High-efficiency PERC and bifacial solar panels [25]	Heliatek GmbH, Germany, OPV and BIPV, Perovskite/Si TSCs [43]	Emcore, USA Triple-junction solar cells and panels [53]
First Solar, USA Si/thin-film and CdTe solar panels [25, 36]	SolarWindow Technologies, USA Transparent PV coatings for windows and other surfaces and perovskite /Si TSCs [44]	Sharp, Japan High-efficiency multi-junction solar cells [54]
SunPower Corporation, USA High-efficiency solar panels Maxeon cell technology [37]	Tandem PV, USA Perovskite /Si TSCs research) [40]	Alta Devices, USA GaAs solar cells [47]
REC Group, Singapore Advanced mc-Si solar modules [38]	Meyer Burger Technology AG, Switzerland heterojunction and tandem solar cells [45]	
Hanwha Q CELLS, South Korea Si, polysilicon Q.ANTUM technology-based module [25]	GCL-Poly Energy Holdings Limited, China Emerging solar technologies [46]	
JA Solar Technology, China mc-Si, pc-Si, PERC technology modules/ panels [25, 39]	Microquanta Semiconductor China Perovskite and OPV [48]	
Risen Energy, China TopCon, mc-Si, half-cut cell and multi-busbar designs module [25]		
TATA Power Solar, India Bifacial Modules [35]		

(Continued)

Table 14.2. (*Continued*)

Terrestrial solar cell/module/panel		Space solar cells/module/panel
Silicon/thin film	Emerging technologies/tandem solar cell	III–V semiconductor
Adani Solar, India TopCon, Mono PERC [35]		
Reliance New Energy Solar PV & Module based on REC Technology [35]		
Waaree Energies, India Bifacial Solar Modules, Mono PERC Solar Modules, Polycrystalline Solar Modules, Flexible Solar Modules [35]		
Vikram Solar, India Bifacial Solar Modules, Mono PERC Solar Modules, Polycrystalline Solar Modules, Flexible Solar Modules [35]		
Emmvee Solar, India TopCon PV Modules [35]		
Goldi Solar, India [35] Monofacial & Bifacial Modules		
Jakson Group, India Monofacial & Bifacial Modules [35]		
GCL-Poly Energy Holdings Limited, China (Si-based technologies) [46]		
Alta Devices, USA (GaAs thin film solar cells) [47]		

technologies with a focus on cost-effective, high-efficiency thin-film technology. SunPower Corporation, USA, now Maxeon, Singapore (since August, 2020) [37] is known for producing high-efficiency solar panels using its proprietary Maxeon cell technology. It has a strong presence in both residential and commercial markets. REC Group, Singapore [38] is a leading European brand of solar panels known for its high-quality, high-efficiency solar products. The company has developed innovative technologies such as the REC Alpha series, which features high power density and excellent performance in real-world conditions. Hanwha Q CELLS, South Korea [25] is a major manufacturer of solar cells and modules, known for its high-efficiency Q.ANTUM technology which enhances cell efficiency by reducing recombination losses, and its products consistently rank among the highest in performance and reliability. JA Solar Technology, China [25] is a prominent manufacturer of high-performance solar cells and modules, with a strong global distribution network. JA Solar has made significant advancements in PERC technology and produces a range of high-efficiency c-Si and pc-Si modules. China Coal Energy Group ordered 4 GW of solar modules at $0.10/W, including 3 GW TOPCon and 1 GW PERC panels from JinkoSolar, Tongwei, JA Solar, and DAS Solar [39]. Risen Energy, China [25] is a leading manufacturer of solar PV products and has a strong focus on high-efficiency solar modules. The company has developed innovative technologies such as half-cut cell and multi-busbar designs to improve module efficiency and performance.

14.2.4.2 Emerging solar cell industries

Emerging solar cell technologies are pushing the boundaries of efficiency, cost-effectiveness, and application versatility. Several industries and companies are at the forefront of developing and commercializing these next-generation solar technologies, such as perovskite solar cells, organic photovoltaics (OPVs) including polymer solar cells, and tandem solar cells.

Tandem PV, founded in 2016, develops perovskite-silicon thin-film solar modules, currently achieving 28% efficiency and expected to surpass 30% by late 2025 [40]. Oxford PV, UK [41] is a leader in the development of PSCs. Oxford PV is focused on creating high-efficiency perovskite/silicon tandem solar cells that can surpass the performance of traditional silicon solar cells. Saule Technologies, Poland [42] is pioneering inkjet-printed flexible and transparent perovskite solar panels. Heliatek GmbH, Germany, [43] specializes in OPVs. Their products are lightweight, flexible, and can be integrated into building materials for BIPV. SolarWindow Technologies, USA [44] focuses on transparent, electricity-generating coatings for windows and other surfaces. Its technology aims to transform ordinary windows into solar power generators. While First Solar, USA [25, 36] is known for its thin-film CdTe solar cells, the company is also investing in the next-generation PV such as advanced thin-film and tandem cell technologies to improve the efficiency and reduce the cost. Meyer Burger Technology AG, Switzerland [45] is a technology leader in the solar industry, focusing on heterojunction and tandem solar cell technologies. GCL-Poly Energy Holdings Limited, China [46], a major player in the polysilicon and wafer production market, is also investing in emerging solar technologies to enhance efficiency and lowering the costs. Alta Devices, USA [47], a subsidiary of Hanergy, focuses on developing high-efficiency GaAs solar cells for both terrestrial and space applications. Microquanta Semiconductor, China [48] is involved in the development of advanced solar cell technologies, including perovskite and organic photovoltaics. Raynergy Tek, Taiwan [49] specializes in organic photovoltaic materials and solutions, aiming to bring flexible and lightweight solar products to market.

14.2.4.3 Space solar cells industries

Industries specializing in space solar cells are crucial for powering satellites, space stations, and other extra-terrestrial missions. These companies focus on developing highly efficient, lightweight, and durable solar cells that are capable of withstanding the harsh conditions of space.

Spectrolab, USA [50] is a leading manufacturer of space solar cells and panels. The company is renowned for its high-efficiency MJSCs, which are widely used in satellites and other space applications. SolAero, New Mexico, USA/Rocket Lab, Long Beach, California [51] is a prominent provider of high-efficiency solar cells, panels, and arrays for space and terrestrial applications. SolAero has a strong presence in the space industry, supplying solar cells for numerous space missions. Azur Space Solar Power GmbH, Germany [52] specializes in the development and production of MJSCs for space applications. The company is known for its high-efficiency GaAs-based MJSCs products and strong research capabilities. Emcore, USA [53] provides advanced solar cells and solar panel assemblies for space missions. Its products are known for high efficiency and reliability under the extreme conditions of space. Sharp, Japan [54] is a global electronics company specializing in high-efficiency solar cells for space applications, and has achieved a 33.66% power conversion efficiency for its silicon tandem solar cell. Sharp has supplied solar cells for several high-profile space missions. Examples of leading industries producing thin-film, silicon, and emerging technology-based terrestrial solar cells and III–V semiconductors-based space solar cells are listed in table 14.2.

14.3 Potential improvements in solar cell for future applications

Looking towards the future, there are several specific areas where improvements in solar cell technology could greatly impact their applications and effectiveness; some of them are discussed here. Enhancing the efficiency of solar cells under low-light conditions, such as during cloudy days or in urban environments with partial shading, would make photovoltaics more reliable and consistent [55]. Developing solar cells that can be seamlessly integrated into building materials like glass, concrete, or even textiles would enable widespread adoption without compromising architectural aesthetics. Innovations that lower the manufacturing costs of solar cells while maintaining or improving efficiency will help across the residential, commercial, and industrial sectors. Minimizing the environmental impact of solar cell production through the use of eco-friendly materials, improved recycling methods, and sustainable manufacturing processes is crucial for long-term viability [56]. Developing solar cells that are more durable and have longer lifespans will reduce maintenance costs and ensure reliable performance over decades. Innovations in manufacturing processes, such as roll-to-roll printing, nanostructuring, or 3D printing of solar cells, can lead to more efficient production and further cost reductions [57]. Continued research into new materials, such as perovskites or organic semiconductors, could unlock higher efficiencies and new functionalities in solar cell technology. Exploring the next-generation emerging concepts such as quantum dot solar cells, nanotechnology-based approaches, and new hybrid solar cell architectures are some future applications [58]. By focusing on these areas for improvement, solar cell technology can continue its trajectory towards greater efficiency, lower costs, enhanced durability, and expanded applications, thereby contributing to a more sustainable and resilient energy future.

14.3.1 Lab-scale to prototyping to commercialization

The journey from lab-scale development to prototyping and eventual commercialization of solar cells involves several critical stages that require meticulous research, testing, optimization, and validation [59–61].

14.3.1.1 Lab-scale research and development

Various materials (e.g., silicon, perovskite, organic compounds), for their photovoltaic properties, efficiency, and stability, continue to be explored for research. Small-scale solar cells are fabricated using deposition techniques (e.g., chemical vapor deposition, sputtering) and patterning processes (e.g., photolithography). PV cells undergo rigorous testing to evaluate efficiency, light absorption spectra, electrical properties (e.g., current–voltage curves, lifetime measurements), and durability under controlled lab conditions. Fabrication processes, material compositions, and device architectures are optimized based on experimental data to improve performance.

14.3.1.2 Prototyping and scale-up

Successful lab-scale solar cell prototypes having promising efficiency and stability are developed. Scalable manufacturing techniques are explored to replicate lab-scale processes at larger volumes without compromising performance. Initial pilot production runs are conducted to validate scalability, consistency in performance, and cost-effectiveness of manufacturing processes. Prototypes are integrated into module configurations suitable for larger-scale testing and validation in simulated operational environments.

14.3.1.3 Commercialization

Prototypes undergo extensive validation testing to meet industry standards and obtain certifications required for commercial deployment. Commercial production facilities are established or expanded to meet market demand, with a focus on optimizing yield, quality control, and production efficiency. Solar cell products are introduced to the market through partnerships with distributors, installation companies, and direct sales channels. Continued monitoring of field performance and customer feedback informs iterative improvements in product design, manufacturing processes, and reliability. Manufacturers leverage economies of scale, technological advancements, and supply chain optimizations to drive down costs and enhance competitiveness. Ensuring that materials maintain performance over extended periods in real-world conditions, including temperature variations, humidity, and UV exposure, is critical for transition from large-scale prototyping to commercialization.

Successful commercialization of solar cells involves collaboration across various interdisciplinary teams from material scientists and engineers to manufacturing experts and market strategists—to effectively translate innovative research into practical solutions that contribute to global energy sustainability goals [59–61].

14.3.2 Solar cell thermal management

Thermal management in solar cells is crucial for maintaining optimal operating temperatures, maximizing efficiency, and ensuring long-term reliability. Effective thermal management strategies help dissipate excess heat generated during solar cell operation, which can otherwise reduce efficiency and lifespan [62, 63]. Thermal fluctuations affect the electrical characteristics of solar cells, leading to variations in output power. Solar cell efficiency decreases with increasing temperature due to reduced open-circuit voltage and increased resistive losses [64, 65]. In addition, high temperatures accelerate degradation mechanisms in solar cells, such as thermal stress-induced microcracks, material diffusion, and increased corrosion rates. Passive or active heat sinks attached to the back of solar cells are used to increase the surface area for heat dissipation [66]. Other approaches for thermal management of solar modules include the following: integration of phase change materials that absorb and release heat during phase transitions to maintain stable temperatures [67]; circulation of cooling fluids (e.g., water or glycol) through integrated channels in solar panel designs to remove heat [68]; incorporation of ventilation channels or spacing between solar cells to facilitate natural convection cooling [69]. Selection of encapsulation materials with high thermal conductivity and durability to protect solar cells is important while facilitating heat transfer [70]. Optimal module tilt and orientation to maximize natural airflow [71], and coatings with high emissivity are some of the ways to reduce solar cell heating [72].

By implementing effective thermal management strategies, solar cell manufacturers and system integrators can enhance energy conversion efficiency, improve reliability, and ensure sustainable performance of solar energy systems in various applications while enhancing the lifespan of solar panels.

14.3.3 Advancements and innovations in contact replacements

In solar cells, metallic contacts are made at the front and rear of the device for efficient collection of charge carriers generated by sunlight to produce electrical current. Front contacts are typically made of silver (Ag) or aluminium (Al). In the case of Si, grid lines are patterned on the front side of solar cells to collect current generated by sunlight. Narrower lines are formed to maximize light absorption by minimizing shading and reflection losses. Rear contacts are usually made of aluminium or other conductive metals; these contacts cover the entire back surface of the solar cell to collect the charge carriers and transfer current. In some cases, selective emitter contacts are diffused or deposited on the back surface to improve carrier collection efficiency [73]. Printing techniques such as screen printing and inkjet printing are used to deposit conductive materials e.g. silver nanoparticles or conductive pastes as alternatives to traditional metal grid lines [74]; copper (Cu) and nickel (Ni) have been explored as replacements for Ag due to their lower cost and comparable conductivity [75]; challenges exist in stability and adhesion. Electroplating or electroless plating [76] of metals are utilized to form robust and low-resistance contacts to enhance durability and performance. Graphene is also proposed as a conductive material for contacts due to its high conductivity and mechanical flexibility [77]. Integration of nanowires or nanomaterials can also be used to improve charge carrier collection efficiency and reduce contact resistance [78].

Exploration of contacts which can withstand long-term exposure to environmental stresses (e.g., temperature variations, humidity) is crucial for the next-generation solar cell technologies such as tandem solar cells and perovskite-silicon hybrids [79, 80].

References

[1] Chapin D M, Fuller C S and Pearson G L 1954 A new silicon p-n junction photocell for converting solar radiation into electrical power *J. Appl. Phys.* **25** 676–7

[2] *Vanguard I the World's Oldest Satellite Still in Orbit* 2 May 2003 https://web.archive.org/web/20150321054447/http://code8100.nrl.navy.mil/about/heritage/vanguard.htm

[3] Luque A and Hegedus S 2011 *Handbook of Photovoltaic Science and Engineering* (New York: Wiley)

[4] Nayak P K *et al* 2019 Photovoltaic solar cell technologies: analysing the state of the art *Nat. Rev. Mater.* **4** 269–85

[5] Kant N and Singh P 2022 Review of next generation photovoltaic solar cell technology and comparative materialistic development *Mater. Today Proc.* **56** 3460–70

[6] Almosni S *et al* 2018 Material challenges for solar cells in the twenty-first century: directions in emerging technologies *Sci. Technol. Adv. Mater.* **19** 336–69

[7] Pastuszak J and Węgierek P 2022 Photovoltaic cell generations and current research directions for their development *Materials* **15** 5542

[8] Richter A, Hermle M and Glunz S W 2013 Reassessment of the limiting efficiency for crystalline silicon solar cells *IEEE J. Photovolt.* **3** 1184–91

[9] Suman P, Sharma and Goyal P 2020 Evolution of PV technology from conventional to nano-materials *Mater. Today Proc.* **28** 1593–7

[10] NREL 2024 NREL's Best Research-Cell Efficiency Chart 2024 https://www.nrel.gov/pv/cell-efficiency.html

[11] Preu R *et al* 2020 Passivated emitter and rear cell—devices, technology, and modeling *Appl. Phys. Rev.* **7** 041315

[12] Kashyap S *et al* 2020 Comprehensive study on the recent development of PERC solar cell *2020 47th IEEE Photovoltaic Specialists Conf. (PVSC)*

[13] Pakala P S A *et al* 2022 Comprehensive study on heterojunction solar cell *Machine Learning, Advances in Computing, Renewable Energy and Communication* (Singapore: Springer Singapore)

[14] Chuchvaga N *et al* 2023 Development of hetero-junction silicon solar cells with intrinsic thin layer: a review *Coatings* **13** 796

[15] NREL 2024 Champion Photovoltaic Module Efficiency Chart 2024 https://www.nrel.gov/pv/module-efficiency.html

[16] Singh B P, Goyal S K and Kumar P 2021 Solar PV cell materials and technologies: analyzing the recent developments *Mater. Today Proc.* **43** 2843–9

[17] Kuczyńska-Łażewska A, Klugmann-Radziemska E and Witkowska A 2021 Recovery of valuable materials and methods for their management when recycling thin-film CdTe photovoltaic modules *Materials* **14** 7836

[18] Dunlap-Shohl W A *et al* 2018 Synthetic approaches for halide perovskite thin films *Chem. Rev.* **119** 3193–295

[19] Khalid S *et al* 2021 Third-generation solar cells *Emerging Nanotechnologies for Renewable Energy* ed W Ahmed, M Booth and E Nourafkan (Amsterdam: Elsevier) ch 1 pp 3–35

[20] Wu C *et al* 2020 Multifunctional nanostructured materials for next generation photovoltaics *Nano Energy* **70** 104480

[21] Dambhare M V, Butey B and Moharil S 2021 Solar photovoltaic technology: a review of different types of solar cells and its future trends *J. Phys. Conf. Ser* **1913** 012053

[22] Topaz Solar Farm Wikipedia, The Free Encyclopedia https://en.wikipedia.org/w/index.php?title=Topaz_Solar_Farm&oldid=1269988126 (accessed 7 March 2025)

[23] *Copper Mountain Solar 1 Photovoltaic Plant, Nevada* 2012 https://www.energymonitor.ai/projects/copper-mountain-solar-1-photovoltaic-plant-nevada/

[24] Burgaleta J, Arias S and Ramirez D 2011 Gemasolar, the first tower thermosolar commercial plant with molten salt storage *SolarPaces Int. Conf. (Marrakech)*

[25] Power Technology 2016 The world's biggest solar photovoltaic cell manufacturers https://www.power-technology.com/features/featurethe-worlds-biggest-solar-photovoltaic-cell-manufacturers-4863800/?cf-view&cf-closed

[26] LONGi 2024 Wikipedia, The Free Encyclopedia. https://en.wikipedia.org/w/index.php?title=LONGi&oldid=1249092836 (accessed 7 March 2025)

[27] Gill T 2024 *The world's biggest solar farms* https://www.theecoexperts.co.uk/solar-panels/biggest-solar-farms

[28] Masters N S A J 2019 *Morocco in the fast lane with world's largest concentrated solar farm* https://edition.cnn.com/2019/02/06/motorsport/morocco-solar-farm-formula-e-spt-intl/index.html (accessed 7 March 2025)

[29] Whitlock R 2021 Sun Cable announces Global Expert Team to deliver the Australia-Asia PowerLink Project https://www.hatch.com/about-us/news-and-media/2021/10/sun-cable-announces-global-expert-team-to-deliver-the-australia-asia-powerlink-project

[30] Woolway R I *et al* 2024 Decarbonization potential of floating solar photovoltaics on lakes worldwide *Nat. Water* **2** 566–76

[31] *Fraunhofer Institute for Solar Energy Systems ISE* (2024) https://ise.fraunhofer.de/en/business-areas/photovoltaics-materials-cells-and-modules/silicon-solar-cells-and-modules.html

[32] *IEA—International Energy Agency* https://www.iea.org/data-and-statistics/charts/renewable-electricity-capacity-additions-by-technology-and-segment-2016-2028

[33] IEA-Renewable electricity capacity growth by country/region, main case, 2017–2030 https://www.iea.org/data-and-statistics/charts/renewable-electricity-capacity-growth-by-country-region-main-case-2017-2030

[34] de Giovanni A 2023 RECAI 62-Renewable Energy Country Attractiveness Index (RECAI) https://www.ey.com/en_gl/insights/energy-resources/are-the-global-winds-of-change-sending-offshore-in-a-new-direction

[35] Rumage J 2024 20 Solar companies in India to know *Built In* https://builtin.com/articles/solar-companies-in-india

[36] Roselund C 2016 First Solar sets new cadmium telluride thin-film cell efficiency record at 22.1% *pv Mag.* https://www.pv-magazine.com/2016/02/23/first-solar-sets-new-cadmium-telluride-thin-film-cell-efficiency-record-at-22-1_100023341/
[37] Solar Magazine 2022 SunPower solar panels: the leading option in the U.S. solar industry *Solar Mag.* https://solarmagazine.com/solar-panels/sunpower-solar-panels/
[38] PR Newswire 2017 REC group emerges as the largest European solar panel manufacturer in India and brings its latest innovative product range at 11th edition of Renewable Energy India Expo https://www.prnewswire.com/in/news-releases/rec-group-emerges-as-the-largest-european-solar-panel-manufacturer-in-india-and-brings-its-latest-innovative-product-range-at-11th-edition-of-renewable-energy-india-expo-645259313.html
[39] Shaw V 2024 Chinese PV industry brief: China coal secures 4 GW of modules at $0.10/W *pv Mag.* https://www.pv-magazine.com/2024/11/29/chinese-pv-industry-brief-china-coal-secures-4-gw-of-modules-at-0-10-w/
[40] Fischer A 2025 Tandem PV moves toward perovskite manufacturing with $50 million funding *pv Mag.* https://www.pv-magazine.com/2025/03/06/tandem-pv-moves-toward-perovskite-manufacturing-with-50-million-funding/
[41] Jowett P 2024 Oxford PV starts commercial distribution of perovskite solar modules *pv Mag.* https://www.pv-magazine.com/2024/09/05/oxford-pv-starts-commercial-distribution-of-perovskite-solar-modules/
[42] Spaes J 2021 Saule Technologies opens perovskite solar cell factory in Poland *pv Mag.* https://www.pv-magazine.com/2021/06/17/saules-technologies-opens-perovskite-solar-cell-factory-in-poland/
[43] Overton G 2018 Heliatek completes largest organic photovoltaics BIPV installation to date *Laser Focus World* https://www.laserfocusworld.com/detectors-imaging/article/16571605/heliatek-completes-largest-organic-photovoltaics-bipv-installation-to-date
[44] SolarWindow Technologies, Inc. 2020 SolarWindow first-ever: electricity-generating flexible glass using high-speed manufacturing process https://www.globenewswire.com/news-release/2020/11/23/2131850/0/en/SolarWindow-First-Ever-Electricity-Generating-Flexible-Glass-Using-High-Speed-Manufacturing-Process.html
[45] Meyer Burger Wikipedia The Free Encyclopedia, https://en.wikipedia.org/w/index.php?title=Meyer_Burger&oldid=1275001614
[46] Ng E 2024 China's GCL investing US$98 million in next-generation solar technology to revive sector South *China Morning Post Publishers* https://www.scmp.com/business/china-business/article/3273455/chinas-gcl-investing-us98-million-next-generation-solar-technology-revive-sector?module=perpetual_scroll_0&pgtype=article
[47] Alta Devices Wikipedia, The Free Encyclopedia, https://en.wikipedia.org/w/index.php?title=Alta_Devices&oldid=1189432914
[48] Shaw V and Hall M 2022 Chinese PV industry brief: microquanta builds 12 MW ground-mounted project with perovskite solar modules *pv Mag.* **February 2022** https://www.pv-magazine.com/2022/02/18/chinese-pv-industry-brief-microquanta-builds-12-mw-ground-mounted-project-with-perovskite-solar-modules/
[49] Bellini E 2020 Large-area organic PV module with 9.5% efficiency *pv Mag.* https://www.pv-magazine.com/2020/09/08/large-area-organic-pv-module-with-9-5-efficiency/
[50] Law DC *et al* 2012 Recent progress of Spectrolab high-efficiency space solar cells *38th IEEE Photovoltaic Specialists Conf.*

[51] Bellini E 2021 US aerospace manufacturer Rocket Lab takes over space solar cell maker Solaero *pv Mag.* **December 2021** https://www.pv-magazine.com/2021/12/16/us-aerospace-manufacturer-rocket-lab-takes-over-space-solar-cell-maker-solaero/

[52] Khorenko V *et al* 2017 BOL and EOL characterization of Azur 3G Lilt solar cells for ESA Juice Mission *E3S Web of Conf.* **16** 03011

[53] Sharps P R *et al* 2002 Ultra high-efficiency advanced triple-junction (ATJ) solar cell production at Emcore photovoltaics *IECEC '02. 2002 37th Intersociety Energy Conversion Engineering Conf.*

[54] Bellini E 2023 Sharp claims 33.66% efficiency for silicon tandem solar cell *pv Mag.* https://www.pv-magazine.com/2023/11/01/sharp-claims-33-66-efficiency-for-perovskite-silicon-tandem-solar-cell/

[55] Oni A M *et al* 2024 A comprehensive evaluation of solar cell technologies, associated loss mechanisms, and efficiency enhancement strategies for photovoltaic cells *Energy Rep.* **11** 3345–66

[56] Alhodaib A *et al* 2024 Sustainable coatings for green solar photovoltaic cells: performance and environmental impact of recyclable biomass digestate polymers *Sci. Rep.* **14** 11221

[57] Weerasinghe H C *et al* 2024 The first demonstration of entirely roll-to-roll fabricated perovskite solar cell modules under ambient room conditions *Nat. Commun.* **15** 1656

[58] Chen J *et al* 2021 Emerging perovskite quantum dot solar cells: feasible approaches to boost performance *Energy Environ. Sci.* **14** 224–61

[59] Wang F *et al* 2022 Recent progress of scalable perovskite solar cells and modules *Energy Rev.* **1** 100010

[60] Martulli A *et al* 2023 Towards market commercialization: lifecycle economic and environmental evaluation of scalable perovskite solar cells *Prog. Photovolt. Res. Appl.* **31** 180–94

[61] Parvazian E and Watson T 2024 The roll-to-roll revolution to tackle the industrial leap for perovskite solar cells *Nat. Commun.* **15** 3983

[62] Poudel S, Zou A and Maroo S C 2022 Thermal management of photovoltaics using porous nanochannels *Energy Fuels* **36** 4549–56

[63] Du Y 2017 Advanced thermal management of a solar cell by a nano-coated heat pipe plate: a thermal assessment *Energy Convers. Manage.* **134** 70–6

[64] Shaker L M *et al* 2024 Examining the influence of thermal effects on solar cells: a comprehensive review *Sustain. Energy Res.* **11** 6

[65] Singh P and Ravindra N M 2012 Temperature dependence of solar cell performance—an analysis *Sol. Energy Mater. Sol. Cells* **101** 36–45

[66] Krstic M *et al* 2024 Passive cooling of photovoltaic panel by aluminum heat sinks and numerical simulation *Ain Shams Eng. J.* **15** 102330

[67] Yazdani McCord M R and Baniasadi H 2024 Advancements in form-stabilized phase change materials: stabilization mechanisms, multifunctionalities, and applications—a comprehensive review *Mater. Today Energy* **41** 101532

[68] Lotfi M *et al* 2022 Cooling of PV modules by water, ethylene-glycol and their combination *Energy Environ. Eval.* **7** 1047–55

[69] Akrouch M A *et al* 2023 Advancements in cooling techniques for enhanced efficiency of solar photovoltaic panels: a detailed comprehensive review and innovative classification *Energy Built Environ.* **6** 248–76

[70] Chu Q-Q *et al* 2023 Encapsulation: the path to commercialization of stable perovskite solar cells *Matter* **6** 3838–63

[71] Yunus Khan T M *et al* 2020 Optimum location and influence of tilt angle on performance of solar PV panels *J. Therm. Anal. Calorim.* **141** 511–32

[72] Salehi S-D and Kingstedt O 2023 Critical assessment and demonstration of high-emissivity coatings for improved infrared signal quality for Taylor–Quinney coefficient experimentation *Int. J. Impact Eng.* **178** 104593

[73] Bilal B and Najeeb-ud-Din H 2021 Fundamentals of and recent advances in carrier selective passivating contacts for silicon solar cells *J. Electron. Mater.* **50** 3761–72

[74] Zhang J *et al* 2022 Silver nanoparticles for conductive inks: from synthesis and ink formulation to their use in printing technologies *Metals* **12** 234

[75] Unsur V 2024 Implementation of nickel and copper as cost-effective alternative contacts in silicon solar cells *Prog. Photovolt. Res. Appl.* **32** 267–75

[76] Hatt T *et al* 2021 Electroplated copper metal contacts on perovskite solar cells *Sol. RRL* **5** 2100381

[77] Jain P *et al* 2024 Recent advances in graphene-enabled materials for photovoltaic applications: a comprehensive review *ACS Omega* **9** 12403–25

[78] Tala-Ighil R 2015 Nanomaterials in solar cells *Green.Org* https://green.org/2024/01/30/nanotechnology-in-solar-cells-the-future-of-solar-energy/

[79] Arya S and Mahajan P 2023 Future in solar cell technology *Solar Cells: Types and Applications* ed S Arya and P Mahajan (Singapore: Springer Nature Singapore) pp 237–56

[80] Ahmad W *et al* 2024 Revolutionizing photovoltaics: from back-contact silicon to back-contact perovskite solar cells *Mater. Today Electron.* **9** 100106

Chapter 15

Challenges, opportunities and future directions

Various challenges in solar photovoltaics such as supply chain, vertical integration, and polysilicon production are addressed in this chapter, along with possible paths to future developments. Advancements in emerging technologies to reduce production costs for solar photovoltaics are described. The associated incentives, silicon wafer size, glass in solar panels, and workforce development are explored as ways to improve the manufacturing processes as well as the supporting infrastructure. The recycling and lifecycle analysis of solar panels is discussed, including the recycling of thin-film and monocrystalline silicon panels. The market growth and projections for the future of solar panels, as well as the expansion of the recycling industry, are addressed.

15.1 Introduction

Solar photovoltaic (PV) systems are gaining popularity as a source of green and sustainable energy to meet the energy demands across the world. While several forms of renewable energy have evolved in recent years, solar PV stands out from the others. Solar panels consist of multiple solar cells, convert solar energy into electricity, making them quieter and more environment friendly than fossil fuels. The yearly addition of solar capacity increased by 74% in 2023, attaining a record 346 GW (figure 15.1(a)) [1]. Although China was the main driver of this acceleration, 28 nations have added more than one gigawatt of additional solar capacity by 2023, demonstrating the global spread of PV solar panels' spectacular growth [1].

Furthermore, the International Energy Agency (IEA) reports that solar power now produces about 4% of the world's electricity, with global PV capacity rising from 1.4 GW in 2000 to 760 GW in 2020. Additionally, according to *World Energy Outlook 2019*, released by IEA [2], if countries follow specified policies and targets, solar PV could surpass coal and gas as the world's largest source of installed power capacity in the next 20 years (figure 15.1(b)).

doi:10.1088/978-0-7503-5994-8ch15

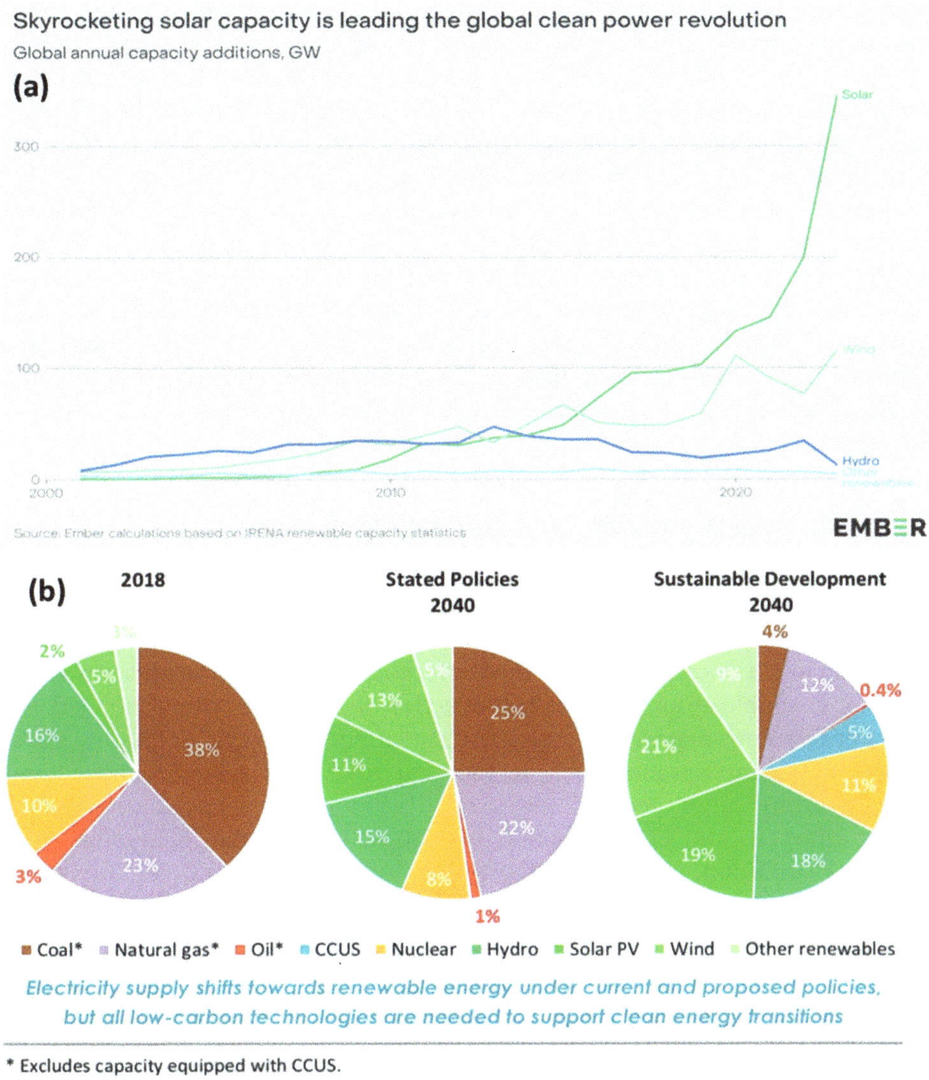

Figure 15.1. (a) Global solar installation. Source: Ember computations based on IRENA renewable capacity statistics. Reproduced from [1]. CC BY 4.0. (b) Global electricity generation by scenario (TWh). The IEA's 'Stated Policies' scenario takes into account current and previously announced policies, whereas the 'Sustainable Development' scenario is fully aligned with the Paris Agreement's goal of 'holding the increase in global average temperature to well below 2 °C above pre-industrial levels and pursuing efforts to limit the temperature increase to 1.5 °C above pre-industrial levels.' Reproduced from [2] CC BY 4.0.

The current PV market is dominated by crystalline silicon and thin-film-based solar panels. Over 95% of the worldwide PV market is made up of silicon cells, which include crystalline silicon (c-Si), multicrystalline (mc-Si) and polycrystalline silicon (pc-Si). These solar panels use silicon as a substrate and have small amounts of valuable metals implanted within it, such as silver (Ag), aluminium (Al), or copper (Cu). Commercial efficiencies of these cells have ranged from 15% to 24%, while the most recent c-Si cells have demonstrated 26.1% efficiency in a laboratory environment [3, 4]. Silicon solar panels are cost effective, long-lasting and modules are predicted to last at least 25 years. Thin film solar panels are based on thin layers of cadmium telluride (CdTe), copper indium gallium selenide (CIGS), and amorphous silicon (a-Si) on a substrate material such as glass, plastic, or metal. While CdTe-based solar panels are the second most common, made with low-cost production processes, their efficiency is not as high as that of silicon panels. However, recent advancements have brought the lower efficiencies of CdTe and CIGS solar cells closer to that of c-Si cells, with efficiencies of 22.1% and 23.4%, respectively [4–6]. In spite of these impressive efficiencies, the PV industry/solar panels still face numerous challenges for their widespread adoption and integration into the global energy system. This chapter will explore various solar PV-related challenges as well as potential future issues and concerns.

15.2 Solar panels—challenges and future directions

Despite the solar panel's sustainable and clean alternative to fossil fuels, there are several limitations in further reducing costs, improving long-term durability, and addressing environmental concerns related to certain materials, such as the toxicity of cadmium in CdTe cells [3, 6]. Furthermore, solar power is dependent on sunlight, so it is not available at night and is less effective on cloudy or rainy/snowy days [7]. This intermittency requires complementary energy storage solutions such as batteries or backup systems such as the grid to ensure a stable power supply. However, current battery technologies are expensive and have limitations in terms of capacity, lifespan, and environmental impact as well as the availability of raw materials. In addition, over time, solar panels lose efficiency and degrade at about 0.5%–1% per year due to degradation caused by exposure to sunlight, weather conditions, and other environmental factors [8]. Furthermore, solar panels require substantial space, to generate significant amounts of electricity, which may not be feasible in densely populated urban areas. Upfront cost of purchasing and installing solar panels can be high, although prices have been decreasing. This includes costs for the panels themselves, inverters, mounting systems, and the associated labor for installation and maintenance. While generally low, maintenance costs can add up over time, particularly for cleaning and occasional repairs. The payback period for solar investments can be long, often 5–10 years, depending on local electricity prices, government incentives, and energy consumption patterns. Proper disposal and recycling of old/used solar panels is critical to prevent environmental contamination. The adoption of solar energy often depends on government policies and incentives, which can vary widely between regions. Uncertain or inconsistent policies can hinder investment and deployment. Integrating solar power into existing electricity grids can be challenging [9, 10], requiring upgrades and changes to the grid infrastructure to handle distributed generation and variable power input. There can be resistance from communities and individuals due to aesthetic concerns, land use issues, or a lack of understanding of the benefits of solar energy. According to recent report [1],

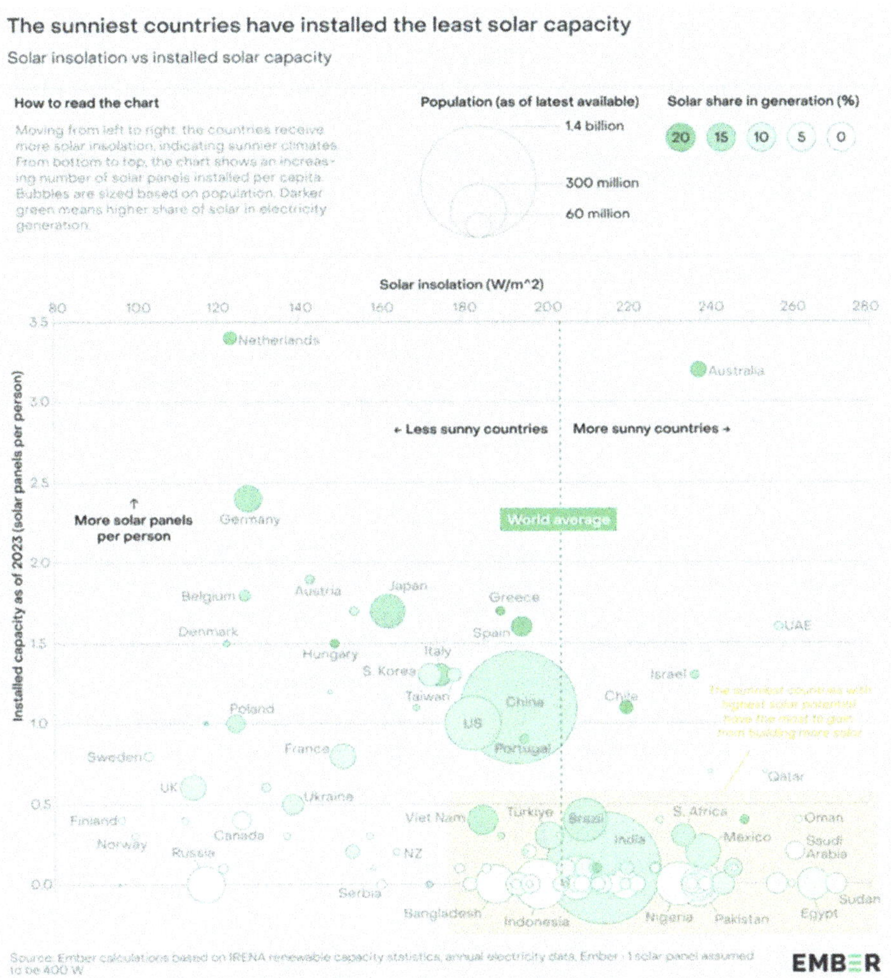

Figure 15.2. Solar insolation against installed solar capacity. The sunniest nations have installed least solar capacity. Source: Ember calculations based on IRENA renewable capacity statistics, yearly electricity data, Ember-1 solar panel considered to be 400 W. Reproduced from [1]. CC BY 4.0.

less solar has been installed in the sunniest countries, as shown in figure 15.2. As of 2023, only 14% of worldwide solar capacity (204 GW) was deployed in markets with solar insolation than the global average. Surprisingly, Japan has 41 times more solar panels per person than Egypt and 13 times more than India, even though a solar panel in these two more sun-filled nations would generate 64% and 32% more electricity, respectively [1].

Despite several challenges, the future of solar PV is promising, with several trends and innovations poised to address current limitations and expand their adoption. According to projections made by the IEA, 95% of the estimated increase in worldwide renewable power capacity by 2028 will come from solar PV and wind energy [11]. A global data and analysis organization has focused on energy transitions; Wood Mackenzie projects that over the next eight years, solar installations will rise at a stable average rate of about 350 GW annually between 2024 and 2032, with peaks of 371 GW in 2025 and 2030 [12].

Some key directions in which solar panel technology and deployment are likely to evolve are discussed in the following section.

15.2.1 Supply chain challenges

The supply chain for PV technologies is complex [13] and can be influenced by various factors, such as geopolitics and tariffs. These factors can impact the availability, cost, and reliability of materials and components that are needed for manufacturing and deploying solar panels. China dominates the global supply chain for PV technologies, manufacturing a significant majority of solar panels and components, including polysilicon, wafers, cells, and modules (figure 15.3) [14]. The strategic control of PV manufacturing by a single country can lead to supply disruptions in the case of geopolitical tensions or trade disputes. Certain materials essential for PV manufacturing, such as rare earth elements and specific metals (e.g., Cd, Te, In, Ga), are geopolitically sensitive. China and a few other countries control the majority of polysilicon production, which is critical for silicon-based solar panels [13]. Political instability in countries that are key suppliers of raw materials or components can disrupt the PV supply chain. For example, unrest in countries producing rare metals can lead to shortages and price spikes.

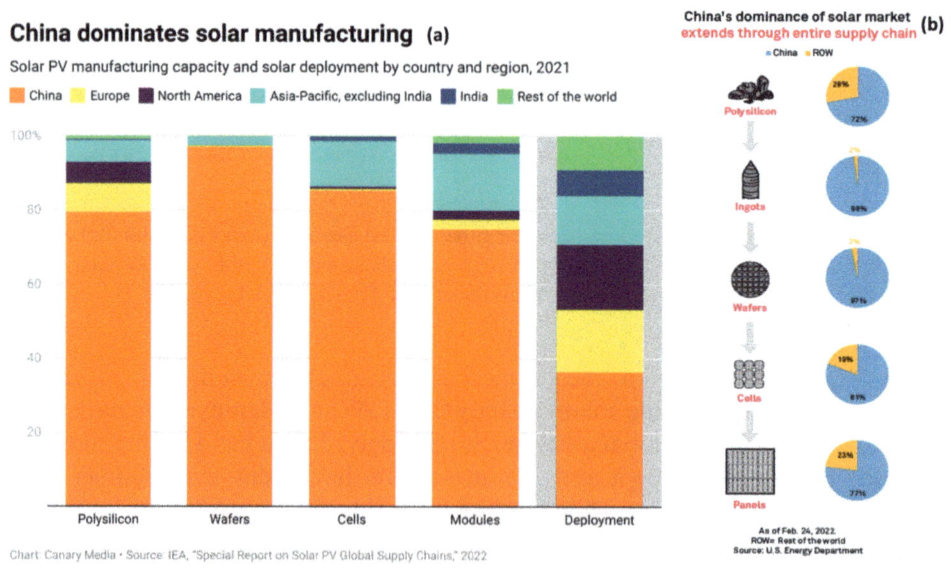

Figure 15.3. Solar manufacturing dominance of China (a) Solar PV manufacturing capacity and solar deployment by country and region, 2021. Reproduced from [13] CC BY 4.0. (b) China accounts for 70%–98% of global production capacity for silicon-based raw materials and solar panel components [15] image credit: US Department of Energy.

In recent years, trade tensions between U.S. and China have led to tariffs on Chinese solar panels and related components, increasing costs for U.S. solar developers and potentially slowing the adoption of solar technology [16]. Tariffs and trade barriers increase the cost of imported solar panels and components, affecting the overall economics of solar projects. In order to avoid tariffs, some companies may seek to diversify their supply chains by sourcing materials and components from other countries, potentially leading to higher costs. Furthermore, global supply chains are vulnerable to logistical challenges, such as shipping delays, port congestion, and transportation bottlenecks. Variability in quality can lead to performance issues and reduced lifespan of solar panels [13]. Lack of standardization in components complicates the integration and maintenance of solar systems. In addition, the extraction and processing of raw materials for PV manufacturing can have significant environmental impacts, including habitat destruction, water pollution, and greenhouse gas emissions. Effective recycling methods are necessary to handle the increasing volume of solar waste [17, 18]. Ensuring fair labour practices and avoiding forced labour in the supply chain is critical.

Diversifying suppliers, sourcing materials from multiple regions, developing local manufacturing capabilities and investing in recycling technologies can reduce dependence on single countries or companies and improve supply chain resilience. Governments can implement policies to support domestic PV manufacturing, including subsidies, tax incentives, and grants.

15.2.2 Vertical integration challenges

Vertical integration in the solar PV industry, where a company controls multiple stages of the supply chain from feedstock to ingot to wafer to solar cell to panel (figure 15.3(b)), can offer significant advantages but also comes with numerous challenges [19, 20]. Establishing facilities for each stage of the supply chain requires substantial capital investment. This includes setting up plants for polysilicon production, ingot casting, wafer slicing, cell manufacturing, and panel assembly [21] (figure 15.3(b)). Ensuring high quality and consistency across all production stages is critical. Defects in any stage can affect the performance and reliability of the final solar panels, necessitating stringent quality control measures. Efficient supply chain management is essential to minimize delays and maintain production schedules. The solar market is subject to fluctuations in demand due to policy changes, economic conditions, and technological advancements. Vertically integrated companies must be able to adapt to these changes to avoid overproduction or shortages. If one segment of the supply chain experiences a downturn, it can affect the entire vertically integrated operation. While vertical integration can lead to efficiencies, it may also reduce flexibility. Companies may find it harder to quickly adopt new technologies or processes if they have already invested heavily in existing infrastructure [20].

Vertical integration in the solar PV industry offers potential benefits such as cost savings, improved quality control, and supply chain efficiencies. However, it also presents significant challenges, including high capital investment, technological complexity, quality assurance, and market dynamics. Companies must carefully weigh these factors and implement strategic measures to mitigate risks and leverage the advantages of vertical integration.

15.2.3 Polysilicon production challenges

Obtaining polysilicon for PV production presents a variety of challenges that span economic, environmental, technological, and geopolitical dimensions [22, 23]. These problems can significantly impact the solar industry's supply chain and overall market dynamics. The production of polysilicon, especially through the Siemens process, is highly energy-intensive, often sourced from non-renewable energy, leading to a substantial carbon footprint [24–26]. The production process can emit hazardous by-products like silicon tetrachloride and greenhouse gases if not properly managed, contributing to environmental pollution. The electricity costs for maintaining the necessary temperatures and reactions are substantial. Setting up and maintaining polysilicon production facilities require significant capital investment in advanced technology and infrastructure. The price of polysilicon is subject to significant fluctuations based on supply and demand dynamics. Factors such as production capacity changes, new market entrants, or regulatory shifts can cause sudden price changes. Enhancing the yield and efficiency of polysilicon production is an ongoing challenge. High purity levels (99.9999% or higher) are necessary for solar-grade polysilicon [27–29]. China dominates the global polysilicon market (figures 15.3 and 15.4), accounting for a significant majority of production.

Investing in local polysilicon production facilities can decrease dependence on imports and enhance supply chain resilience. Developing and adopting more efficient production methods, such as the fluidized bed reactor (FBR) process, can enhance efficiency and lower costs [30, 31]. Using renewable sources and improving energy efficiency in production processes can reduce the carbon footprint and production costs. Developing international trade policies that promote fair and open markets for polysilicon can stabilize supply chains and reduce price volatility.

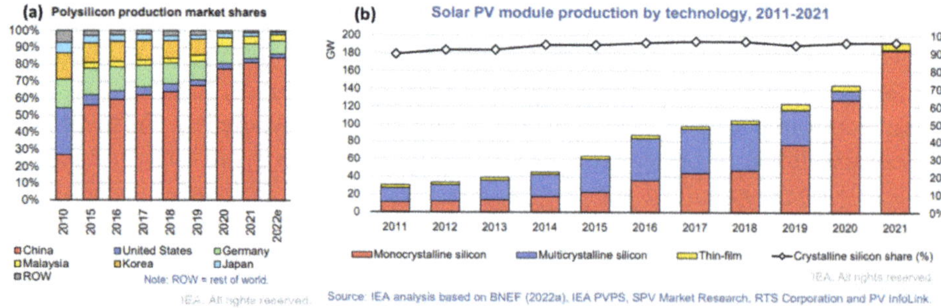

Figure 15.4. (a) Market share of global producers of polysilicon (b) Production of solar photovoltaic modules using thin film, multicrystalline, and monocrystalline silicon solar cells. Crystalline silicon has accounted for almost 95% share of the market. Reproduced from [13] CC BY 4.0.

15.2.4 Methods to reduce cost of manufacturing

Reducing the cost of manufacturing solar cells is critical for making solar energy more competitive with conventional energy sources. Manufacturers employ various strategies and technologies to achieve cost reductions throughout the production process.

15.2.4.1 Advancements in emerging technologies

The recent developments in emerging PV cell materials and designs include perovskite solar cells (PSCs) having efficiency of more than 25% [32], organic PV cells having ~15%–18% efficiency [33] and rapidly increasing efficiency of quantum dot solar cells [34]. As discussed in chapter 14, second, third and fourth generation PV technologies require only a fraction of the semiconductor material compared to the traditional silicon-based cells; hence, they are a suitable option for lowering cost. PSC has been one of the most exciting recent breakthroughs [35, 36] and is comparable to the best silicon-based cells, and researchers are aiming to increase their long-term stability [36]. Perovskites can absorb light across a broad range of the solar spectrum, from the visible to the near-infrared [37, 38] (figure 15.5), enhancing their potential conversion efficiency. By stacking multiple layers of different materials, each capturing a different part of the solar spectrum, multijunction solar cells can achieve very high efficiencies, potentially over 40% [32, 39, 40]. In perovskite silicon tandem solar cells, the perovskite top cell absorbs high-energy photons, while the silicon bottom cell absorbs lower-energy photons, maximizing the utilization of the solar spectrum and increasing efficiency [39, 40]. The material stability of perovskite and its ability to match the solar spectrum, especially in the form of heterojunction solar cells, present both challenges and opportunities [41]. Perovskites can degrade when exposed to moisture, oxygen, light, and heat [42]. This degradation can lead to the formation of non-perovskite phases, reducing their efficiency and lifespan. The migration of ions within the perovskite structure under operational conditions can lead to

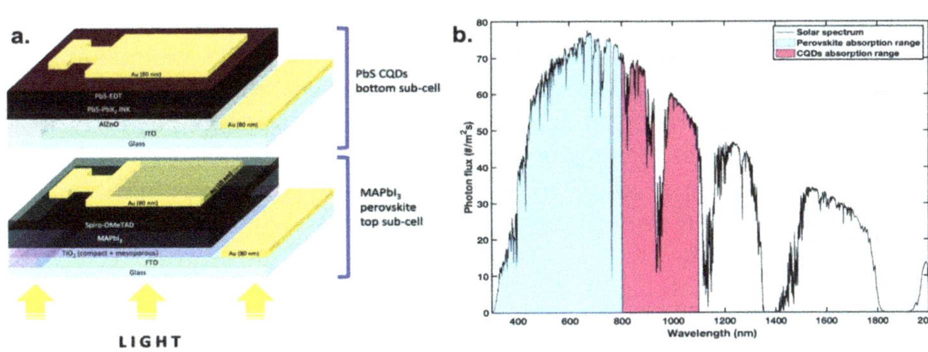

Figure 15.5. Perovskite quantum dot tandem design. (a) Schematic illustration of a 4-terminal (4T) tandem device, consisting of a $MAPbI_3$ perovskite front sub-cell and a lead sulfide (PbS colloidal quantum dot (CQD) back sub-cell. (b) Solar spectrum highlighting the light absorption regions corresponding to both the perovskite and CQD solar cells. Reprinted with permission from [38], copyright 2021 Royal Society of Chemistry. CC BY-NC 3.0.

phase segregation and instability. This necessitates effective encapsulation and barrier technologies. Elevated temperatures accelerate degradation processes; therefore, thermal stability is crucial for the practical deployment of perovskite solar cells [43]. Adjusting the composition of perovskites, such as by incorporating mixed cations (e.g., formamidinium and cesium) or halides (e.g., iodide and bromide), can enhance their stability. Surface passivation techniques reduce defect states and ion migration, improving stability [44]. Developing advanced encapsulation methods to protect perovskite layers from environmental factors can significantly extend their operational lifetime [45]. Significant advancements have been made in the field of PSCs by researchers at the Australian Institute for Bioengineering and Nanotechnology, who modified a nanomaterial and used doping techniques to increase the efficiency and thermal stability of perovskite solar cells [46]. Research into new perovskite compositions that offer better stability and performance is ongoing. This includes exploring double perovskites, lead-free perovskites, and other novel materials.

Nanostructuring perovskite layers can enhance light absorption and improve charge carrier dynamics. Despite being an amazing solar material, silicon can only capture a portion of the sun's total light. Recent study [38] integrates a methylammonium lead iodide (MAPbI3) perovskite solar cell as the front sub-cell and a lead sulfide (PbS) colloidal quantum dot (CQD) as the back sub-cell (figure 15.5(a)), enabling efficient and broad spectral absorption from visible to near-infrared light (figure 15.5(b)). The perovskite absorbs light up to $\sim$800 nm, while the CQDs extend absorption to $\sim$1100 nm. To enhance photon transmission to the back sub-cell, the interlayer is modified by incorporating a semi-transparent gold (Au) electrode on the perovskite cell, followed by a molybdenum (VI) oxide (MoO_3) or ionomer resin interlayer. This study emphasizes the interlayer's key role in connecting PSCs and CQDs for efficient tandem photovoltaics. Enhancing perovskite structures or quantum dot sizes can further boost performance, making these devices a promising solution for next-generation solar technology [38].

The development of scalable manufacturing processes, such as roll-to-roll printing and vapor deposition, is crucial for the commercial viability of PSCs [47]. Focusing on reducing material and production costs while maintaining high efficiency and stability will drive the adoption of perovskite technologies. Perovskites hold immense promise for the future of solar energy due to their high efficiency and tunable bandgap properties. Through ongoing research and innovation in composition engineering, encapsulation, and interface optimization, PSCs can become a mainstream technology, contributing significantly to the global renewable energy landscape.

Advances in materials relating to thin-film solar cells and organic photovoltaics (OPV) are leading to the development of lightweight and flexible solar panels that can be integrated into a variety of surfaces, including building facades, windows, and even clothing [33, 48]. OPV cells degrade rapidly due to breaking of their weak bonds under the influence of high-energy UV photons, leading to a decline in their performance over time. However, the lifespan of organic solar cells has been increased to 30 years by researchers at the University of Michigan [49] by adding buffer layers to stop internal degradation and protective layers to screen against UV rays. This breakthrough enables their use in applications such as power-generating windows [49]. The invention of bifacial panels which capture sunlight on both sides, has increaed their overall energy output. They are particularly effective in areas with high albedo surfaces (reflective ground surfaces) [50]. Building integrated photovoltaics (BIPV) involves the integration of solar panels into building materials, such as roof tiles and facade elements; it is becoming more popular, allowing for seamless integration of solar power into urban environments [51, 52]. Agrivoltaics [53] is a combination of solar panels with agriculture-related activities; this approach uses dual-use land to produce food and generate electricity simultaneously, optimizing land use and creating additional revenue streams for farmers. Installation of solar panels on bodies of water floating solar farms, such as reservoirs and lakes, is a growing trend. Floating solar farms [54] can reduce land use and potentially increase panel efficiency due to the cooling effect of water. Sustainable manufacturing efforts are being implemented today to reduce the environmental impact of solar panel manufacturing, such as using less energy-intensive processes and more eco-friendly materials; such efforts are gaining traction in solar manufacturing throughout the world [55].

Improved energy storage such as solid-state batteries is being developed, and promises higher energy densities, longer lifespans, and improved safety compared to the current lithium-ion batteries. Innovations in large-scale storage technologies, such as flow batteries [56] and advanced compressed air storage [57] are essential for stabilizing power grids with high solar penetration. Energy storage and grid-forming technologies need to be integrated more and more as solar installations proliferate rapidly in order to maintain grid stability and allow for the continued use of renewable energy sources [58]. Continued advancements in manufacturing processes, economies of scale, and material science are expected to further reduce the cost of solar panels and associated components.

15.2.4.2 Improvements in manufacturing processes

In the production of c-Si solar cells, multi-wire sawing and wafer processing techniques have reduced material loss and costs. Roll-to-roll deposition in thin-film manufacturing has enhanced efficiency and scalability while lowering production costs [6, 47]. Implementation of automated manufacturing processes such as robotic handling and precision laser cutting reduces labor costs, improves precision, and increases production throughput [59]. Real-time monitoring and quality control systems minimize defects, reduce rework, and improve overall yield. Artificial intelligence (AI) and machine learning (ML) techniques can be used to optimize production parameters, predict equipment maintenance needs, and improve quality control for higher yields. Advancements in cell architectures (e.g., PERC, TopCon, bifacial cells), anti-reflection coatings, and light-trapping techniques can be utilized to produce efficient solar cells. Optimizing the use of raw materials and consumables reduces costs that are associated with material procurement and waste disposal. Examples of these include exploring cheaper alternatives such as using Cu instead of Ag without compromising performance or reliability or exploring non-toxic and abundant materials for encapsulation and backsheet materials. Chemical vapor deposition, physical vapor deposition, and screen printing techniques can be used for applying conductive pastes. Streamlining supply chain logistics and negotiating favourable terms with suppliers reduces material costs.

Expansion of manufacturing facilities, investments in larger production lines, and standardized module designs are some of the future options. Frameless module designs, integration of mounting hardware during module assembly, and modular component designs for easier maintenance can be explored.

15.2.4.3 Silicon wafer thickness and size

From a materials perspective, reducing the thickness of silicon wafers and scaling from M10 to G12 and beyond, including the shift to rectangular wafers, involves several considerations and potential benefits [60, 61]. Thinner wafers use less silicon, which can significantly reduce material costs and lead to better light absorption and improved efficiency if they are integrated with advanced light-trapping techniques [62, 63]. Thinner wafers are more prone to cracking and breaking during manufacturing and handling [63]; hence, advances in wafer handling technology are necessary to mitigate these risks. Figure 15.6(a) depicts the evolution of wafer size and thickness of silicon [64]. Moving to a larger wafer size e.g., M10 to G12 increases the active area of each wafer, which can lead to higher power output per wafer. Larger wafers reduce the number of wafers required for a given power output, potentially improving manufacturing throughput and reducing costs (figure 15.7). Wafer sizes in the silicon solar PV sector have grown from M2, M4, G1, M6, to M10 and M12 (G12) [65] (figures 15.6 (a) and 15.7). This scaling also allows for more efficient use of the module's frame and overall area. Scaling to larger wafers necessitates changes in the production equipment and processes. The size of square wafers is increasing, ranging from 125 to 166 mm and 182 to 210 mm, as shown in figure 15.6(b) [66]. Prior to 2010, c-Si wafers were dominated by size 125 mm × 125 mm (165 mm silicon ingot diameter) (figure 15.6(a)). In the year 2013 and 2017, following the development of aluminium back surface field solar cell technology, 156 mm (200 mm silicon ingot diameter) 'M0' and 156.75 mm 'M2' wafers dominated the market [67] (figure 15.7) due to their reduced cost per watt. Subsequently, the wafer diameters of 158.75 mm and M6 wafers of 166 mm are developed. The transition to mono PERC (passivated emitter rear contact) technology in 2018 marked the start of rapidly evolving various formats on the market. Currently, M6 (166 mm), M10 (182 mm), and G12 (210 mm) wafers hold significant market shares, with the latter two being the most commonly used (figure 15.7) [65, 67].

The shift to thinner, larger, and rectangular silicon wafers in PV technology holds significant promise for improving efficiency and reducing costs, but it requires addressing challenges related to mechanical stability, manufacturing processes, and material handling. Rectangular wafers can reduce wasted space in solar modules, leading to higher packing density and increased module efficiency [63, 66]. Rectangular shapes simplify the manufacturing process, especially in the assembly of solar modules, potentially reducing production time and costs. Ensuring consistent doping and surface texturing across larger and thinner wafers is crucial for maintaining efficiency and yield. Thinner wafers can be more susceptible to thermal stress [63]. Advanced cooling and thermal management strategies are needed to maintain

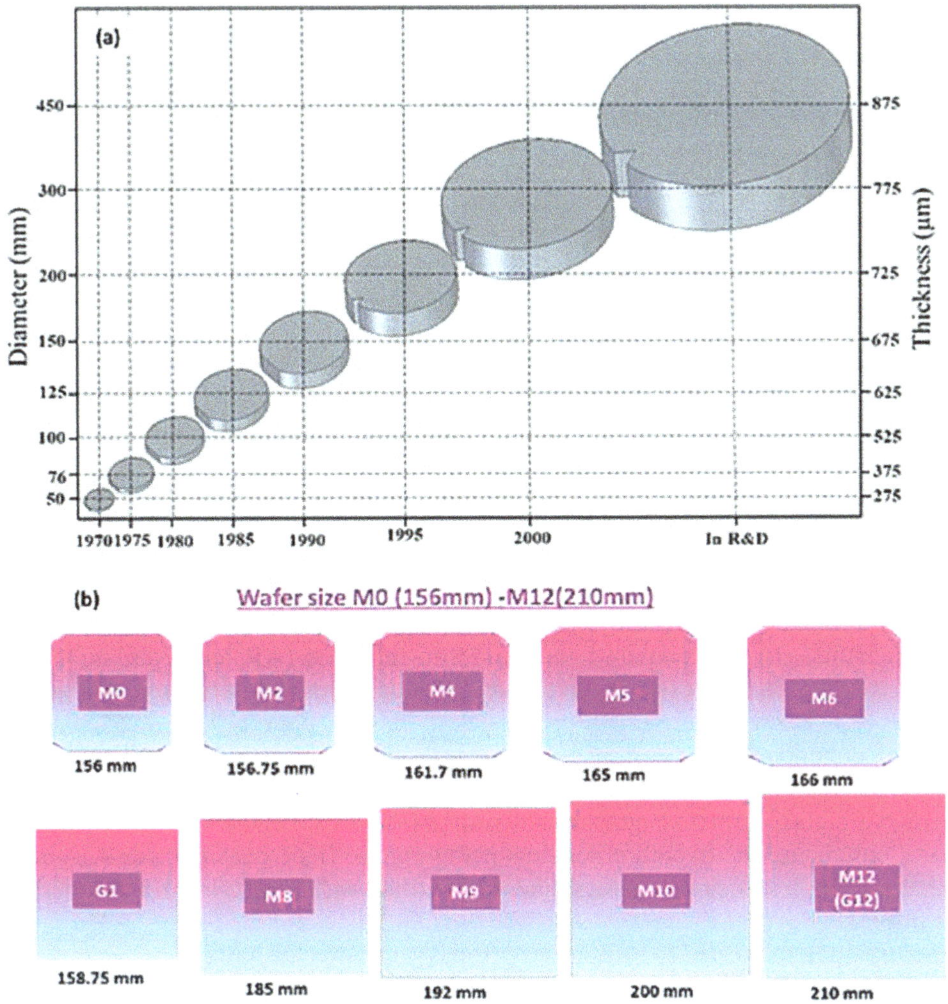

Figure 15.6. (a) Evolution of silicon wafer size and thickness. Reproduced from [64] CC BY 4.0. (b) Schematic representation of wafer size from 156 mm (M0) to 210 mm (M12).

performance and reliability. Diamond wire sawing technique [68] allows for more precise and less wasteful cutting of silicon ingots into thinner wafers. Kerfless wafer technologies [69] such as epitaxial growth on reusable substrates or direct wafer bonding produces thin wafers without traditional sawing, reducing material waste and cost. Advanced passivation and anti-reflective coatings are used to enhance the performance of thinner wafers by reducing surface recombination and improving light absorption.

15.2.5 Glass in solar panels

Glass in solar panels is a critical component, providing protection for PV cells while ensuring long-term durability and efficiency [70]. In order to withstand hail, constant

Figure 15.7. PV module production in relation to the size of wafers and the advancement of cell technology. Reproduced with permission from [67]. Credit: InfoLink Consulting.

interaction with nature, and ensuring 30-year guarantees, several considerations and advancements in materials and design are necessary. Several types of glasses are used in solar panels such as tempered glass and anti-reflective (AR) coated glass [70, 71]. Tempered glass is about four to five times stronger than standard annealed glass. It undergoes a thermal tempering process that induces compressive stresses on the surface, enhancing its resistance to impact and thermal stress. When broken, tempered glass fractures into small, blunt pieces, reducing the risk of injury. AR coatings on glass improve light transmission to the solar cells by reducing reflection losses, enhancing the overall efficiency of the panel. These coatings are designed to withstand environmental factors like UV radiation, moisture, and temperature variations.

Solar panels are tested to meet international standards for hail resistance (e.g., IEC 61215), which involves shooting ice balls of specific sizes and speeds at the glass to simulate hail impact [72]. High-quality tempered glass can withstand hailstones up to 25 mm in diameter impacting at speeds of up to 83 km h^{-1}. The glass and encapsulant materials should resist moisture ingress and chemical degradation. Ethylene-vinyl acetate (EVA) and polyvinyl butyral (PVB) are common encapsulants that provide good protection [73] to the panel. Using a combination of tempered glass layers laminated with a polymer interlayer (PVB or EVA) can further enhance impact resistance and prevent shattering. UV-resistant coatings and materials prevent degradation due to prolonged exposure to sunlight. Hydrophobic or superhydrophilic coatings [74] can help keep the glass surface clean by preventing the buildup of dirt and dust, ensuring optimal light transmission. These coatings must withstand environmental wear and tear over the lifetime of the panel.

Future developments include ion-exchange strengthened glass [75, 76] which undergoes a chemical strengthening process, resulting in higher strength compared to tempered glass. Ultra-thin, flexible glass materials are being explored for applications requiring lightweight and bendable solar panels. Nanocoatings [77] are designed to enhance light absorption and provide superior resistance to scratches, dirt, and further improve panel efficiency. Bifacial panels, which generate power from both sides, often use double-glass configurations [78]. Advances in glass technology for these panels enhance durability and efficiency while providing additional energy yield. Continuous innovations in materials and manufacturing processes are essential to meet the stringent requirements for 30-year guarantees in diverse environmental conditions.

15.2.6 Workforce development

A trained workforce that appreciates the benefits of renewables, particularly solar PV, is crucial for the sustained growth and adoption of RE technologies [79]. Vocational schools and training centres that are focused on PV technologies need be established. These programs should offer hands-on training in solar panel installation, maintenance, and repair. Certification programs for solar installers, technicians, and engineers should be developed and promoted to ensure that workers have the necessary skills and knowledge to perform their jobs effectively [80]. Workshops, seminars and campaigns should be conducted to educate the public about the economic, environmental, and social benefits of solar PV. Collaboration between academic institutions, industry, and government agencies should be encouraged and funding for research and development and incubators need to be established to support startups and entrepreneurs in the solar PV sector.

Building a trained workforce that appreciates the benefits of solar PV involves a multifaceted approach, including comprehensive education and training programs, supportive policies, public awareness campaigns, and strong industry partnerships.

15.2.7 Role of incentives

Incentives play a critical role in the adoption and growth of solar PV technology. They can significantly influence the economics of solar projects, making them more attractive for both residential and commercial users [81, 82]. Financial incentives such as the investment tax credit (ITC) in the United States allows homeowners and businesses to deduct a percentage of the cost of installing a solar energy system from their federal taxes. This has been a major drive of solar adoption. Direct cash incentives are provided by government programs or utilities to offset the initial cost of solar installations. Feed-in tariffs (FITs) and regulatory incentives are the payments made to solar energy producers for the electricity they generate and feed into the grid. This reduces their overall electricity bills and improves the return on investment. Renewable portfolio standards (RPS) mandates that a certain percentage of a utility's energy mix comes from RE sources, often including specific targets for solar PV. This creates a stable market for solar power. Solar renewable energy certificates (SRECs) represent the environmental attributes of solar power. System owners can sell SRECs to utilities that need to meet RPS requirements, providing an additional revenue stream. Options such as solar leases and power purchase agreements (PPAs), where a third party owns the solar system and the customer pays for the electricity produced, often at a lower rate than traditional grid electricity, are some of the incentives.

Incentives significantly lower the initial cost of solar PV systems, making them more affordable for a broader range of consumers. Grants, rebates, and tax credits can reduce the effective cost by 20–50% or more, depending on the specific program and region. By reducing upfront and operational costs, incentives improve the return on investment (ROI) and shorten the payback period for solar installations. Performance-based incentives and net metering ensure that solar system owners receive ongoing financial benefits, enhancing the overall economic case for solar. Financial and regulatory incentives create a favorable environment for innovation in solar technology and related industries. By promoting solar energy adoption, incentives contribute to national and international climate targets. Many governments are enhancing their renewable energy policies, including tax credits, subsidies, and feed-in tariffs, which will drive further adoption of solar technology [83, 84]. In the United States, ITC (Investment Tax Credit) has been playing an important role in solar growth, along with state-specific programs like California's solar initiative and New Jersey's SREC market. In Germany, the country's feed-in tariff program, which guaranteed payments for solar electricity fed into the grid, led to a significant increase in solar installations. In China, government subsidies and favorable policies have made China the world leader in solar PV manufacturing and deployment. More businesses are committing to renewable energy targets, leading to increased investment in solar installations, both on-site and through PPAs.

By implementing the abovementioned methods and strategies, solar cell manufacturers can achieve significant reductions in manufacturing costs, making solar energy more accessible and competitive in the global energy market.

15.3 Recycling of solar panels

Solar panels generate electricity safely and without emitting any pollutants into the atmosphere while in operation. However, when solar panels reach the end of their useful life, they, like any other source of energy, generate waste that must be recycled or disposed of correctly. The number of end of life panels will increase along with the solar PV market [85]. The most widely used solar panels are based on thin films and c-Si. [23]. It has been predicted by 2030 that 8 million metric tons of these panels are expected to reach the end of their useful lifespan, whereas by 2050, this number is expected to increase to 80 million metric tons [86, 87]. It is anticipated that the United States will generate up to one million tons of discarded solar panel waste by 2030 and is predicted to have up to 10 million tons by 2050, making it the second largest country in the world. The US Environmental Protection Agency (EPA) has proposed a rule to include solar panels in the universal waste regulations and is engaged in a new rulemaking initiative to enhance the recycling and management of end-of-life solar panels [88]. Technologies that promise to recover considerably more of the valuable elements from PV cells, while lowering the costs and negative environmental effects of recycling, are being developed. According to NREL, recycling a silicon PV module in the US costs roughly $15–$45, while disposing of it in landfill only costs $1–$5. Consequently, more landfill limitations, combined with less expensive methods that increase the economic worth of PV waste, could help shift the balance in favour of recycling [86, 87]. By 2050, 60 million tons of waste from PV panels would be in landfills due to a lack of recycling procedures. Since all PV cells contain some amount of harmful material, this would make using PV panels a source that is less environment friendly [85]. As the number of decommissioned solar panels grows, efficient recycling methods become increasingly important. Recycling solar panels involves several steps to recover valuable materials and reduce environmental impact. A recent study [89] has introduced a simple salt-etching method for recycling Ag and Si from end-of-life Si solar panels without using toxic mineral acids or causing secondary pollution. The process utilizes molten hydroxide to efficiently etch surface layers, enabling direct Ag separation from Si wafers in just 180 seconds, achieving recovery rates of >99.0% for Ag and >98.0% for Si. Additionally, Cu, Pb, Sn, and Al are recovered through oxidation, alkaline leaching, and electrodeposition. This approach offers a sustainable solution for solar panel recycling, contributing to a circular economy.

Generally, used or damaged solar panels are collected from various sources such as residential, commercial, and industrial installations [89–91]. Collaboration with solar installers, manufacturers, and waste management companies facilitate efficient collection and transport. The panels are then transported to recycling facilities and inspected for condition and type. Functional panels may be refurbished and resold, while non-functional ones are sorted for recycling. Crystalline silicon and thin-film solar panels are separated and components such as glass, metals, and semiconductor materials are identified [89]. Hazardous materials, such as lead in solder is removed to ensure safe processing. Some of the techniques for recycling used solar panels are discussed in the following section.

In mechanical recycling, panels are first manually disassembled to separate the aluminium frame and junction box. The remaining module is crushed into smaller pieces, typically through shredding or milling [17, 89] (figure 15.8(a)). Mechanical processes such as vibration, magnetic separation, and eddy current separation methods are used to segregate materials like glass, plastics, and metals. Mechanical recycling is simple and cost-effective and can recover a high percentage of materials, especially glass and metals. However, it is less effective in recovering high-purity silicon and other valuable materials and produces secondary waste that requires further processing. Another method is thermal recycling, in which panels are subjected to high temperatures in a furnace to break down the encapsulating materials (such as EVA) and separate the silicon cells from other components. After burning off the encapsulating materials, the silicon wafers, metals, and glass are recovered. This method is effective in breaking down and separating materials and recover high-purity silicon [17, 85]. However, it is an energy-intensive process and has potential release of harmful emissions if not properly controlled. In chemical recycling, chemical solutions are used to dissolve the encapsulating materials and separate the silicon cells. Chemical treatments are used to purify the silicon, recover metals like silver, and separate other materials. This method has achieved higher recovery rates for silicon and other valuable materials and also produces high-purity recovered materials [17, 85]. Nevertheless, it involves hazardous chemicals that require careful handling and

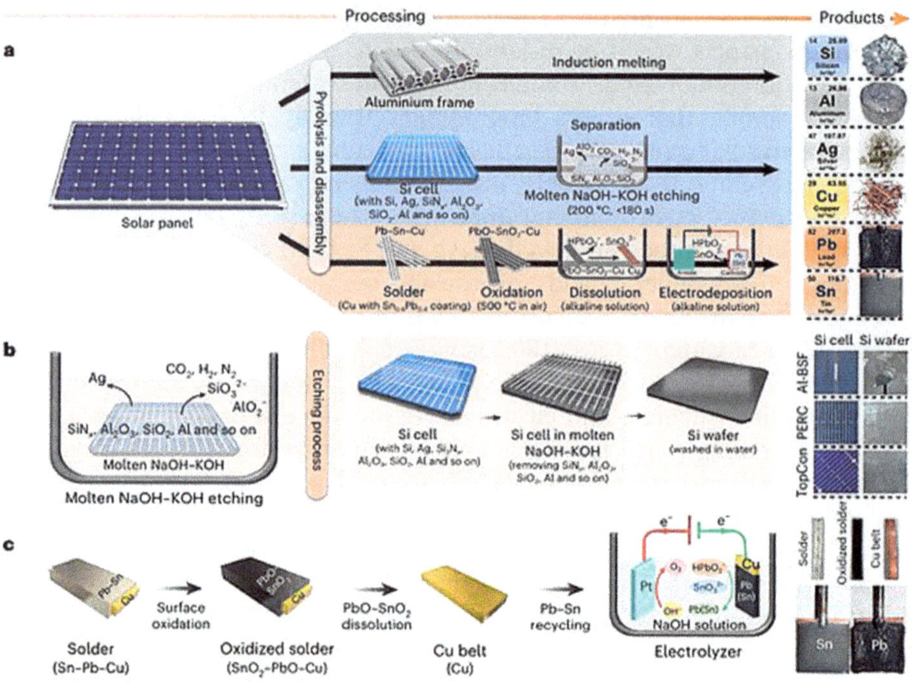

Figure 15.8. Schematic representation of end-of-life PV panel recycling and recycled products. (a) Flowchart illustrating the recycling process of end-of-life PV panels, encompassing initial dismantling, induction melting of Al frames and glass, salt-etching, and solder recycling. The images on the right display the recovered materials. (b) Schematic depiction of the salt-etching process in molten NaOH-KOH. (c) Schematic representation of the solder recycling process, including oxidation, alkali leaching, and electrodeposition. Reprinted with permission from [89], copyright 2024 Springer Nature.

disposal and can be expensive due to the cost of chemicals and the need for specialized equipment. Electrochemical method is used to selectively dissolve and recover metals from the panels and is helpful in purifying silicon and other materials. Although this method is effective in recovering metals like silver and copper and produces high-purity materials, it requires specialized equipment and processes and can be costly and complex. A hybrid method employs a combination of mechanical, thermal, chemical, and electrochemical processes to maximize material recovery. For example, mechanical separation followed by chemical etching or thermal treatment to recover high-purity materials is a good option. This maximizes material recovery and purity and can be tailored to specific types of solar panels and materials. However, it is more complex and costly compared to single-method approaches and requires coordination of multiple processes and technologies. Emerging technologies such as laser technology [90] are used to selectively remove and recover materials from solar panels. This method is promising in reducing the use of chemicals and improving material recovery rates. Research is ongoing into the use of biological agents (e.g., microorganisms) to break down and separate materials in an environmentally friendly way.

15.3.1 Recycling of silicon solar panels

The methods described above are employed in recycling of silicon panels, beginning with disassembling the device and then separating the aluminium and glass components [17, 85] [89, 90]. While all external metal components are used to create new cell frames, 95% of the glass can be reused. The remaining components are heated to 500 °C in a thermal processing facility to ease binding between cell elements [91]. The encapsulating plastic evaporates due to the intense heat, leaving the silicon cells, which are then prepared for additional processing. Even this plastic is not thrown away; rather, it serves as a heat source for extra thermal processing. The green hardware is physically divided following the heat treatment. About 80% of this is easily reusable, while the rest is polished. Acid is used to etch away silicon wafers. The silicon material is recycled at a rate of 85% when broken wafers are melted and used to create new silicon modules [91].

15.3.2 Recycling of thin-film solar panels

Thin-film solar panels differ from traditional silicon-based panels and require specific recycling processes due to their unique materials and structure [17, 85, 90]. In comparison to Si PV panel, thin-film-based panels are processed more drastically. The first step is to put them through a shredder. A hammermill is then used to ascertain that all of the particles are no bigger than 4–5 mm, at which point the lamination holding the internal components together breaks and can be removed. In contrast to Si PV panels, the residual material is made up of both liquid and solid components. These are separated by a revolving screw, which causes the solid components to continue rotating inside a tube, whereas the liquid drips into a container [91]. Mechanical techniques such as flotation, magnetic separation, and density separation are used to isolate different materials. Purity is ensured in liquids via precipitation and dewatering. The final product is subjected to metal processing in order to fully separate the semiconductor components [91]. Chemical recycling methods are used to separate the semiconductor materials (CdTe, CIGS) from the glass and other components. CdTe panels typically involve nitric acid leaching to recover cadmium and tellurium. CIGS panels use a combination of acids and solvents to extract copper, indium, gallium, and selenium. Chemical processes are used to further purify the recovered metals for reuse in manufacturing new solar panels or other products. On average, 95% of the semiconductor material is reused. Solid matter is contaminated with interlayer materials, which are lighter in mass and can be eliminated by vibrating the surface. Finally, the material is rinsed [91]. The glass, which makes up a significant portion of the panel, is cleaned and recycled. It can be reused in new panels or other glass products [91]. Certain thin-film materials, such as cadmium, are toxic and require careful handling and processing to prevent environmental contamination.

The cost of recycling processes versus the value of recovered materials can impact the economic feasibility of recycling programs. Compliance with environmental regulations and standards is crucial for safe and effective recycling practices. Advances in recycling technologies are needed to improve recovery rates and reduce costs. The development of automated systems for disassembly and material separation will increase efficiency and reduce labour costs.

15.4 Solar panel recycling market, growth and forecast

The solar panel recycling market is an emerging sector driven by the increasing deployment of solar energy systems and the subsequent need to manage end-of-life panels sustainably. Researchers are working hard to create recycling techniques that can profitably extract the majority of a solar panel's components. Certain nations have enacted design regulations to guarantee PV panel recycling, which will come into force in 2025 [91]. Although the solar recycling sector is still in its early stages, according to a new Rystad Energy analysis, recyclable elements from solar panels may have a market value of over $2.7 billion by 2030 [92]. Governments and environmental agencies are implementing stringent regulations to manage waste from solar panels. Companies and countries are aiming to reduce their carbon footprint and promote circular economy practices, driving the demand for recycling solutions.

Many waste management facilities in the European Union only collect bulk items such as aluminium frames and glass covers, which account for more than 80% of the mass of silicon panels, despite laws requiring PV recycling [87]. Even though the residual mass includes elements like Si, Cu, and Ag which together make up two-thirds of the monetary worth of the ingredients used to make a silicon panel, it is frequently burned [87]. Europe is a leading region in solar panel recycling, with strong regulatory support and established recycling infrastructure. The global solar panel recycling market is witnessing significant growth. In 2022, the market was valued at £250 million. It is projected to reach £1.29 billion by 2028. Recycle solar technologies, ILM Highland, and H&H Pro are a few well-known companies in the UK that provide recycling services [91]. EU funds have supported a number of European initiatives aimed at enhancing the recycling of solar panels that have reached the end of their useful life [92]. The Photorama Consortium has developed a new delamination technique that may make it possible to salvage glass in its whole and use it in new photovoltaic modules. Another initiative, ReProSolar, is constructing a trial recycling factory that will utilize a mild base to extract silver from solar cells, allowing full wires to be removed [92]. Countries like Germany and France are at the forefront in this endeavor. There are no federal regulations requiring PV recycling in the United States, and NREL reports that less than 10% of the country's decommissioned panels get recycled [87]. The United States and Canada are witnessing growing investments in solar panel recycling technologies and facilities. Researchers at Arizona State University (ASU) are developing a novel recycling method that effectively extracts Si and Ag from solar panels using chemicals [92]. The US Department of Energy awarded the ASU team funding for the research, which they anticipate will result in the construction of a pilot recycling plant. A number of novel recycling procedures are being tested by the American recycling firm TG Companies. One such method would use a hot steel blade to separate the silicon cells of solar panels from the glass sheets of the panel. Starting in 2025, a law in Washington is anticipated to encourage solar recycling [92].

China, Japan, and India are significant markets due to their large solar energy capacities and increasing focus on sustainable practices. Major players in the market include First Solar, Veolia, Reiling GmbH & Co. KG, and NPC Incorporated. By 2050, photovoltaic recycling is expected to generate around £11 billion in recoverable value and more green employment possibilities [89, 91]. In Australia, scientists are developing a novel method for removing silicon from solar panels and converting it into nano-silicon, which is intended for use in lithium-ion batteries [92]. The nano-silicon will be used to make high-energy anodes, which will be extremely useful in lowering the cost of battery materials [92].

Innovations in recycling technologies are improving efficiency and reducing costs. Government policies and incentives are encouraging investments in recycling infrastructure and research. Companies are forming partnerships to develop comprehensive recycling solutions and expand their market reach. Increased awareness about the environmental benefits of recycling solar panels is driving demand. In a recent report [93], global solar panel recycling market, segmented by different recycling processes (thermal, mechanical, and laser) from 2019 to 2030, with the market value measured in millions of dollars ($M) [93] (figure 15.9). The recycling market is projected to surpass $1 billion by 2030, driven by increasing demand for sustainable waste management. Mechanical recycling dominates the market, while thermal and laser methods show gradual growth. Post-2026, the market is expected to experience an accelerated expansion, with rising adoption of advanced recycling techniques.

Efficiently separating and recovering materials from complex panel structures is technologically challenging. The market is fragmented with various players employing different technologies and processes, leading to inconsistencies in recycling outcomes.

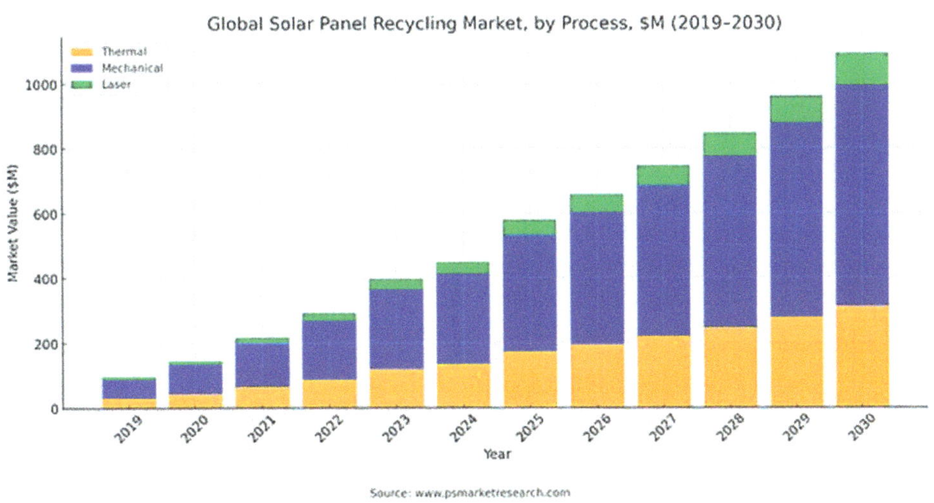

Figure 15.9. Global solar panel recycling market, by process thermal, mechanical and laser (2019–2030). Reproduced with permission from [93]. Credit: P&S Intelligence: https://www.psmarketresearch.com/.

The solar panel recycling market is expected to grow significantly over the next decade, driven by the increasing volume of decommissioned panels and advancements in recycling technologies. The market is poised to create economic opportunities through the recovery and resale of valuable materials, as well as the creation of green jobs in the recycling sector. Effective recycling can substantially reduce the environmental impact of solar panels, conserving resources and minimizing waste. The recently passed Inflation Reduction Act (IRA), in the United States, has allocated funding for solar PV recycling programs, which will also encourage growth over the next decade. Solar recycling companies are already emerging, seeking to recycle and sell the materials and keep panels out of landfills [92].

With all of these technological breakthroughs underway, a cost-effective solar recycling technology might hit the market just as the first significant wave of solar panels near the end of their useful life. There will probably be more solar panels constructed of recyclable material in the future [92]. Reusing valuable materials from solar panels that have reached the end of their useful life could help reduce supply chain bottlenecks in the future and, eventually, increase the sustainability of the whole sector. The solar panel recycling market is at a nascent stage but is set to expand rapidly due to the growing need for sustainable management of solar panel waste. Technological innovations, supportive policies, and increased awareness will be crucial in driving the market forward and ensuring the long-term sustainability of the solar energy sector.

15.5 Lifecycle analysis of solar panels

Lifecycle analysis (LCA) of solar panels assesses their environmental impact throughout their entire lifecycle, from raw material extraction to manufacturing, use phase, and end-of-life disposal or recycling [94, 95]. An overview of the key stages and considerations involved in the lifecycle analysis of solar panels is discussed in this section.

Primary materials include Si for silicon panels, CdTe, CIGS for thin-film panels, metals (Ag, Al, Cu), glass, plastics, and rare earth elements (in some cases). Mining and processing of raw materials can lead to energy consumption, water use, land disturbance, and emissions of greenhouse gases (GHGs) and pollutants. Panel production involves refining raw materials, manufacturing silicon wafers, depositing thin-films, assembling panels, and adding frames and encapsulants. Energy consumption, water use, emissions of GHGs and air pollutants, waste generation, and potential occupational health and safety hazards [94, 95] are some of the concerns. Transport includes shipping of raw materials, components, and finished panels to manufacturing facilities and distribution centres. Energy and materials are used during the installation of solar panels on rooftops, ground mounts, or integrated into buildings. Solar panels generate electricity without direct emissions during operation, reducing reliance on fossil fuels. Cleaning, monitoring, and occasional repairs or replacements are required during the operational lifetime.

Lifecycle analysis helps stakeholders make informed decisions about solar panel technologies, identifying opportunities to reduce environmental impacts and enhance sustainability throughout their lifecycle. Ongoing research and collaboration is essential to advancing the understanding and implementation of sustainable practices in the solar energy sector.

15.6 Conclusion

Potential advancements in the field of solar photovoltaics, as well as a number of challenges including supply chain, vertical integration, polysilicon production, and recycling are described in this chapter. Innovations in developing technology to lower solar photovoltaic production costs are discussed. Manufacturing process optimization is investigated in relation to incentives, silicon wafer thickness, glass in solar panels, and workforce development. The lifecycle analysis and methods for recycling of solar panels are explored. Future market growth and cost estimates for solar panel recycling are discussed.

References

[1] Rangelova K 2024 2023's record solar surge explained in six charts https://ember-energy.org/latest-insights/2023s-record-solar-surge-explained-in-six-charts/#:~:text=Solar%20skyrocketed%20in%202023.,built%20in%202023%20was%20solar.

[2] Patel S 2019 *IEA World Energy Outlook: Solar Capacity Surges Past Coal and Gas by 2040* (*Power*)

[3] Marques Lameirinhas R A, Torres J P N and de Melo Cunha J P 2022 A Photovoltaic technology review: history, fundamentals and applications *Energies* **15** 1823

[4] NREL 2024 NREL's Best Research-Cell Efficiency Chart 2024 https://www.nrel.gov/pv/cell-efficiency.html

[5] NREL 2024 Champion Photovoltaic Module Efficiency Chart 2024 https://www.nrel.gov/pv/module-efficiency.html

[6] Ali O 2024 *Solar Innovation: Exploring the Future Possibilities of Photovoltaic Technology* (AZoCleantech)

[7] Wu C, Zhang X-P and Sterling M 2022 Solar power generation intermittency and aggregation *Sci. Rep.* **12** 1363

[8] Solar Magazine 2022 Solar panel degradation: what is it and why should you care? *Solar Mag.* **October 2022** https://solarmagazine.com/solar-panels/solar-panel-degradation/

[9] Nwaigwe K N, Mutabilwa P and Dintwa E 2019 An overview of solar power (PV systems) integration into electricity grids *Mater. Sci. Energy Technol.* **2** 629–33

[10] Shafiullah M, Ahmed S D and Al-Sulaiman F A 2022 Grid integration challenges and solution strategies for solar pv systems: a review *IEEE Access* **10** 52233–57

[11] IEA 2024 *Renewables 2023* (Paris: IEA)

[12] Jowett P 2024 WoodMac predicts strong yet flat global PV growth through to 2032 *PV Mag. Global* https://www.pv-magazine.com/2024/01/24/woodmac-predicts-strong-yet-flat-global-pv-growth-through-to-2032/

[13] IEA 2022 Global Supply Chains An IEA Special Report https://www.iea.org/reports/solar-pv-global-supply-chains

[14] Wood Mackenzie 2023 China to hold over 80% of global solar manufacturing capacity from 2023–26 https://www.woodmac.com/press-releases/china-dominance-on-global-solar-supply-chain/

[15] Copley M 23 May, 2022 A decade into tariffs, US solar manufacturing is still deep in Asia's shadow https://www.spglobal.com/market-intelligence/en/news-insights/articles/2022/5/a-decade-into-tariffs-us-solar-manufacturing-is-still-deep-in-asia-s-shadow-70236202
[16] Fang M M 2020 A crisis or an opportunity? The trade war between the US and China in the Solar PV Sector *J. World Trade* **54** 103–26
[17] Vinayagamoorthi R *et al* 2024 Recycling of end of life photovoltaic solar panels and recovery of valuable components: a comprehensive review and experimental validation *J. Environ. Chem. Eng.* **12** 111715
[18] Farrell C C *et al* 2020 Technical challenges and opportunities in realising a circular economy for waste photovoltaic modules *Renew. Sustain. Energy Rev.* **128** 109911
[19] Zhang F and Gallagher K S 2016 Innovation and technology transfer through global value chains: evidence from China's PV industry *Energy Policy* **94** 191–203
[20] Reker S, Schneider J and Gerhards C 2022 Integration of vertical solar power plants into a future German energy system *Smart Energy* **7** 100083
[21] Saga T 2010 Advances in crystalline silicon solar cell technology for industrial mass production *NPG Asia Mater.* **2** 96–102
[22] Shiradkar N *et al* 2022 Recent developments in solar manufacturing in India *Solar Compass* **1** 100009
[23] Ballif C *et al* 2022 Status and perspectives of crystalline silicon photovoltaics in research and industry *Nat. Rev. Mater.* **7** 597–616
[24] Drouiche N *et al* 2015 Hidden values in kerf slurry waste recovery of high purity silicon *Renew. Sustain. Energy Rev.* **52** 393–9
[25] Xakalashe B and Tangstad M 2011 Silicon processing: from quartz to crystalline silicon solar cells (Unpublished)
[26] Dazhou Y 2017 Siemens process *Handbook of Photovoltaic Silicon* ed D Yang (Berlin, Heidelberg: Springer) pp 1–32
[27] Yu X and Yang D 2017 Growth of crystalline silicon for solar cells: Czochralski Si *Handbook of Photovoltaic Silicon* ed D Yang (Berlin, Heidelberg: Springer) pp 1–45
[28] Liu X, Payra P and Wan Y 2019 Polysilicon and its characterization methods *Handbook of Photovoltaic Silicon* ed D Yang (Berlin, Heidelberg: Springer) pp 9–36
[29] Yadav S, Chattopadhyay K and Singh C V 2017 Solar grade silicon production: a review of kinetic, thermodynamic and fluid dynamics based continuum scale modeling *Renew. Sustain. Energy Rev.* **78** 1288–314
[30] Filtvedt W O *et al* 2010 Development of fluidized bed reactors for silicon production *Sol. Energy Mater. Sol. Cells* **94** 1980–95
[31] Jiang L *et al* 2019 Fluidized bed process with silane *Handbook of Photovoltaic Silicon* ed D Yang (Berlin, Heidelberg: Springer) pp 69–108
[32] Bati A S R *et al* 2023 Next-generation applications for integrated perovskite solar cells *Commun. Mater.* **4** 2
[33] Solak E K and Irmak E 2023 Advances in organic photovoltaic cells: a comprehensive review of materials, technologies, and performance *RSC Adv.* **13** 12244–69
[34] Aqoma H *et al* 2024 Alkyl ammonium iodide-based ligand exchange strategy for high-efficiency organic-cation perovskite quantum dot solar cells *Nat. Energy* **9** 324–32
[35] Noman M, Khan Z and Jan S T 2024 A comprehensive review on the advancements and challenges in perovskite solar cell technology *RSC Adv.* **14** 5085–131

[36] Pastuszak J and Węgierek P 2022 Photovoltaic cell generations and current research directions for their development *Materials* **15** 5542

[37] Kothandaraman R K *et al* 2020 Near-infrared-transparent perovskite solar cells and perovskite-based tandem photovoltaics *Small Methods* **4** 2000395

[38] Andruszkiewicz A *et al* 2021 Perovskite and quantum dot tandem solar cells with interlayer modification for improved optical semitransparency and stability *Nanoscale* **13** 6234–40

[39] Jin Y *et al* 2024 Efficient and stable monolithic perovskite/silicon tandem solar cells enabled by contact-resistance-tunable indium tin oxide interlayer *Adv. Mater.* **36** 2404010

[40] Shi Y, Berry J J and Zhang F 2024 Perovskite/silicon tandem solar cells: insights and outlooks *ACS Energy Lett.* **9** 1305–30

[41] Kahandal S S *et al* 2024 Perovskite solar cells: fundamental aspects, stability challenges, and future prospects *Prog. Solid State Chem.* **74** 100463

[42] Kore B P, Jamshidi M and Gardner J M 2024 The impact of moisture on the stability and degradation of perovskites in solar cells *Mater. Adv.* **5** 2200–17

[43] Zhou Y *et al* 2021 Improving thermal stability of perovskite solar cells by suppressing ion migration using copolymer grain encapsulation *Chem. Mater.* **33** 6120–35

[44] Noman M, Khan A H H and Jan S T 2024 Interface engineering and defect passivation for enhanced hole extraction, ion migration, and optimal charge dynamics in both lead-based and lead-free perovskite solar cells *Sci. Rep.* **14** 5449

[45] Mariani P *et al* 2024 Low-temperature strain-free encapsulation for perovskite solar cells and modules passing multifaceted accelerated ageing tests *Nat. Commun.* **15** 4552

[46] Bati A *et al* 2021 Cesium-doped $Ti_3C_2T_x$ MXene for efficient and thermally stable perovskite solar cells *Cell Rep. Phys. Sci.* **2** 100598

[47] Wang F *et al* 2022 Recent progress of scalable perovskite solar cells and modules *Energy Rev.* **1** 100010

[48] Li M and He F 2024 Organic solar cells developments: what's next? *Next Energy* **2** 100085

[49] Li Y *et al* 2021 Non-fullerene acceptor organic photovoltaics with intrinsic operational lifetimes over 30 years *Nat. Commun.* **12** 5419

[50] Gu W *et al* 2020 A comprehensive review and outlook of bifacial photovoltaic (BPV) technology *Energy Convers. Manage.* **223** 113283

[51] Biyik E *et al* 2017 A key review of building integrated photovoltaic (BIPV) systems *Eng. Sci. Technol., Int. J.* **20** 833–58

[52] Pillai D S, Shabunko V and Krishna A 2022 A comprehensive review on building integrated photovoltaic systems: emphasis to technological advancements, outdoor testing, and predictive maintenance *Renew. Sustain. Energy Rev.* **156** 111946

[53] Widmer J *et al* 2024 Agrivoltaics, a promising new tool for electricity and food production: a systematic review *Renew. Sustain. Energy Rev.* **192** 114277

[54] Almeida R *et al* 2022 Floating solar power could help fight climate change—let's get it right *Nature* **606** 246–9

[55] Tawalbeh M *et al* 2021 Environmental impacts of solar photovoltaic systems: a critical review of recent progress and future outlook *Sci. Total Environ.* **759** 143528

[56] Zhang L *et al* 2022 Emerging chemistries and molecular designs for flow batteries *Nat. Rev. Chem.* **6** 524–43

[57] Zhang X *et al* 2024 Advanced compressed air energy storage systems: fundamentals and applications *Engineering* **34** 246–69

[58] Rana M M *et al* 2023 Applications of energy storage systems in power grids with and without renewable energy integration—a comprehensive review *Energy Storage* **68** 107811
[59] Weerasinghe H C *et al* 2024 The first demonstration of entirely roll-to-roll fabricated perovskite solar cell modules under ambient room conditions *Nat. Commun.* **15** 1656
[60] Hu X 2021 Research on silicon wafer manufacturing process and physical properties testing using high-purity polysilicon *J. Phys. Conf. Ser.* **2083** 022050
[61] Stallhofer P 2011 Why are silicon wafers as thick as they are? *Ultra-thin Chip Technology and Applications* ed J Burghartz (New York: Springer) pp 3–12
[62] Dong Z and Lin Y 2020 Ultra-thin wafer technology and applications: a review *Mater. Sci. Semicond. Process.* **105** 104681
[63] Wu T *et al* 2024 Free-standing ultrathin silicon wafers and solar cells through edges reinforcement *Nat. Commun.* **15** 3843
[64] Pal P *et al* 2021 High speed silicon wet anisotropic etching for applications in bulk micromachining: a review *Micro Nano Syst. Lett.* **9** 4
[65] *Bigger Wafers, Half-Cut Technology, Multi Bus-Bar (MBB) Make Higher Power Solar Panel* (2021) https://dsneg.com/info/bigger-wafers-half-cut-technology-multi-bus-62643057.html
[66] Brian 2023 New trend in PV cells: rectangular silicon wafers (182R & 210R) https://maysunsolar.com/new-trend-in-pv-cells-rectangular-silicon-wafers-182r-210r/
[67] Tu A 2023 Wafer formats continue to evolve *PV Magazine* https://www.pv-magazine.com/2023/08/17/wafer-formats-continue-to-evolve/
[68] Li A *et al* 2023 Recent advances in precision diamond wire sawing monocrystalline silicon *Micromachines* **14** 1512
[69] Lee H-S *et al* 2018 Enhanced efficiency of crystalline Si solar cells based on kerfless-thin wafers with nanohole arrays *Sci. Rep.* **8** 3504
[70] Belançon M P *et al* 2023 Glassy materials for silicon-based solar panels: present and future *J. Non-Cryst. Solids* **619** 122548
[71] Law A M, Jones L O and Walls J M 2023 The performance and durability of Anti-reflection coatings for solar module cover glass—a review *Sol. Energy* **261** 85–95
[72] Friesen T 2013 Hail testing of PV modules: results of a round robin for hail grain quality determination and testing results of different module designs *28th European Photovoltaic Solar Energy Conference (Paris)*
[73] Gaddam S K, Pothu R and Boddula R 2021 Advanced polymer encapsulates for photovoltaic devices—a review *J. Materiomics* **7** 920–8
[74] Rajbahadur Y N K *et al* 2024 Evaluation of hydrophobic/hydrophilic and antireflective coatings for photovoltaic panels *J. Coat. Technol. Res.* (in press) https://doi.org/10.1007/s11998-024-00929-0
[75] Allsopp B L *et al* 2020 Towards improved cover glasses for photovoltaic devices *Prog. Photovolt. Res. Appl.* **28** 1187–206
[76] Chen H *et al* 2023 Re-structuring of glass surface by ion post-embedding for transmission improvement and application in solar panels *Sol. Energy* **252** 373–9
[77] Chala G T *et al* 2024 Effects of nanocoating on the performance of photovoltaic solar panels in Al Seeb, Oman *Energies* **17** 2871
[78] Riedel-Lyngskær N *et al* 2022 The effect of spectral albedo in bifacial photovoltaic performance *Sol. Energy* **231** 921–35

[79] Hassan Q *et al* 2023 A review of hybrid renewable energy systems: solar and wind-powered solutions: challenges, opportunities, and policy implications *Res. Eng.* **20** 101621
[80] Barki D *et al* 2020 Innovative skill development techniques for solar power plants and solar PV job creation in India *2020 47th IEEE Photovoltaic Specialists Conf. (PVSC)*
[81] Kılıç U and Kekezoğlu B 2022 A review of solar photovoltaic incentives and policy: selected countries and Turkey *Ain Shams Eng. J.* **13** 101669
[82] Silveira B *et al* 2024 Incentives for photovoltaic energy generation: a comparative analysis of policies in Spain, Germany, and Brazil *Energy Strategy Rev.* **54** 101415
[83] Timilsina G R, Kurdgelashvili L and Narbel P A 2012 Solar energy: markets, economics and policies *Renew. Sustain. Energy Rev.* **16** 449–65
[84] Nijsse F J M M *et al* 2023 The momentum of the solar energy transition *Nat. Commun.* **14** 6542
[85] Chowdhury M S *et al* 2020 An overview of solar photovoltaic panels' end-of-life material recycling *Energy Strategy Rev.* **27** 100431
[86] Walzberg J, Carpenter A and Heath G A 2021 Role of the social factors in success of solar photovoltaic reuse and recycle programmes *Nat. Energy* **6** 913–24
[87] Peplow M 2022 Solar panels face recycling challenge *ACS Cent. Sci.* **8** 299–302
[88] *EPA Announced Plans to Propose New Universal Waste Regulations* (Nov. 16, 2023) https://fedcenter.gov/Announcements/index.cfm?id=41566
[89] Gao S *et al* 2024 Recycling of silicon solar panels through a salt-etching approach *Nature Sustain* **7** 920–30
[90] Divya A *et al* 2023 Review on recycling of solar modules/panels *Sol. Energy Mater. Sol. Cells* **253** 112151
[91] Vekony A T 2024 The opportunities of solar panel recycling https://www.greenmatch.co.uk/blog/2017/10/the-opportunities-of-solar-panel-recycling
[92] Bluesystem Energy Solutions 2023 Solar Panel Recycling: New Methods Look Promising https://www.bluestemenergysolutions.com/solar-panel-recycling-new-methods-look-promising/
[93] *Solar Panel Recycling Overview* (2024) https://www.psmarketresearch.com/market-analysis/solar-panel-recycling-market
[94] Gerbinet S, Belboom S and Léonard A 2014 Life Cycle Analysis (LCA) of photovoltaic panels: a review *Renew. Sustain. Energy Rev.* **38** 747–53
[95] Prabhu V S, Shrivastava S and Mukhopadhyay K 2022 Life cycle assessment of solar photovoltaic in India: a circular economy approach *Circ. Econ. Sustain.* **2** 507–34

www.ingramcontent.com/pod-product-compliance
Ingram Content Group UK Ltd.
Pitfield, Milton Keynes, MK11 3LW, UK
UKHW051906100926
8556IPUK00005B/778

9 780750 359924